3rd International Conference on Advanced Technologies in Design, Mechanical and Aeronautical Engineering (ATDMAE 2019)

IOP Conference Series: Materials Science and Engineering
Volume 616

Shanghai, China
5 – 7 July 2019

ISBN: 978-1-5108-9823-3
ISSN: 1755-8981

Printed from e-media with permission by:

Curran Associates, Inc.
57 Morehouse Lane
Red Hook, NY 12571

Some format issues inherent in the e-media version may also appear in this print version.

Institute of Physics
Dirac House, Temple Back
Bristol BS1 6BE UK

Phone: 44 1 17 929 7481
Fax: 44 1 17 920 0979

techtracking@iop.org

Additional copies of this publication are available from:

Curran Associates, Inc.
57 Morehouse Lane
Red Hook, NY 12571 USA
Phone: 845-758-0400
Fax: 845-758-2633
Email: curran@proceedings.com
Web: www.proceedings.com

Table of contents

Volume 616

3rd International Conference on Advanced Technologies in Design, Mechanical and Aeronautical Engineering (ATDMAE 2019)

5–7 July 2019, Shanghai, China

Accepted papers received: 19 August 2019
Published online: 16 October 2019

Preface

Papers

Aerospace Engineering

IOP Conf. Series: Materials Science and Engineering **616** (2019) 011001 doi:10.1088/1757-899X/616/1/011001

PREFACE

3rd International Conference on Advanced Technologies in Design, Mechanical and Aeronautical Engineering (ATDMAE 2019) was held in Shanghai, China from July 5 to 7, 2019. ATDMAE 2019 was co-organized by Nazarbayev University and Hong Kong Society of Mechanical Engineers (HKSME). The conference provides a useful and wide platform both for display the latest research and for exchange of research results and thoughts in Design, Mechanical and Aeronautical Engineering and other topics. The participants of the conference were from almost every part of the world, with background of either academia or industry, even well-known enterprise. The success and prosperity of the conference is reflected high level of the papers received.

The proceedings are a compilation of the accepted papers and represent an interesting outcome of the conference. This book covers 3 chapters: Aerospace Engineering; Mechanical Engineering; Design Engineering.

We would like to acknowledge all of those who supported ATDMAE 2019. Each individual and institutional help were very important for the success of this conference. Especially we would like to thank the organizing committee for their valuable advices in the organization and helpful peer review of the papers.

We sincerely hope that ATDMAE 2019 will be a forum for excellent discussions that will put forward new ideas and promote collaborative researches. We are sure that the proceedings will serve as an important research source of references and the knowledge, which will lead to not only scientific and engineering progress but also other new products and processes.

Prof. Yong Zhao, Nazarbayev University, Kazakhstan

IOP Conf. Series: Materials Science and Engineering **616** (2019) 011002 doi:10.1088/1757-899X/616/1/011002

COMMITTEES

Conference Chairman

Prof. Yong Zhao, Nazarbayev University, Kazakhstan

Conference Chair

Prof. Xiaohui Su, Dalian University of Technology, China

Conference Program Committee

Prof. Zijian Wang, University of Kansas, USA

Prof. Tongming Zhou, The University of Western Australia, Australia

Prof. Bin Wang, Brunel University London, UK

Dr. Yang Wenming, National University of Singapore, Singapore

Organizing Committee

Prof. Gopalan Jagadeesh, Indian Institute of Science, India

International Technical Commmitte

Prof. Weihua Liu, Nanjing University of Aeronautics and Astronautics, China (Chair)

Prof. TEH Soo Lee, Nazarbayev University, Kazakhstan

Prof. S.Thanigaiarasu, Madras Institute of Technology, India

Prof. Kuei-Yuan Chan, National Taiwan University, Taiwan

Prof. K.M.Parammasivam, Madras Institute of Technology, India

Prof. B. T. N. Sridhar, Anna University, India

Dr. Zhauyt Algazy, Satbayev University, Kazakhstan

Dr. Zhangming Wu, Cardiff University, UK

Dr. Redouane Zitoune, Clément Ader Institute of Toulouse University, France

Dr. Pierre Kerfriden, Cardiff University, UK

Dr. GUELAILIA Ahmed, Satellite Development Center, Algerian Space Agency, Algeria

Peer review statement

All papers published in this volume of *IOP Conference Series: Materials Science and Engineering* have been peer reviewed through processes administered by the proceedings Editors. Reviews were conducted by expert referees to the professional and scientific standards expected of a proceedings journal published by IOP Publishing.

IOP Conf. Series: Materials Science and Engineering **616** (2019) 012001 doi:10.1088/1757-899X/616/1/012001

Numerical Simulations for the Tip-Jet Drive Rotor Based Upon Momentum-Source Method

CUI Fei[1], ZHUANG Jian[1] and WANG Bo[2]

[1]Innovation & Research Institute of HIWING Technology Academy, Beijing 100074, China
[2]Science and Technology on Rotorcraft Aeromechanics Laboratory Nanjing University of Aeronautics and Astronautics, Nanjing 210016, China

E-mail: 381343120@qq.com

Abstract. Based upon a momentum-source method, a numerical simulation method has been developed to calculate the tip-jet drive rotor/fuselage flowfield. The momentum-source term is used to describe the effect of rotor on its flowfield. The mass-source term is used to describe the tip-jet. According to the characteristics of rotor/fuselage flowfield and requirements of momentum-source and mass-source method, a hybrid mesh generation approach is developed. In the present method, on the basis of the known movement, geometric and aerodynamic characteristics on blade section, a scheme containing momentum-source and mass-source term is adopted for the solution of N-S equations. By the present method, the downwash flowfields of rotor and rotor/fuselage are simulated, and some meaningful conclusions are obtained.

1. Introduction

The jet-driven rotor generates a reaction force to drive the rotor to rotate through the high-speed jet of the wing tip. Compared with the shaft-driven rotor, the heavy gear transmission system and the anti-torque system are eliminated, which effectively reduces the complexity of the system, and is simple and easy to use[1-2]. With the increasing demand for vertical take-off and landing and high-speed cruise aircraft, a new concept of rotorcraft with stagnable rotor, such as X-50A in the United States[3], has emerged. The rotor of this type of aircraft has two major characteristics. First, the rotor is driven by jet, which can not only meet the power demand of vertical take-off and landing, but also meet the thrust demand of high-speed leveling in a single power situation. Second, the rotor can rotate at high speed to provide tension for vertical take-off and landing, and can be locked into a fixed wing to provide lift for high subsonic speed cruise.

In order to reduce the computational cost and improve the efficiency of simulated rotor downwash flow field, some researchers have made many attempts. At the end of the last century, Rajagopalan[4] and Chaffin[5] calculated the rotor flow field by using a paddle instead of a rotor, and made significant progress.

Based on the principle of momentum source[6-8], the aerodynamic characteristics of CRW(Canard Rotor/Wing) are numerically simulated by CFD method.

2. Principle and geometric characteristics of tip-jet drive rotor

A typical tip-jet rotor system consists of an air intake, an engine, a main nozzle, a main shaft duct, jet pipes, tip-jet nozzles, and a rotor/wing, as shown in Figure 1. The jet enters the main shaft pipe from the tail nozzle of the engine, and is divided into two blades that enter the rotation through the "Y"-shaped pipe at the hub. The main nozzle prevents the back pressure from being too high and causes the engine to surge.

The jet pipe is located inside the blade, as shown in Figure 2. At the tip of the blade, the cross-sectional shape of the pipe changes with the shape of the airfoil, as shown in Figure 3.

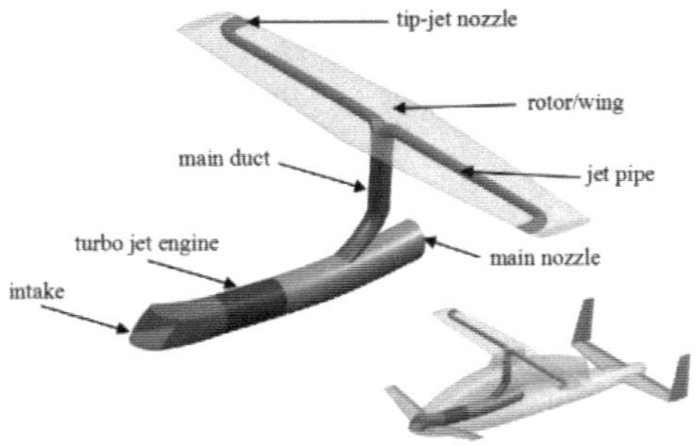

Figure 1. Typical tip-jet rotor system.

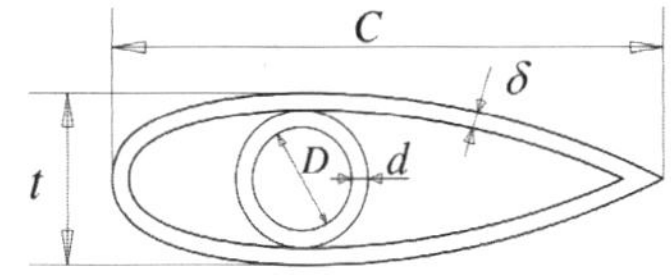

Figure 2. Blade cross section at hub.

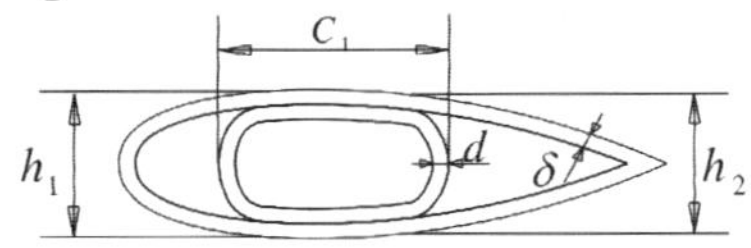

Figure 3. Blade cross section at tip.

3. Momentum-source method

By projecting the lift and the drag at (r,ϕ,n) in the blade profile coordinate system to the n, ϕ direction, respectively, the tensile force f_n perpendicular to the rotational direction and the rotational drag f_ϕ in the rotational direction can be obtained.

$$f_n = d\mathrm{Y}\cos\beta - d\mathrm{X}\sin\beta \tag{1}$$

$$f_\phi = -d\mathrm{Y}\sin\beta - d\mathrm{X}\cos\beta \tag{2}$$

The resultant force $\vec{F}$ of the tensile force f_n and the resistance f_ϕ and the radial force (which can be ignored here) is the force of the surrounding airflow on the blade micro-segment.

4. Mass-source method

Under the premise of not considering the generation of internal gas, transportation mode and whether it has the property of high temperature and high pressure, the driving mode of the tip jet can be expressed by the cross section of the propeller disk that continuously emits a certain mass of gas in synchronization with the blade:

Jet function = mass source item + mass source item corresponding to the momentum source item

Equivalent algorithm for mass flow $\dot{m}$: The mass of gas ejected per unit time of the nozzle is evenly distributed to the volume of space swept by the nozzle per unit time.

5. Characteristics of flow field in the hovering state

5.1. Computational grid and conditions

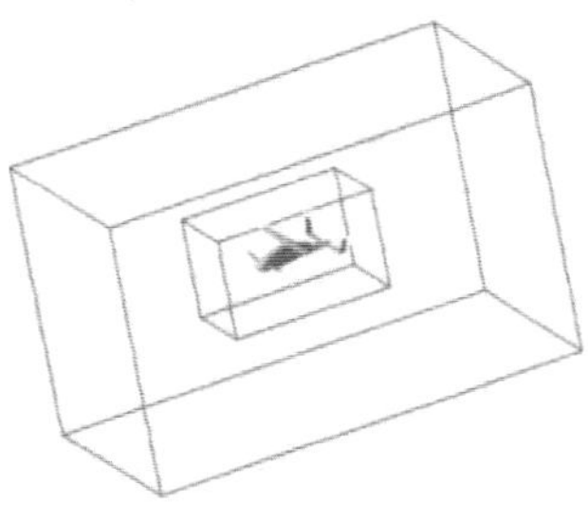

Figure 4. Selection of flow field calculation domain for CRW.

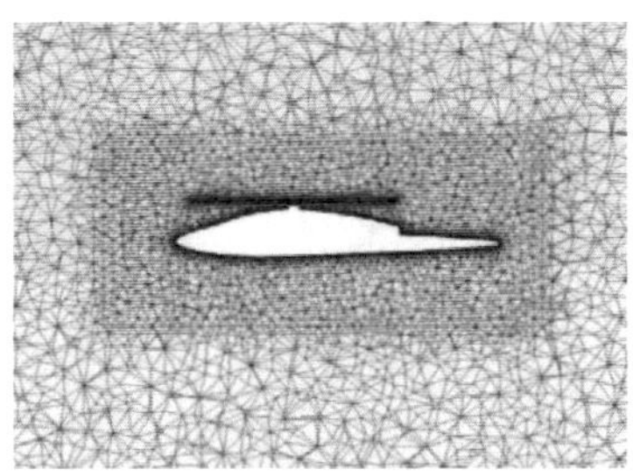

Figure 5. A cross-sectional view of the flow field grid in the hovering state of CRW.

The selection of computational region should adapt to the characteristics of flow field, taking the hovering state as an example, select cuboid of 9R*9R*14R (R is the radius of rotor) as the calculation domain, as shown in Figure 4. Figure 5 is a cross-sectional view of the flow field grid in the hovering state of the CRW aircraft.

5.2. The flow field in hovering state

Figure 6 shows the flowfield of the fuselage in hovering state. The black arrow represents the flow line inside the surface. The far-field airflow above the fuselage has a downward acceleration trend when approaching the paddle. The downwash airflow shrinks into the paddle plate and flows downward around both sides of the fuselage when encountering the fuselage. After bypassing the fuselage, the airflow continues to accelerate downward. The light yellow part on both sides of the fuselage in the figure is the tip vortex.

Figure 7 shows the velocity vector of the airflow in the longitudinal direction of the fuselage. It can be seen that in the longitudinal direction of the fuselage, with the center of the hub as the boundary, the downwash flow in front of the fuselage hits the back of the fuselage and flows up and down along the front fuselage surface respectively. The downward airflow leaves the fuselage at the nose and flows downwards, while the upward airflow flows along the wall to the hub, and then flows vertically upward along the hub wall to form the upward airflow. Similarly, due to the influence of the tip vortex, the airflow behind the fuselage is sucked by the paddle, which is also divided into two parts along the wall surface upward and downward. The upward airflow to the hub forms the upward airflow along the wall of the hub, and the downward airflow flows along the fuselage to the tail, which is gradually upstream under the influence of the tail tip vortex.

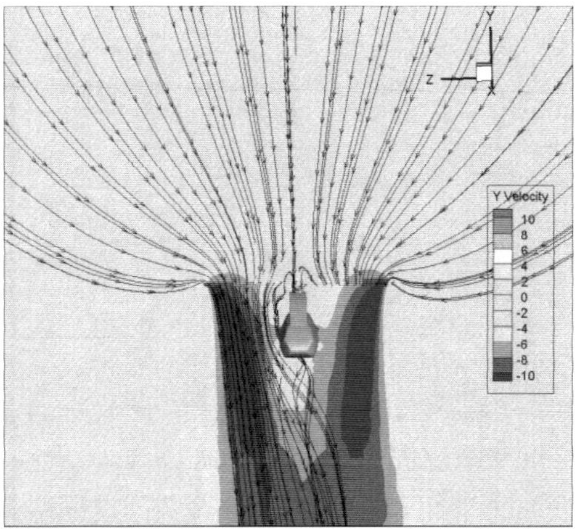

Figure 6. Velocity distribution of downwash flow field of CRW aircraft in hovering state.

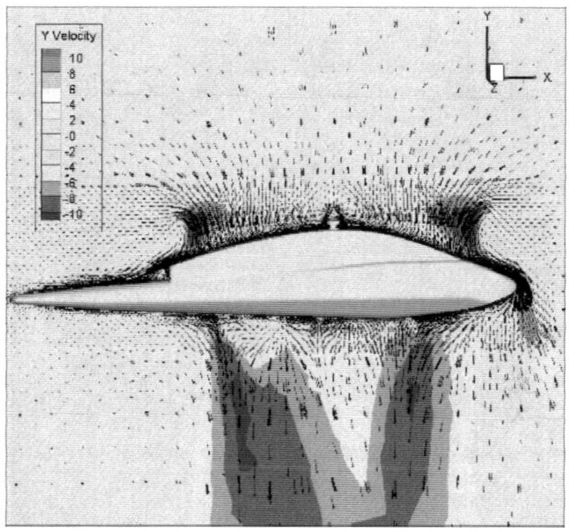

Figure 7. The velocity vector of the fuselage surface.

6. Analysis with/without mass jet

Aiming at the forward flight state of CRW aircraft, the flow field with/without mass jet is compared and calculated. Among them, the forward flying speed of the CRW aircraft is 42m/s, the rotor speed is 1000r/min, the angle of attack is 0°, the collective pitch is 10°, the wing tip jet velocity is 400m/s, the wing tip jet flow rate is 2.5kg/s, and the flow temperature is 723K.

The following data are derived from the calculation results of two groups of flow fields: a) only added momentum source term (not simulated tip jet); b) added both momentum source term and mass source term (simulated tip jet). Figure 8 and Figure 9 shows the comparison of the density distributions with and without mass jet. It is assumed that the distance below and above the flow field space is 20H, and the central position in the y direction is at the point where h/H =0.

IOP Conf. Series: Materials Science and Engineering **616** (2019) 012001 doi:10.1088/1757-899X/616/1/012001

By comparing the results of group a) and group b), it can be seen that the density contour is significantly different. First, in group a), the density of the rotor area is significantly higher than that of the non-rotor area. And due to the influence of the CRW fuselage, the density of the outer ring portion of the paddle is higher than other area of the paddle. But the phenomenon gradually fades with the increasing vertical distance, the density on the entire paddle gradually becomes uniform. On the contrary, group b) with the mass source added has a reasonable density distribution. A ring is formed around the fuselage in the area with enhanced density corresponding to the area with jet tip injection. With the increasing of vertical distance, the influence of fuselage gradually weakens, and a ring is formed in the area with increased density under the influence of jet flow. Second, the upper limit of the density of group b) is much larger than that of group a). It can be analyzed that because of the introduction of the jet the density of the rotor nozzle area will increase sharply, and the increasing trend will continue to spread around the rotor, making the surrounding area also increases. The density of the so-called "ring" area actually reaches the corresponding density of the group a) value.

In addition, due to the influence of fuselage (vertical tail, canard wing), under the action of jet flow, the distribution of density cloud shows the phenomenon of left- right asymmetry.

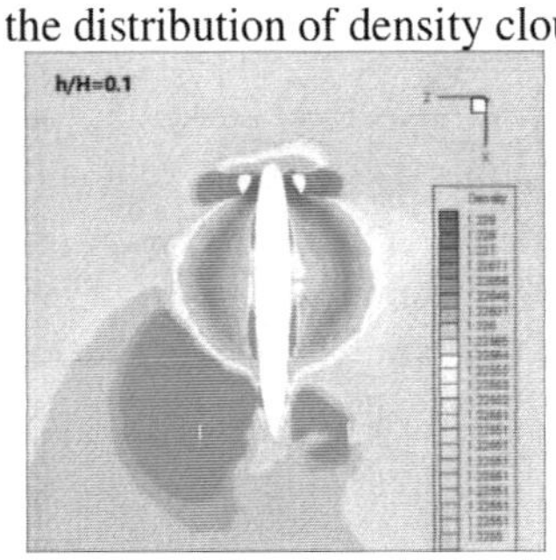

(a)h/H=0.1(momentum-source).

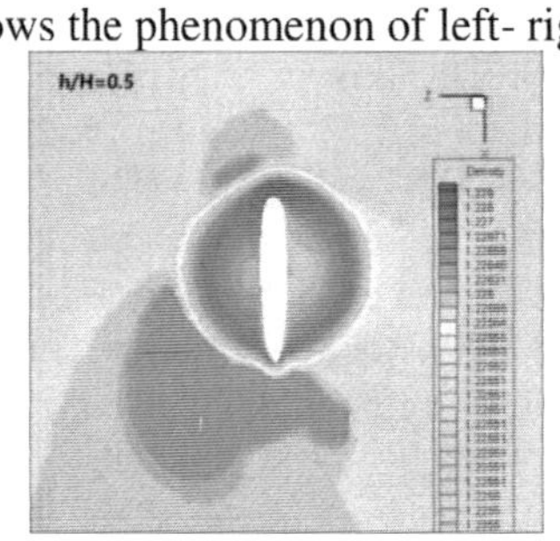

(c)h/H=0.5(momentum-source).

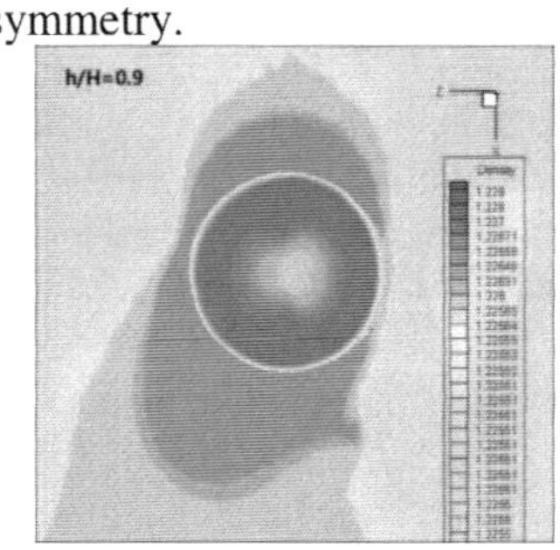

(e)h/H=0.9(momentum-source).

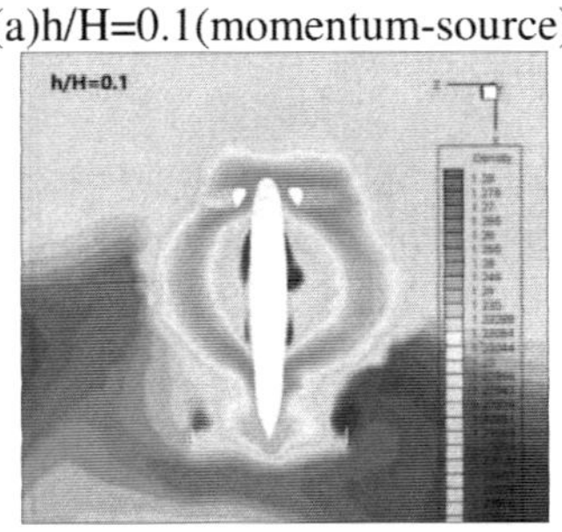

(b) h/H=0.1(mass-source term) .

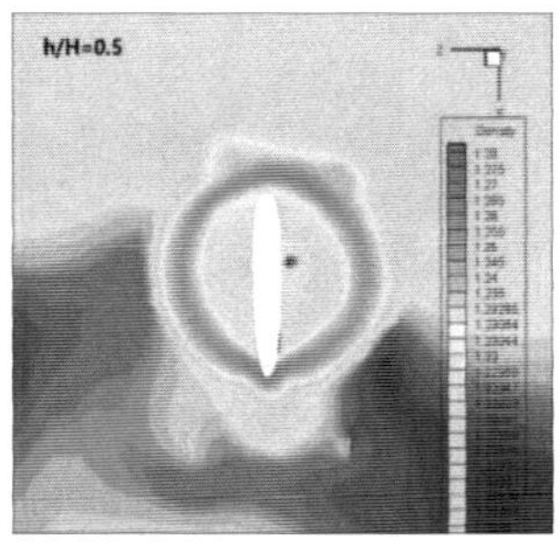

(d)h/H=0.5(mass-source term)

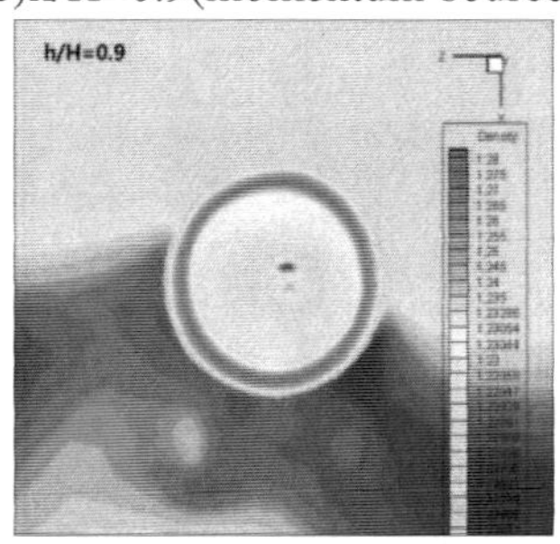

(f) h/H=0.9(mass-source term).

Figure 8. Comparison of flow field density distribution with/ without tip-jet mass source.

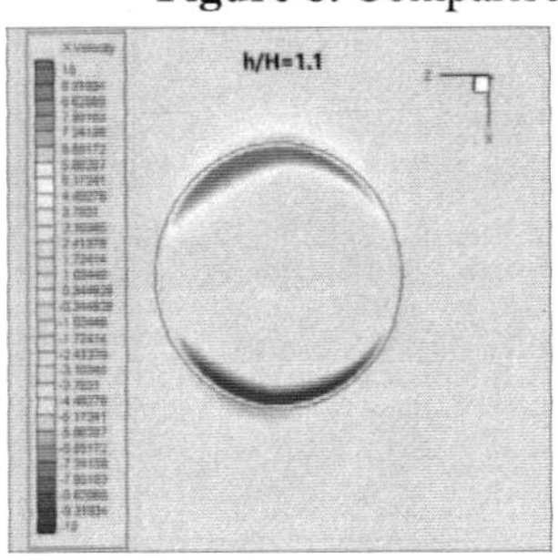

(a)X velocity(momentum-source).

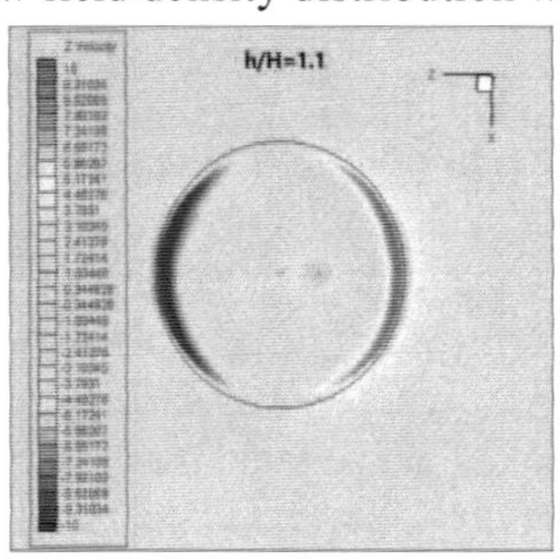

(c)Z velocity(momentum-source).

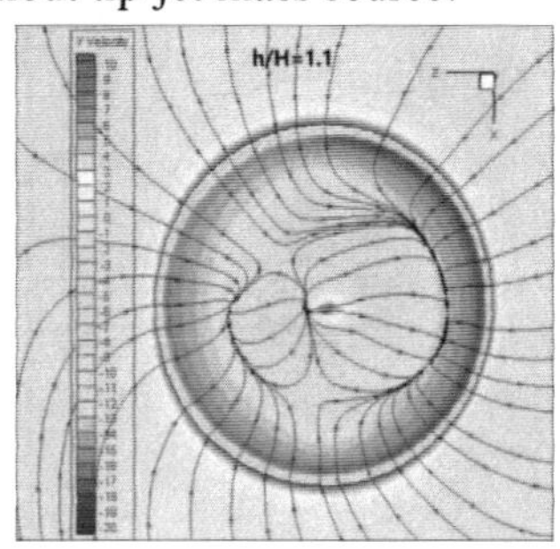

(e)Y velocity(momentum-source).

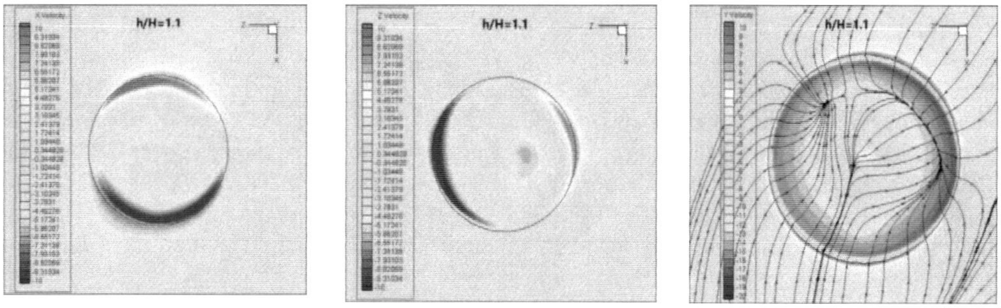

(b)X velocity(mass-source). (d)Z velocity(mass-source). (f) Y velocity (mass-source)

Figure 9. Comparison of velocity component and flow direction distribution of nozzle source with/ without tip-jet mass source.(h/H=1.1)

In Figure 10, the pressure distribution cloud diagram of fuselage surface in three different states (rotation only, jet only and rotation plus jet) are listed respectively. The four maps in each column represent the pressure distribution at different viewing angles of the fuselage in the same state. Comparing (a) rotation only and (b) jet only state, we can see the effect of the induction speed caused by the rotation. In (a), the red area indicates the high pressure area, and the blue area indicates the low pressure area. The high pressure area is mainly concentrated near the canard wing and near fuselage, the front part of the fuselage and the middle part of the fuselage slant back before the tail, and the distribution of left and right sides are asymmetric. Group (b) just simulates the effect of the tip jet. The induction speed generated in the direction perpendicular to the paddle is too small, causing no significant fluctuation of the pressure. The entire surface is in the same color block area set by the current cloud image, and the pressure change can only be seen when the numerical range of the cloud image is reduced. Group (c) simulates the combination of rotation and jet, the area of high pressure expand, and the value of the pressure increases.

In addition, it is not difficult to see that the place where the high pressure region appears is corresponding to the tip end region where the rotor down washing speed is the largest. Due to the maximum speed of the tip of the blade, the induced velocity is also correspondingly the largest, which results in the formation of a high-pressure region on the surface of the fuselage. It can also be seen from the figure that the disturbance of rotor down-wash flow on the pressure of CRW tail is very small.

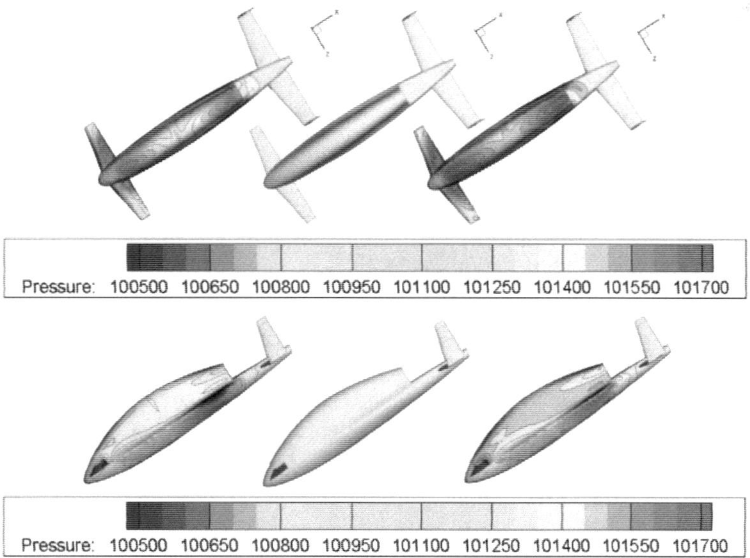

(a) Rotation only. (b) Jet only. (c) The combined of rotation and jet.

Figure 10. Comparison of pressure distribution on fuselage surface in different states.

IOP Conf. Series: Materials Science and Engineering **616** (2019) 012001 doi:10.1088/1757-899X/616/1/012001

In Figure 11 the pressure distribution of the flow field section in three different states (rotation only, jet only and rotation plus jet) are listed respectively. The cross sections shown are in the same spatial position and are located on the symmetrical plane of the fuselage. It can be seen from the figure that the disturbance of rotor downwash flow on the tail fin is indeed relatively small, which is consistent with the previous conclusion. The low pressure area near the paddle is mainly concentrated in the area near the tip of the propeller, and there are few low pressure areas in the central area of the paddle. Similarly, the high pressure area under the paddle plate is mainly concentrated at the tip, and the distribution law of the rotor-induced speed is reflected from the side, which is consistent with the actual situation.

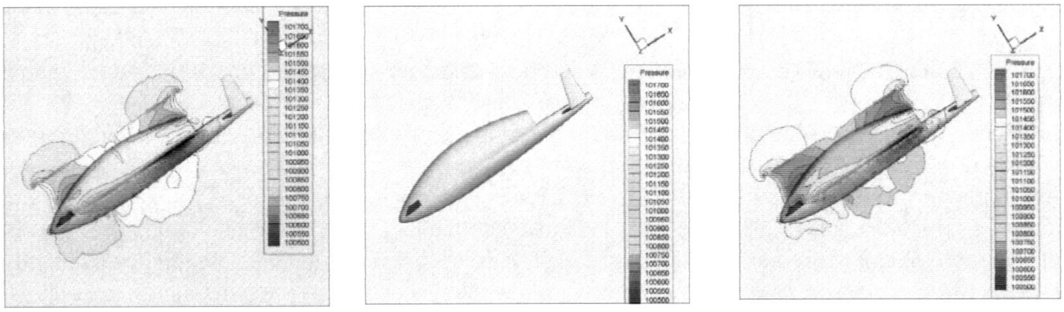

(a) Rotation only. (b) Jet only. (c)The combined of rotation and jet.

Figure 11. Comparison of pressure distribution on the symmetrical plan

7. Conclusion

Based on the momentum source principle, the aerodynamic characteristics of CRW are simulated by CFD method. Firstly, the CFD numerical simulation method is described, and the flow field simulation flow charts are given. Then, the flow field characteristics in the hovering state are given. It can be seen that the distribution trend generated by momentum source is basically reasonable. Finally, the density distribution, velocity distribution and streamlines comparison of the fuselage under the action of momentum source only and the interaction of momentum source and mass source are given.

References

[1] Clark A. Mitchell, Barbara J. Vogel. The Canard Rotor Wing(CRW) Aircraft-A New Way to Fly[R]. *AIAA* 2003-**2517**, 2003.

[2] J.W. Lee, K.S. Jeon, M.Kim, et al. Rotor Deisign for the Performance Optimization of Canard Rotor/Wing Aircraft[C], *The 2nd International Basic Research Conference on Rotorcraft Technology*, Nanjing, China, 2005:322-335.

[3] J.W. Rutherford, S.M. Bass, S.D. Larsen. Canard Rotor/Wing: A Revolutionary High-speed Rotorcraft Concept[R]. *AIAA* **93-1175**, 1993

[4] Rajagopalan R. G.. A procedure for rotor performance, flowfield and interference - A perspective[R]. *AIAA*-2000-116. 2000.

[5] M. S. Chaffin, J. D. Berry, S. D. Lsarsen. Navier-Stokes Simulation of a Rotor Using a Distributed Pressure Disk Method, *51st American Helicopter Society International Annual Forum* 1995. **112-136**

[6] Rajagopalan R. G, Mathur S. R. Three dimensional analysis of a rotor in forward flight. *AIAA* **89-1815**, 1989.

[7] Rajagopalan. R. G. A procedure for rotor performance, flowfield and interference- A perspective[R] *AIAA*-2000-**116**, 2000.

[8] Leishman J. G. Principles of helicopter aerodynamics. *Cambrideg Aerospace Series,* Cambridge University Press, 2000.

A Comparative Analysis of Different Turbulence Models for Simulating Complex Turbulent Separated Flows over Cubic Geometries

Nurgeldy Praliyev[1], Aydarkhan Sarsen[1], Nazira Kaishubayeva[1], Yong Zhao[1,*], Sai Cheong Fok[1] and Soo Lee Teh[1]

[1]Department of Mechanical and Aerospace Engineering, Nazarbayev University, Astana, Kazakhstan, 010000

[*]Corresponding author's E-mail: yong.zhao@nu.edu.kz

Abstract. This study is aimed to compare the performance of four Reynolds Averaged Navier Stokes (RANS) turbulence models in simulating complex flow over two- and three-dimensional (2D and 3D) cubic geometries using experimental measurements for model validation. The four turbulence models were k-Epsilon (k-ε), k-Omega (k-ω), Shear Stress Transport (SST) and BSL Reynolds Stress (RSM) models. The model validation was performed by comparative analysis with the experimental values of pressure coefficient around the cubic geometries. It was found that the SST and BSL RSM models performed better than the others in capturing the complex separated flows and strong shear in the boundary layer. The good agreement with empirical data can be attributed to the inclusion of transport effects of turbulence and the anisotropic nature in the formulations of the models.

1. Introduction

The fluid flow around bluff bodies is of great interest in engineering. Computational Fluid Dynamics (CFD) has become a useful tool in the study of wind around buildings and for the improvement of building design [1, 2, 3]. Poor wind engineering analysis of buildings may lead to issues during their construction and operation. For instance, in 2018, quite a few buildings in Astana (Kazakhstan) were damaged because of strong wind gusts, which reached the speed of 25 m/s. Later it was found that these building structures did not match local authority construction standards, according to which the environmental conditions of the local area must be considered [4]. Therefore, it is important to test buildings design under different environmental conditions (including strong wind, heavy rain etc.) using engineering tools, such as CFD, especially for areas such as Astana, where average wind speed exceeds 14 m/s, and occasionally reaches 23 m/s [5, 6]. According to the statistical study by Koppen [7], the majority of buildings in Astana are of rectangular and cube shapes. For this reason, it was decided in this study to investigate turbulent flow over a cubic block using different turbulence models, namely, Shear Stress Transport (SST), k-Epsilon (k-ε), k-Omega (k-ω) and BSL Reynolds Stress Model (RSM). The models were chosen since turbulent flow over a bluff body such as a cube has massive flow separation zones, which are difficult to predict accurately with the Reynolds-Averaged Navier-Stokes (RANS) turbulence simulation. In addition, the study of flow characteristics around bluff bodies such as cubes is always a significant topic in academic research as well as in engineering applications [8].

There are a number of studies on flow around cubes, though they are limited to experimental analysis. Certain flow parameters including static pressure coefficient have been measured over the cube surfaces in those empirical studies. One of the mostly cited paper in this area is the study of turbulent boundary layer flow over a cube by Castro et al. [9], who measured surface pressure coefficient of the cube at different Reynolds numbers. Authors concluded that the effects of Reynolds number (Re) existed for smooth laminar upstream flow, whereas for fully developed turbulent boundary layer the effect disappeared. Furthermore, no effects of Reynolds number have been found when Reynolds Number was greater than 4000.

Regarding the comparison of large eddy simulation for two- and three-dimensional computations for two-dimensional geometry, few papers have shown numerical results for the static pressure coefficient and compared it with that of the three-dimensional analysis for the same geometry. However, there is still doubt whether two-dimensional analysis is adequately representative of the flow in reality to replace three-dimensional analysis in order to avoid expensive three-dimensional computations and measurements. Murakami and Mochida [10] analysed unsteady flow over a two-dimensional square cylinder and compared the results of both 2D and 3D computations. Numerical results for 2D cube showed several discrepancies in comparison with those of 3D cube. Several researchers attributed the deviations to different boundary conditions (where 3D models allow for more realistic BCs), different meshing and boundary-layer development of turbulent flow. Nevertheless, it was noticed that differences in results can be significantly reduced by employing proper turbulent models. They concluded that the standard k-ε model fails in reproducing the vortex shedding, while the BLS RSM could do so rather well in comparison with other models.

Experimental findings by Lee, Y. T et al. [11] have been used by many researchers for validation of turbulence models for air flow over rectangular buildings with various aspect ratios and wind directions. It was found that the k-epsilon model results are in overall agreement with the experimental findings, but sometimes it does not capture certain flow features, for instance, in the regions close to the leading edge. In addition, the surface pressure variation is highly dependent upon the approaching wind direction, and the pressure coefficient is different at various positions on the top face of the three-dimensional rectangular geometry.

Most flows are three-dimensional, turbulent and transient, but it is very computationally expensive to conduct numerical studies on three-dimensional cases with the most realistic conditions. For that reason, the majority of previous studies were limited to two-dimensions [12]. For the current work, however, it was decided to conduct numerical analysis on both two- and three-dimensional cube geometries and make a brief comparison study. The main aim of the work was to compare four turbulence models, namely SST, k-ε, k-ω, BSL RSM and to identify the most accurate one based on its prediction of the flow around the cube both in 2D and 3D. The simulations were based on the experimental setup of Lee, Y. T et al. [10], in which a cube of 80x80x80 mm was tested in the wind tunnel of 320(h) x 560(w) x 1120(l) mm. More detailed information regarding the experimental setup and dimensions is given in the main body.

The current paper is divided into the following sections. First, the governing equations are discussed in the second section. Then, boundary and initial conditions of the numerical computation are presented in the third section. The fourth section covers information on the computational domain and the mesh verification study. Validation of the model and numerical results for the selected turbulence models are presented in the fifth section. The concluding remarks of the key findings are given in the final section.

2. Governing equations

The governing equations were derived based on the following assumptions: the air flow under study is statistically steady, incompressible and turbulent, for which the Reynolds-Averaged Navier-Stokes (RANS) equations are given below:

$$\frac{\partial}{\partial x_i}(\rho u_i) = 0 \qquad (1)$$

$$\frac{\partial}{\partial x_j}(\rho u_i u_j) = -\frac{\partial p}{\partial x_i} + \frac{\partial}{\partial x_j}(\tau_{ij} + {\tau_{ij}}^R) + S_i\,;\ i = 1,2 \tag{2}$$

with

$$\tau_{ij} = \mu\left(\frac{\partial u_i}{\partial x_j} + \frac{\partial u_j}{\partial x_i} - \frac{2}{3}\sigma_{ij}\frac{\partial u_k}{\partial x_k}\right) \tag{3}$$

The Boussinesq turbulent eddy viscosity model was employed for the two-equation turbulence models:

$${\tau_{ij}}^R = \mu_t\left(\frac{\partial u_i}{\partial x_j} + \frac{\partial u_j}{\partial x_i} - \frac{2}{3}\sigma_{ij}\frac{\partial u_k}{\partial x_k}\right) - \frac{2}{3}\rho k\delta_{ij} \tag{4}$$

k-ε equations

The Turbulence Kinetic Energy Equation is as follows:

$$\frac{\partial}{\partial x_i}(\rho u_i k) = \frac{\partial}{\partial x_i}\left(\left(\mu + \frac{\mu_t}{\sigma_k}\right)\frac{\partial k}{\partial x_i}\right) + S_k \tag{5}$$

The Specific Dissipation Rate Equation is given below:

$$\frac{\partial}{\partial x_i}(\rho u_i \varepsilon) = \frac{\partial}{\partial x_i}\left(\left(\mu + \frac{\mu_t}{\sigma_\varepsilon}\right)\frac{\partial e}{\partial x_i}\right) + S_e \tag{6}$$

where, the source terms were defined as

$$S_k = {\tau_{ij}}^R\frac{\partial u_i}{\partial x_j} - \rho\varepsilon + \mu_t P_B \tag{7}$$

$$S_\varepsilon = C_{\varepsilon 1}\frac{\varepsilon}{k}\left(f_1{\tau_{ij}}^R\frac{\partial u_i}{\partial x_j} + \mu_i C_B P_B\right) - C_{\varepsilon 2}f_2\frac{\rho\varepsilon^2}{k} \tag{8}$$

$$P_B = -\frac{g_i}{\sigma_B}\frac{1}{\rho}\frac{\partial \rho}{\partial x_i} \tag{9}$$

with $g_i = (g_x, g_y)$ the gravity vector, the constant σ_B=0.9 and constant C_B = 1 when $P_B > 0$, and 0 otherwise. Whereas

$$f_1 = 1 + \left(\frac{0.05}{f_\mu}\right)^2 \tag{10}$$

$$f_2 = 1 - e^{-{R_T}^2} \tag{11}$$

$$f_\mu = \left[1 - e^{-0.025R_y}\right]^2 * \left(1 + \frac{20.5}{R_T}\right) \tag{12}$$

$$R_T = \frac{\rho k^2}{\mu\varepsilon} \tag{13}$$

$$R_y = \frac{\rho\sqrt{k}y}{\mu} \tag{14}$$

The values of constants C_μ, $C_{\varepsilon 1}$, $C_{\varepsilon 2}$, σ_k, σ_ε were set as:
C_μ=0.09, $C_{\varepsilon 1}$=1.44, $C_{\varepsilon 2}$=1.92, σ_k=1, σ_ε=1.3

SST k-ω equations

The Turbulence Kinetic Energy is written as

$$\frac{\partial k}{\partial t} + U_j\frac{\partial k}{\partial x_j} = P_k - \beta^*\omega k + \frac{\partial}{\partial x_j}\left((v + \sigma_k v_T)\frac{\partial k}{\partial x_j}\right) \tag{15}$$

The Specific Dissipation Rate is

$$\frac{\partial(p\omega)}{\partial t} + \frac{\partial(\rho u_j\omega)}{\partial x_j} = \alpha S^2 - \beta^*\omega^2 + \frac{\partial}{\partial x_j}\left((v + \sigma_\omega v_T)\frac{\partial \omega}{\partial x_j}\right) + 2(1 - F_1)\frac{\sigma_{w2}}{\omega}\frac{\partial k}{\partial x_i}\frac{\partial w}{\partial x_i} \tag{16}$$

where the terms are

$$F_1 = \tanh\{\{\min\left[\max\left(\frac{\sqrt{k}}{\beta^*\omega y}, \frac{500v}{y^2\omega}\right), \frac{4\sigma_{\omega 2}k}{CD_{k\omega}y^2}\right]\}^4\} \quad (17)$$

$$F_2 = \tanh\left[\left[\max\left(\frac{2\sqrt{k}}{\beta^*\omega y}, \frac{500v}{y^2\omega}\right)\right]^2\right] \quad (18)$$

$$P_k = \min\left(\tau_{ij}\frac{\partial U_i}{\partial x_j}, 10\beta^* k\omega\right) \quad (19)$$

$$\mathrm{CDk\omega} = \max\left(2\rho\sigma_{w2}\frac{1}{\omega}\frac{\partial k}{\partial x_i}\frac{\partial \omega}{\partial x_i}, 10^{-10}\right) \quad (20)$$

$$\phi = \phi_1 F_1 + \phi_2(1 - F_1) \quad (21)$$

And the constant coefficients are

$$\alpha_1 = \frac{5}{9}, \ \alpha_2 = 0.44$$

$$\beta_1 = \frac{3}{40}, \ \beta_2 = 0.0828$$

$$\beta^* = \frac{9}{100}$$

$$\sigma_{k1} = 0.85, \sigma_{k2} = 1$$

$$\sigma_{\omega 1} = 0.5, \sigma_{\omega 2} = 0.856$$

BSL Reynolds Stress Model

The Turbulent Kinetic Energy equation is

$$\frac{\partial(pw)}{\partial t} + \partial(Ukpw) = \frac{\alpha_3 w}{k}P_k + P_{wb} - \beta \ \ p\omega^2 + \frac{\partial}{\partial x_j}\left(\left(\mu + \mu\frac{t}{\sigma_{w3}}\right)\frac{\partial w}{\partial x_j}\right) + \frac{(1-F)2p1}{\sigma_{w3}}\frac{\partial k}{\partial x}\frac{\partial w}{\partial x} \quad (22)$$

The constant coefficients are

$$\alpha_2 = 0.44$$

$$\beta_2 = 0.0828$$

$$\beta^* = \frac{9}{100}$$

$$\sigma_1 = 1, \sigma_2 = 0.856$$

$$F_1 = \tanh\{\{\min\left[\max\left(\frac{\sqrt{k}}{\beta^*\omega y}, \frac{500v}{y^2\omega}\right), \frac{4pk}{CD_{k\omega\sigma_{k-e}}y^2}\right]\}^4\} \quad (23)$$

$$CDk\omega = \max\left(2\rho\frac{1}{\omega}\frac{\partial k}{\partial x_i}\frac{\partial \omega}{\partial x_i}, 10^{-10}\right) \tag{24}$$

Reynolds stress Model equation are given below:

$$\frac{\partial}{\partial t}\left(\rho\overline{u_i u_j}\right) + \frac{\partial}{\partial x_k}\left(\rho u_k \overline{u_i' u_j'}\right) = -\frac{\partial}{\partial x_k}\left[\rho\overline{u_i' u_j' u_k'} + \overline{p'\left(\delta_{kj}u_i' + \delta_{ik}u_j'\right)}\right] + \frac{\partial}{\partial x_k}\left[\mu\frac{\partial}{\partial x_k}\left(\overline{u_i' u_j'}\right)\right] - \rho\left(\overline{u_i' u_k'}\frac{\partial u_j}{\partial x_k} + \overline{u_j' u_k'}\frac{\partial u_i}{\partial x_k}\right) + \overline{p'\left(\frac{\partial u_i'}{\partial x_j} + \frac{\partial u_j'}{\partial x_i}\right)} - 2\mu\overline{\frac{\partial u_i'}{\partial x_k}\frac{\partial u_j'}{\partial x_k}} - 2\rho\Omega_k\left(\overline{u_j' u_m'}\epsilon_{ikm} + \overline{u_i' u_m'}\epsilon_{jkm}\right) \tag{25}$$

3. Numerical model setup

Ansys CFX was employed in the numerical simulations. The finite volume method with second-order accuracy was used to solve all the governing equations. The numerical calculations were conducted on the cube with 80x80x80 mm dimensions placed within the wind tunnel of 1120 mm in length and 320 mm in height. The mean inlet velocity at cube height was calculated from the Reynolds number of 4.6×10^4, which resulted in 8.67 m/s. Air at 20° Celsius and 1 atm enters the wind tunnel through the inlet in the left and exists from the outlet as can be seen in Figure 1. Figure 1 also demonstrates the boundary conditions including inlet velocity and relative pressure.

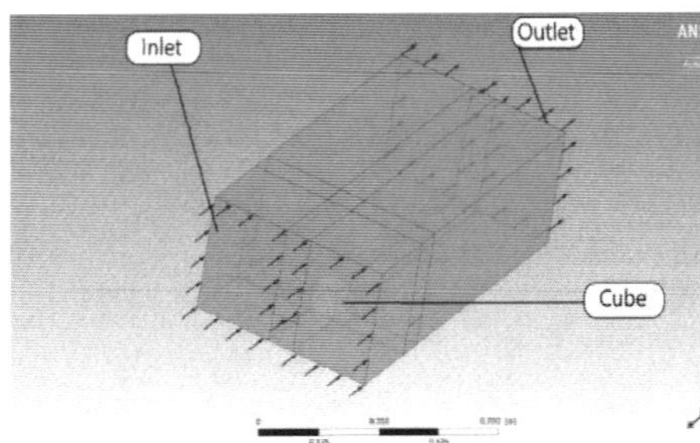

Figure 1. Boundaries and interfaces

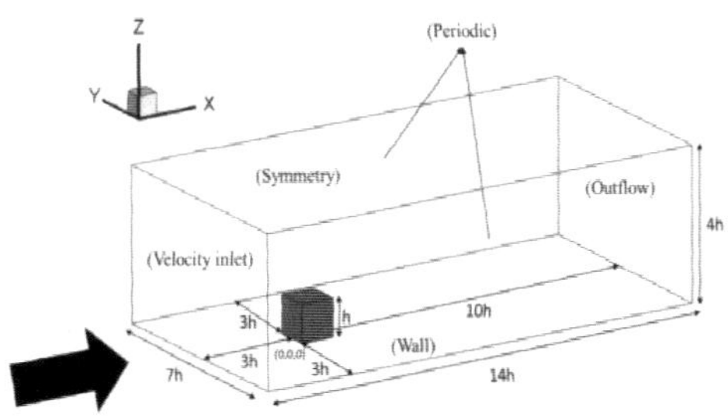

Figure 2. Computational domain and boundary conditions [11]

Figure 2 illustrates the experimental boundary conditions, the outflow (mass flow conservation for the outlet conditions) and symmetry conditions for the three-dimensional cube. Important setup specifications of the model are listed in Table 1.

Table 1. Setup specifications of the model [11]

Specification	Value
Distance from the inlet to the cube (left face)	240 mm
Distance from the cube to the outlet (right face)	800 mm
Cube height	80 mm

Inlet velocity	10 m/s
Outlet relative pressure	0 Pa
Reynolds number	46 000
Kinematic viscosity	$15x10^{-6}m^2/s$
Air density	$1.225\ g/m^3$

4. Mesh convergence study

The governing equations were discretized and solved using the finite volume method. Meshes were created using sweep meshing and inflation layers. The mesh convergence study was conducted using progressively denser meshes until the difference in the reattachment point of less than 1% was reached.

4.1. Y-plus

The y-plus value is a dimensionless distance from the wall. In turbulence modelling, it is an important parameter for Low Reynolds-Number (LRN) turbulence models [13]. Here the y-plus value was set below 3 in order to obtain accurate results in resolving the turbulent boundary layer. The initial mesh near to cube front face contained 700,012 nodes and 717,016 elements, and the y-plus value was around 7. Further mesh refinement was focused at the front face of the cube and the inflation method allowed to reduce the y-plus value from 7 to nearly 3. Finally, the elements size around the geometry were further reduced to reach the expected value. The final maximum y-value value was 1.4 at the front-edge of the cube. The resultant mesh contained 1,097,137 nodes and 1,031,610 elements. The result for y-plus is shown in Figure 3.

4.2. Mesh verification

Mesh verification analysis was conducted with 50% growth in the element number. As was mentioned earlier, less than 1% difference in reattachment point downstream of the cube was reached during the mesh verification process. Table 2 contains results of the mesh verification in several iterations with their errors and reattachment points.

Figure 3. Y-plus along the cube surfaces

Table 2. Mesh verification

Iteration number	Reattachment points (m)	Error (%)	Elements	Nodes
1	1.84622	-	1,031,610	1,097,137
2	1.87646	1.61	2,470,177	2,588,366
3	1.89292	0.87	2,718,309	2,846,333
4	1.89471	0.09	3,309,530	3,453,597

From Table 2, it can be seen that the mesh convergence was achieved in the third iteration, when the relative error in the reattachment point was only 0.87%. Therefore, the mesh with 2,846,333 nodes and 2,718,309 elements was used for the further computational analysis. The resultant mesh around the cube is illustrated in Figure 4.

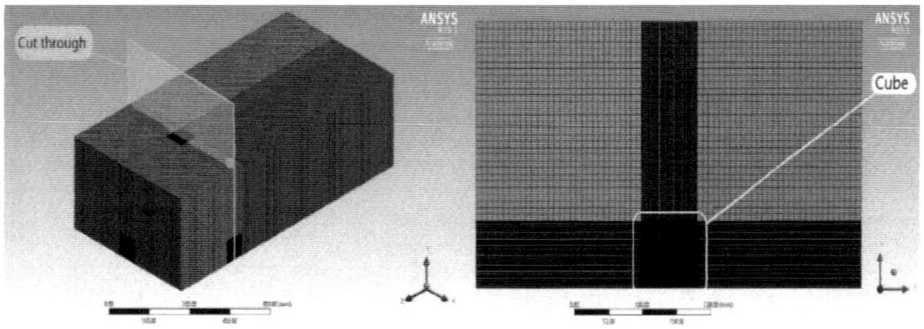

Figure 4. The resultant mesh and mesh density around the cube

5. Validation of the models

For the validation of the models, the experimental data and numerical results were compared. The comparison is done by using pressure coefficient values. To calculate the pressure coefficient the following definition was used:

$$C_p = \frac{P - P_r}{0.5 * \rho * U_h^2} \tag{26}$$

Where U_h is the free stream velocity and P_r is the mean static pressure upstream of the body. Obtained results were plotted in Figure 5, in which the horizontal axis represents the dimensionless length of the cube. The graph is divided in to three sections, which are 0-1, 1-2 and 2-3 and represent the front, top, and back sides of the cube, respectively. Figure 5 also shows the flow direction and the 0-1, 1-2, 2-3 flow paths.

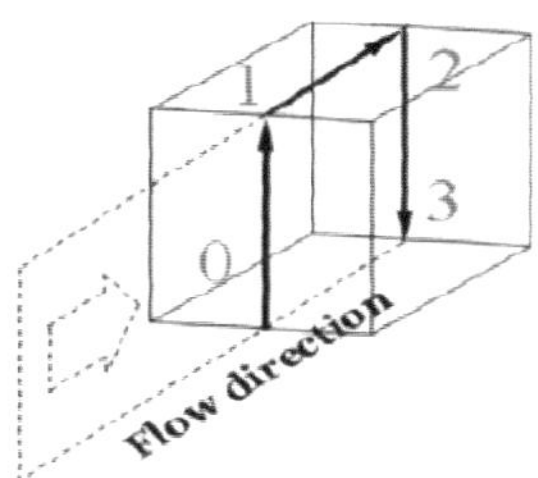

Figure 5. Flow over the central section of the cube [11]

The numerical results agree quite well for the front surface of the cube (0-1), where all the turbulence models show nearly the same results. According to the experimental data for the top surface of the cube, the pressure first decreases after the separation point and increases through the left length. However, as the flow reaches the top surface significant deviations between numerical and experimental results occur. As can be observed in Figure 6, the k-ω model has the least accurate data for the top surface. The k-ε model is more accurate than k-ω, but still has a noticeable difference. The SST model shows a more accurate trend, but the BSL Reynolds model shows the best numerical results for 1-2 region. According to empirical data, the pressure distribution on the back side of the cube should remain constant with slight increase in the pressure coefficient. For this region The BSL Reynolds model shows similar behaviour for the beginning of the region 2-3 but starts deviating towards the end. The k-ε and k-ω models show a similar trend, but the k-ε model's values were higher than the experimental data, while k-ω produced lower values. The SST model results lie between k-ε and k-ω and are closer to the experimental data than the other models. Here the SST model shows better performance because the model uses the combination of the k-ε and k-ω models. Overall, for the

IOP Conf. Series: Materials Science and Engineering **616** (2019) 012002 doi:10.1088/1757-899X/616/1/012002

back side of the cube all the models show certain deviation, which might be due to the complex fluid flow that includes vortices, separation, and reattachment of the flow, and the viscosity effects. To sum up, for the 3-D case the BSL Reynolds and SST models show accurate results in comparison with the k-ε and k-ω models.

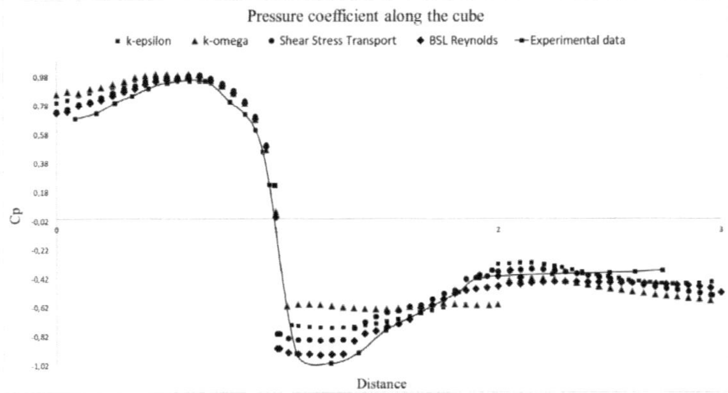

Figure 6. Pressure distribution along the cube surface

The 2-D case of the flow around the cube was also studied in this work. The numerical results for the 2-D case were generated using the same boundary conditions and dimensions and were compared with the 3-D case. Figure 7 shows the comparison, in which the surface pressure distribution is also divided into three sections that are the front, top, and back side of the cube. For the front side the 3-D numerical analysis show more accurate results, while the 2-D numerical values remained constant along the distance. Moreover, at the top surface, the results deviate nearly by factor of 3. The back side of the cube also shows large variations. Overall, the trends of the curves for both cases are similar, but the actual values differ considerably. The reason is that in the 2-D case the air flows in one direction only – along the top surface of the cube, while in the 3-D case the flow also tends to travel around the sides of the cube as shown in Figure . According to [14] for the 3-D finite shaped geometries, the flow will be developed steadily over all sides and Bernoulli's equation can be used to determine the pressure. However, in the 2-D case at the top surface flow separation starts suddenly and the flow recirculation in the wake region is much stronger, where viscous effects become more significant, thus resulting in very steep decrease and increase in pressure and much lower pressure level there than the 3D case.

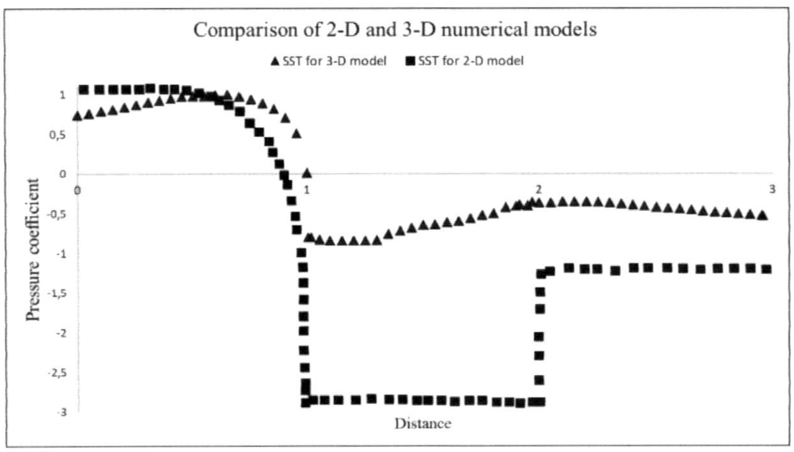

Figure 7. Pressure coefficient around the 2-D and 3-D cubes

Finally, according to the obtained numerical results, the SST and BSL Reynolds models can predict the flow more accurately than the others, but in general, the BSL Reynolds model has the least discrepancy. Here the pressure and velocity contours, and streamlines have been plotted for this model (Figures 8-9). Figure (a) and (b) show the pressure and velocity contours around the cube from the top. There is stagnation at the front surface of the cube, where pressure increases but velocity approaches zero. Moreover, Figure (b) shows flow separation around the sides of the cube and formation of the wake region. The vortices formed in the wake region can be observed in Figure . From the streamlines, it can be noticed that after the stagnation point the fluid starts accelerating, which causes recirculation of the flow and creation of vortices. A clearer view of the recirculating flow is shown from the bottom side of the cube in Figure 8.

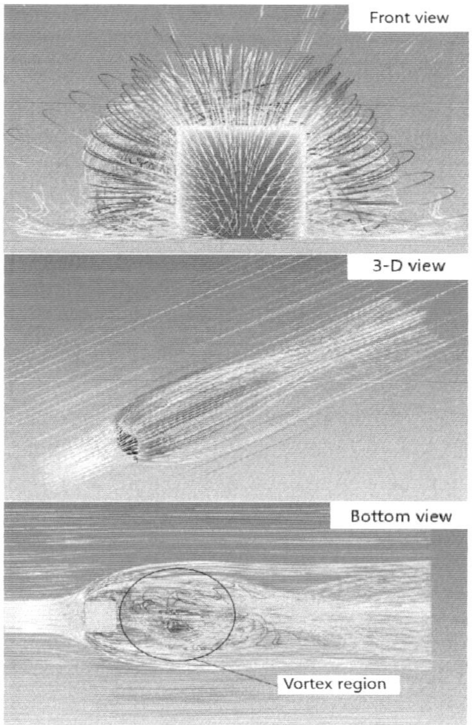

Figure 8. Streamlines of the flow around the cube in the 3D simulations

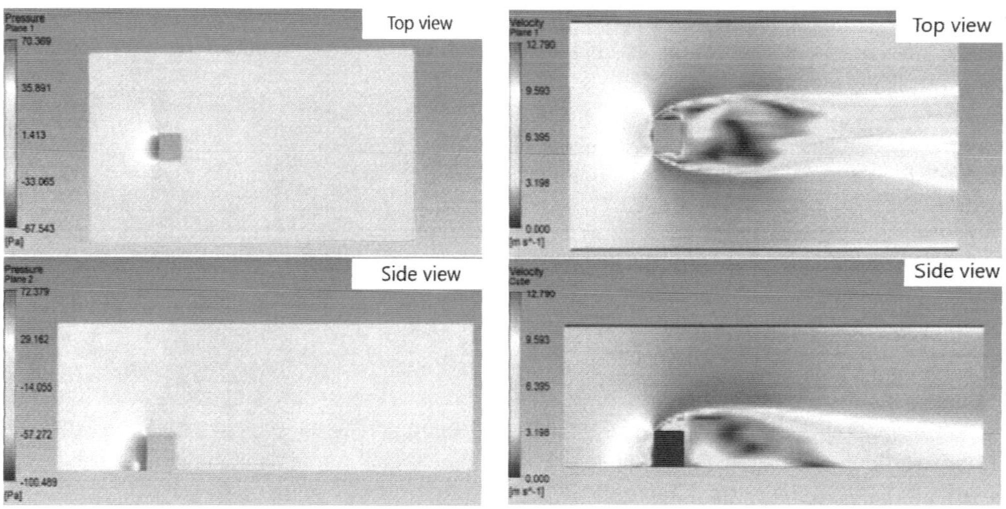

Figure 9. (a) pressure contours | (b) velocity contours

6. Conclusion

Numerical simulations were performed on the turbulent flow around 3-D cube geometry by using the Ansys CFX software. The main aim of the study was to compare four turbulence models, namely k-ε, k-ω, SST, and RSM models and to identify the most efficient turbulence model that accurately describes the flow around the cube. During the investigation, the values of the pressure coefficient (Cp) on the front, top, and back sides of the cube were obtained and compared with experimental data.

Numerical analysis demonstrates that the SST and RSM turbulent models produce more accurate results for separated flows with circulating vortices, than the k-ε and k-ω models. Furthermore, the RSM has less deviation from the empirical data than the SST model. However, it necessary to consider the fact that the RSM needs to solve seven equations, while SST solves only two, which affects the cost of the computations. Therefore, if there is not enough computational power available, SST model can be used instead of the RSM. For the 2-D and 3-D cases the pressure coefficient values were quite different. It was because the actual flow is fully 3D and turbulent with flow separations around the cube.

References

[1] Heidarzadeh, H., Farhadi, M., & Sedighi, K. (2012). Convective heat transfer over a wall mounted cube using large eddy simulation. CFD Letters, 4(2), 80-92.

[2] Lim, H. C., Thomas, T. G., & Castro, I. P. (2009). Flow around a cube in a turbulent boundary layer: LES and experiment. Journal of Wind Engineering and Industrial Aerodynamics, 97(2), 96-109.

[3] Lim., H. C., Castro, I.P., (2006). Turbulent Boundary Layer Flow Over a Cube. European Research Community on Flow, Turbulence and Combustion, Ercoftact databases. Main page.

[4] Kiislicina, Yu. (2018, January 11). High speed hurricane damaged buildings including Universities in Astana. Radio Sputnik, Retrieved from https://ria.ru/world/20180111/1512424874.html

[5] Oliver, K. (2018, March 5). Wind statics for Astana including 2018 year. Windfinde, Retrieved from https://www.windfinder.com/windstatistics/astana

[6] Zhumanova, D. (2018, January 11). Critical weather condition after strong hurricane. Commercial television channel KTK, Retrieved from https://www.ktk.kz/ru/news/video/2018/01/11/87774

[7] Köppen, B. (2013). The production of a new Eurasian capital on the Kazakh steppe: architecture, urban design, and identity in Astana. Nationalities Papers, 41(4), 590-605.

[8] Abohela, I., Hamza, N., & Dudek, S. (2013). Effect of roof shape, wind direction, building height and urban configuration on the energy yield and positioning of roof mounted wind turbines. Renewable Energy, 50, 1106-1118.

[9] Castro, I.P. & Robins, A.G. (1977). The flow around a surface mounted cube in uniform and turbulent streams. J. Fluid Mech., 79, 307-355.

[10] Murakami, S., & Mochida, A. (1995). On turbulent vortex shedding flow past 2D square cylinder predicted by CFD. Journal of Wind Engineering and Industrial Aerodynamics, 54, 191-211.

[11] Lee, Y. T., Boo, S. I., Lim, H. C., & Misutani, K. (2016). Pressure distribution on rectangular buildings with changes in aspect ratio and wind direction. Wind and Structures, 23(5), 465-483.

[12] Fimbres-Weihs, G. A., & Wiley, D. E. (2010). Review of 3D CFD modeling of flow and mass transfer in narrow spacer-filled channels in membrane modules. Chemical Engineering and Processing: Process Intensification, 49(7), 759-781.

[13] Ariff, M., Salim, S. M., & Cheah, S. C. (2009, December). Wall y+ approach for dealing with turbulent flow over a surface mounted cube: Part 1—Low Reynolds number. In Seventh International Conference on CFD in the Minerals and Process Industries (pp. 1-6).

[14] Gao, Y. and Chow, W.K., 2005. Numerical studies on air flow around a cube. Journal of Wind Engineering and Industrial Aerodynamics, 93(2), pp.115-135.

IOP Conf. Series: Materials Science and Engineering **616** (2019) 012004 doi:10.1088/1757-899X/616/1/012004

Research on Machining Fixture of Near-Net-Shaped Jet Engine Blade

Dongbo Wu [1], Yunhai Chen [2], Xuesong Zhen[2*] and Kaiyao Zhang [3]

[1] State Key Laboratory of Tribology, Beijing Key Lab of Precision/Ultra-Precision Manufacturing Equipment and Control, Department of Mechanical Engineering, Tsinghua University, Beijing, China 100084, P.R. China.
[2] BEIJING POLYTECHNIC, Beijing, China 100176, P.R. China.
[3] Yantai University. Yantai, Shandong 264005, P. R.China

[*]Corresponding author's e-mail: 100754@bpi.edu.cn

Abstract. This paper focus on the design and verification of the machining fixture on the background of the adaptive processing technology. Firstly, the positioning and clamping scheme of near-net-shaped jet engine blades are introduced and analyzed, especially the new materials (Polyetheretherketone (PEEK-GF30)) and new structure are applied to fixture, and corresponding fixture is designed, manufactured and applied to engineering. Secondly, the relationship between the blade clamping deformation and clamping force is analyzed by finite element method (FEA). Thirdly, blade cutting deformation is analyzed by cutting experiment. The results show that this new material and structure fixture combined with CNC adaptive processing technology can achieve the precision machining of near-net-shaped jet engine blades Tenon root.

1. Introduction

Blades are the key components of the contemporary aero-engine, and structure design and machining quality of the blades play a vital role in the engine performance [1, 2]. The new generation of aero-engine blade is usually manufactured by precision forging process. Quality of the blade surface by precision forging is guaranteed by the forging die, which doesn't need the secondary processing after forming. However, leading edge and tailing edge can not be formed by precision forging [3]. Because of forging deformation, clamping deformation and so on, leading edge and tailing edge of the blades can not be manufactured according to the theoretical model, and the CNC machining procedures can not be directly shared in the mass production. In order to solve the problems of connection department between the actual surface of the blade and leading edge and tailing edge of the theoretical model, forming of leading edge and tailing edge of the blades, and NC machining program translation in mass production of precision forging blades and so on [4].

A new generation of aero-engine blades uses precision forging to make blanks. Precision forging is a new process with little cutting-free machining developed on the basis of ordinary die forging [5]. The precision forging process ensures the continuity of the metal inside the blade, increases the strength and load carrying capacity of the blade, and improves the performance and life of the blade. The quality of the blade surface of precision forging is guaranteed by the forging die, and the forming precision is high. The shape of the blade body after forming does not need secondary processing [6].

However, the variable radius of the blade is not smooth and the forging edge cannot be formed by precision forging. Must be reprocessed [7].

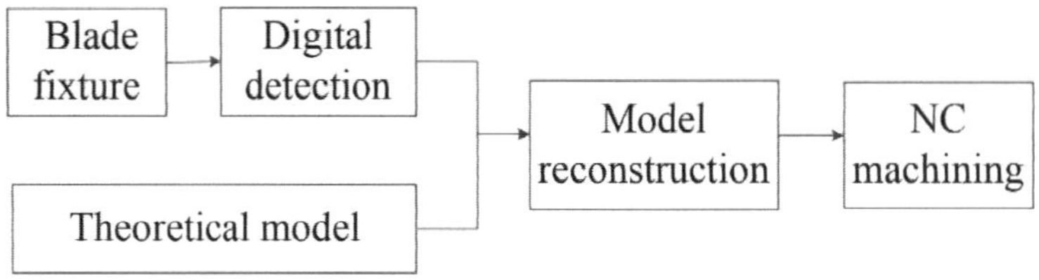

Figure 1. The flow chart of adaptive NC machining technology

Adaptive CNC machining technology is integration of digital detection, digital detection model reconstruction, and NC machining. And the flow chart is shown in Figure 1. The adaptive NC machining technology [8, 9] is based on the design model of the parts to be machined and the corresponding machining program. According to the measured results of the machining area, the orientation change and shape deviation of the part model are obtained, and the actual machining is adaptively generated on the basis of the nominal machining program [10, 11]. The processing code of the area and guarantees a smooth transition to the results of the previous process. This technology is a system engineering technology, which not only covers CNC machining programming technology and surface modeling technology in CAD/CAM [12], but also involves several key technologies such as digital detection, reverse engineering, workpiece clamping positioning, etc. An important part of technological research and development [13, 14].

2. Description of positioning and clamping scheme

In structure, there are nine positioning (or assistant positioning) heads to contact with the body of blade, which is a typical multi-point positioning scheme. Specifically, there are 3 auxiliary positioning points based on the original 6-point positioning of the fixture (see Figure 2).

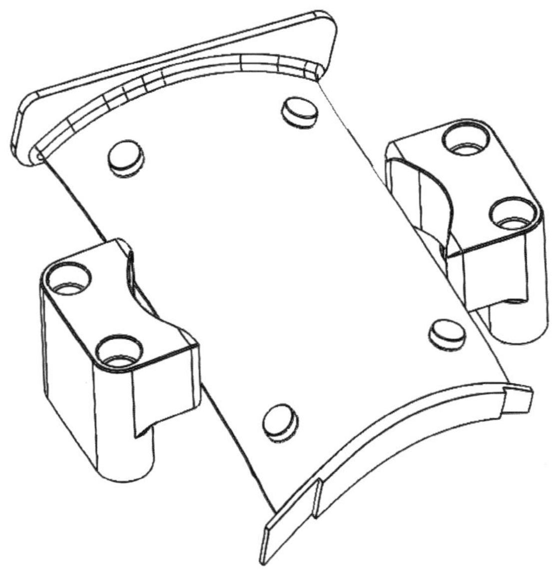

Figure 2. Positioning and clamping scheme of blade

In the fixture, the main structural components are made of steel, and the clamping heads and positioning heads are made of Polyetheretherketone (PEEK-GF30) material, and the main material of blade is Ti-6Al-4V, and the material properties of blade and fixture components are shown in Table 1.

Table 1 Material properties of blade and fixture components

Material	Young's modulus	Poisson ratio	Density
Ti-6Al-4V	110 GPa	0.36	4.40 kg/m^3
PEEK-GF30	7 GPa	0.38	1.51 kg/m^3
STEEL	220 GPa	0.3	7.80 kg/m^3

3. Analysis of clamping force and deformation

Figure 3 is influence curve of single clamping force on deformation, and the other clamping forces are 500N in this case. It can be seen from Figure 3 (a) that deformation maximum gradually decreases with the increase of F1. Until F1 is more than 800N, deformation maximum will increase with the increase of F1. It can be seen from Figure. 3 (b) that deformation maximum gradually decreases with the increase of F2. It can be seen from Figure 3 (c) that deformation maximum gradually decreases with the increase of F3. Until F3 is more than 500N, deformation maximum will increase with the increase of F3. It can be seen from Figure 3 (d) that deformation maximum gradually increases with the increase of F4. When F4 is less than 500N, the increase rate is relatively flat, and when the force is greater than 500N, the increase rate is obvious.

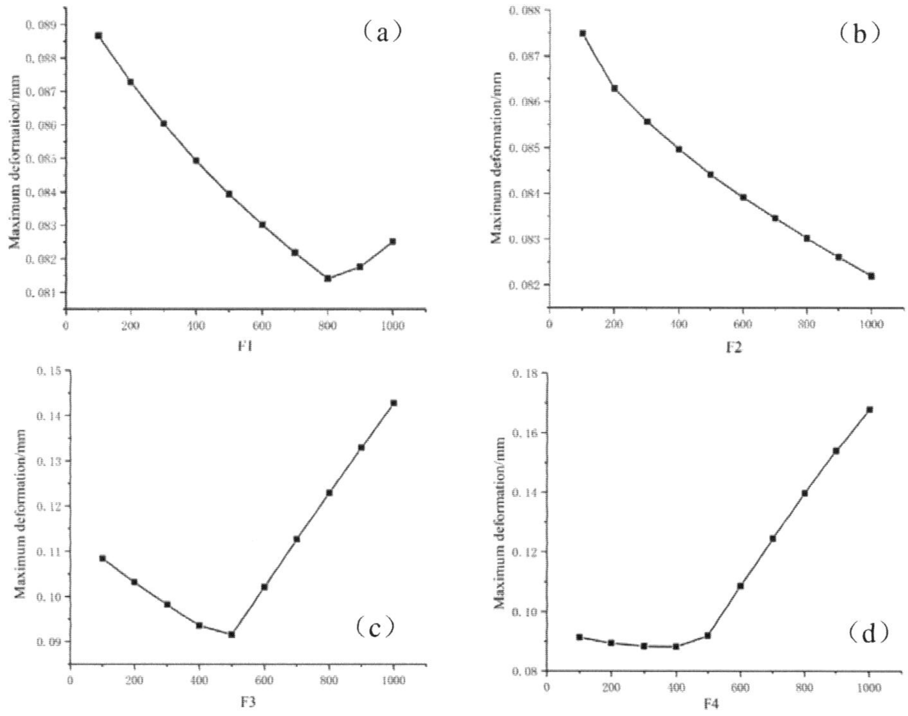

Figure 3. Influence curve of single clamping force on deformation. a Influence curve of F1 on deformation maximum; b. Influence curve of F2 on deformation maximum; c. Influence curve of F3 on deformation maximum; d. Influence curve of F4 on deformation maximum

Experiment is arranged in order to verify the correctness of finite element calculation results. In this experiment, the clamping force is measured by film force sensor, and maximum deformation is obtained by electrical vortex sensor, and corresponding curves between the maximum deformation and clamping force is shown in Figure 4. It can be seen from Figure 4 that the finite element calculation results are basically consistent with the experimental test results when the clamping force is greater than 400N, and the deviation between the finite element calculation result and the experimental test result is relatively large when the clamping force is less than 400N. This is because the contact area is changing with the changing of clamping force.

Blade is directly contacted by the positioning element. Firstly, the blade profile error is large because of blade complex surface characteristic and precision forging processing. Secondly, in order to make the clamping and positioning elements effectively contact to blade surface, positioning and clamping device must be fitted blade surface shape. However, the positioning and clamping elements and blade surface contact surface area are not complete because of the machining process error of actual clamping element.

Therefore, the contact process is not in accordance with the ideal surface contact because of the positioning elements, clamping elements and blade machining errors. Experiment is arranged in order to identify the change of contact areas. A layer of oil was applied to the surface of the blade to observe the change of oil coating area after clamping force in this experiment. It can be seen from Figure 4, the contact area is as an "8" word area when clamping force is 500N, and the contact area is a "0" word area when the clamping force is 1000N. So it can be concluded that the positioning and clamping elements and blade surface contact surface area gradually increases with the increase of clamping force, contact area is starting point contact, after the "8" word area contact, to the final "0" word area contact.

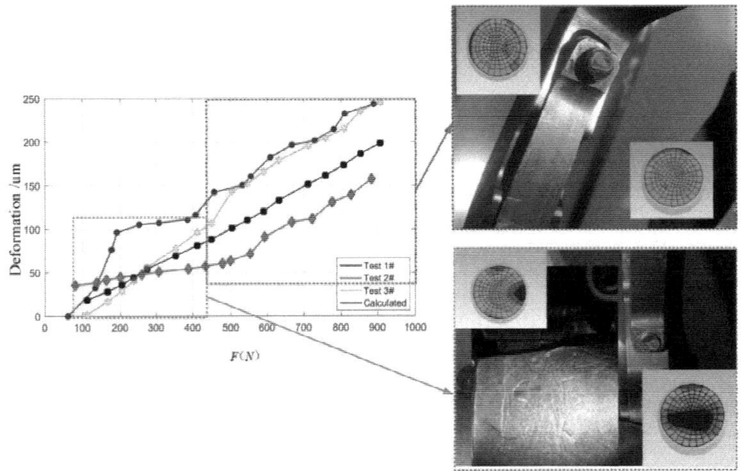

Figure 4. Results comparison of calculation and experiment

Therefore, it is necessary to reset the contact area according to the different contact area, which leads to the decrease of the calculation precision of finite element. The results of finite element calculation are acceptable considering the influence of contact area, which also indicates that the finite element model established above is reliable.

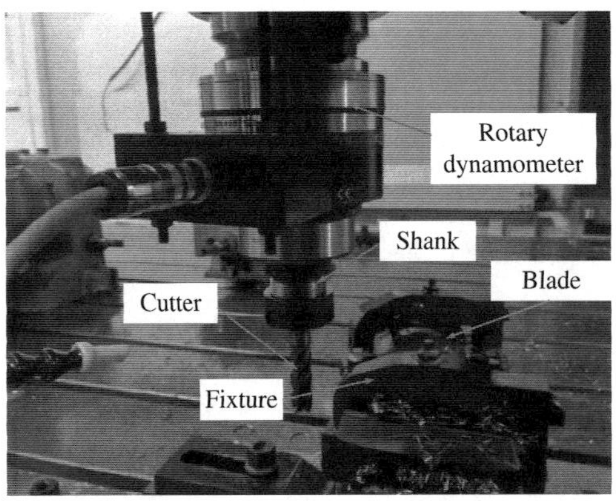

Figure 5. Cutting process

IOP Conf. Series: Materials Science and Engineering 616 (2019) 012004 doi:10.1088/1757-899X/616/1/012004

4. Analysis of cutting experiment and cutting deformation

Cutting experiment, displacement response measurement and the application of fixture is carried out, and the test setup is shown in Figure 5. The cutting process parameters are shown in Table 2.

Table 2. Machining parameters

	Rotation speed	Feed rate	Cutting depth
Roughing milling	800 rpm	0.3mm/min	5mm
Finishing milling	1200 rpm	0.2mm per tooth	0.2mm

As shown in Figure 6, it is the change of the blade cutting force RMS of the roughing and finishing under the corresponding process parameters. It can be seen that the roughing cutting force is much larger than the finishing cutting force, because the roughing mainly improves the blade cutting efficiency. Cutting efficiency is dominant, so a large amount of cutting material removal is required, resulting in a large depth of cut. Therefore, the cutting force is large.

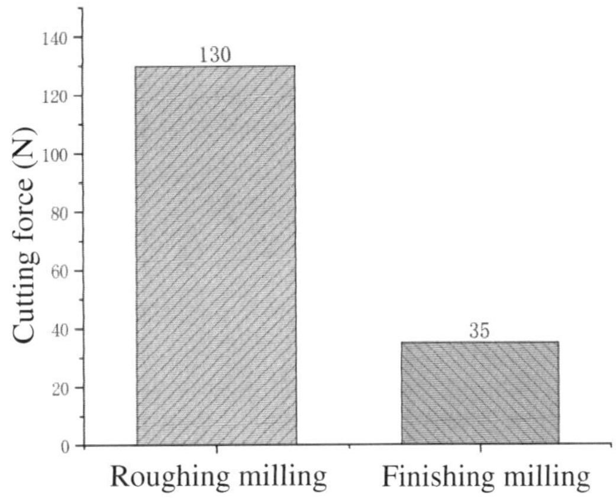

Figure 6. Cutting force test result

As shown in Figure 7, it is the change of the cutting deformation of the roughing and finishing blades under the corresponding process parameters. It can be seen that the cutting deformation of the roughing blades is much larger than the finishing cutting deformation, because the roughing cutting force is much larger than the finished cutting force

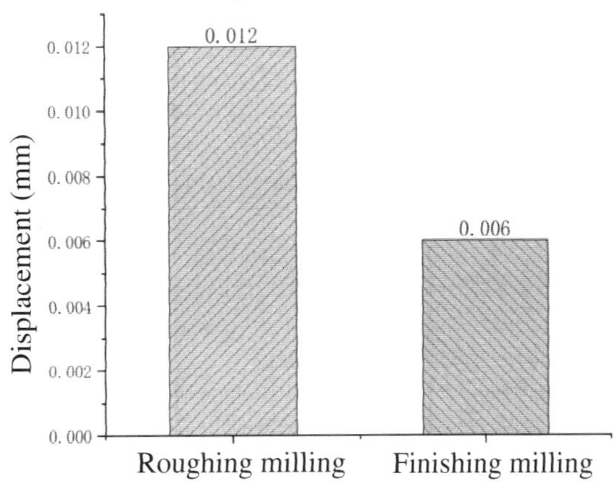

Figure 7. Displacement result of blade

However, these cutting deformations are acceptable for comparison with machining accuracy requirements, and the machining deformation is small. This also shows that the proposed fixture meets the process requirements and can be used in the manufacture of blades in the context of adaptive CNC machining technology.

IOP Conf. Series: Materials Science and Engineering **616** (2019) 012004 doi:10.1088/1757-899X/616/1/012004

5. Conclusion

In this work, machining fixture in adaptive CNC machining process of near-net-shaped thin-walled blade were investigated by theoretical analysis, FEA and engineering experiment test. The results can be summarized as follows:

1) The results of FEA are acceptable considering the influence of contact area, which also indicates that the established FEM is reliable.
2) The proposed fixture and process can realize high-precision blade manufacturing, and can be used in the industrial production of blades

Acknowledgements

The author would like to acknowledge the support and contributions of our colleagues in Xi'an Aero-Engine (Group) Ltd. This research is supported in part by Xi'an Aero-Engine (Group) Ltd.

References

[1] Lin X, Wu D, Yang B, Wu G, Shan X, Xiao Q, Hu L, Yu J (2017) Research on the mechanism of milling surface waviness formation in thin-walled blades. The International Journal of Advanced Manufacturing Technology 93 (5-8):2459-2470. doi:10.1007/s00170-017-0669-3

[2] Hui W, Baohai W, Xiaoqiang L (2014) Advanced machining technology of new generation commercial aeroengine blade[J]. Aeronautical Manufacturing Technology (20):26~31 (in Chinese)

[3] Lin X, Wu D, Shan X, Wu G, Cui T, Zhang Y, Hu L, Yu J (2018) Study on flexible CNC polishing process and surface integrity of blade. Journal of Mechanical Science and Technology 32(6):2735-2746

[4] XJ L, WH W, CW S (2009) Research on the new manufacturing process of aeroengine blade. Aviation Precision Manufacturing Technology (45(05)):262-274

[5] Marini D, Cunningham D, Corney JR (2018) Near net shape manufacturing of metal: A review of approaches and their evolutions. Proc IMechE Part B: J Engineering Manufacture Vol. 232(4) (650–669):650–669

[6] Katz R, Srivatsan V, Patil L (2011) Closed-loop machining cell for turbine blades. The International Journal of Advanced Manufacturing Technology 55 (9):869-881. doi:10.1007/s00170-010-3138-9

[7] Zhang Z, Zhang D, Luo M, Wu B (2016) Research of Machining Vibration Restraint Method for Compressor Blade. Procedia CIRP 56:133-136. doi:10.1016/j.procir.2016.10.042

[8] Lin X, Yue C, Wang Z, Yan G, Yuan G, Zhang X (2015) Front and rear edge model reconstruction of precision forged blades for adaptive machining. Journal of Aeronautics 36 (05):1695-1703

[9] Zhang D, Zhang Y, Wu B, Li S, Zhang S (2008) Application of Adaptive Processing Technology in blisk Manufacturing. Aeronautical Manufacturing Technology 13:51-55

[10] Hui W, Zhou M, Jie Y, Wang W, Huang B, Rong Y (2016) Numerical modeling and analysis of positioning error in aero-engine blade conformal machining. Computer Integrated Manufacturing System 22 (09):2118-2126

[11] D LY, H GP (2004) Free-form surface inspection techniques state of the art review[J]. Computer-Aided Design 36(13):1395-1417

[12] H XZ, Y WM, X . A LZ (2003) A computer-aided probing strategy for workpiece localization. IEEE International Conference on Robotics & Automation (3941-3946)

[13] Gao J, Chen X, Yilmaz O, Gindy N (2008) An integrated adaptive repair solution for complex aerospace components through geometry reconstruction. The International Journal of Advanced Manufacturing Technology 36 (11):1170-1179. doi:10.1007/s00170-006-0923-6

[14] A EH, A B, K KG (2004) Integrated inspection and machining for maximum conformance to design tolerances [J]. CIRP Annals - Manufacturing Technology 53(1):411-416

IOP Conf. Series: Materials Science and Engineering **616** (2019) 012005 doi:10.1088/1757-899X/616/1/012005

Numerical Investigations on High Lift Force Generation of 3D Dragonfly Wing during Hovering Motions

Xiaohui Su[1], Kaixuan Zhang[1], Juan Zheng[1], Jiantao Zhang[1], Yong Zhao[2]

[1]School of Hydraulic Engineering, Dalian University of Technology, Dalian 116024, China

[2]College of Engineering, Nazarbayev University, 53 Kabanbaybatyr Ave., Astana, 010000, Republic of Kazakhstan

[*]Corresponding author e-mail: zkx@mail.dlut.edu.cn

Abstract. In the paper, a novel numerical method is present to numerically investigate life force generation mechanisms of 3D dragonfly wing during hovering motions. The dynamic mesh which descripts the motion of 3D wing and gird change in inner computational domain is generated by an ALE model we developed. Then the mesh is transferred at each time step into ANSYS CFX using Junction Box Routine. The simulation results show that a high lift force generated by implementing one specified advanced flapping action during hovering motion in which the translation time period of downstroke is shortened.

1. Introduction

The dragonfly flight mode existing in nature has long been the inspiration source for human beings to design Micro Air Vehicles (MAVs) which would be widely used in the field of national defense and civil field. Great progress literature reviews [1-4] summarized the aerodynamics of insect flight and the lift generation mechanism of rigid wing.

In this paper, the aerodynamic performance and energy consumption of dragonfly's wings during hovering flight are simulated with the geometrical model used in reference [5] which is simplified based on a real geometric parameters of Aeschna juncea hindwings [6,7] in 1972.

Moreover, a new rotation mode only by changing the beginning time of the down stroke reversal is present and the aerodynamic performance of the new rotation motion is numerically investigated.

2. Geometry of Hindwing and Kinematics

2.1 Geometry of dragonfly hindwing

The hindwing model is described as a rigid plate with a span length of R, 4.6 *cm*, average chord length of c, 1.12 *cm*, and a thickness of 1% of average chord length, about 0.0112 *cm*. The symmetric plane is located at 0.61 *cm* away from the wing root, and the rotation axis of the Angle of Attack (AOA) is 0.28 *cm* away from the leading edge c/4. The simplified geometric model of the dragonfly's hindwing is plot in Figure 1.

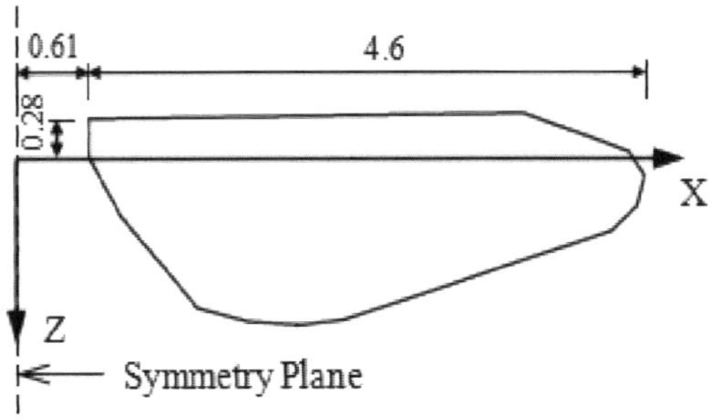

Figure 1. Simplified geometric model of the dragonfly hindwing

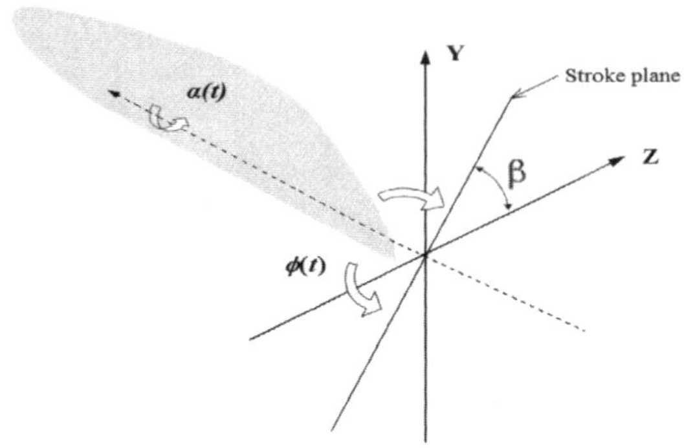

Figure 2. The sketch of the flapping motion of wing

2.2 The traditional kinematics of flapping motion

The flapping function of dragonfly wing during hovering flight is the same as the formula used in reference [5], as shown in Figure 2. It assumes that flapping plane and horizontal plane have an inclined angle β, 52°, deviation angle, 0°. Thus, only two parameters, flapping angle and AOA, control the movement of dragonfly wing. Flapping frequency is set to 36 *Hz*.

The flapping angle, $\Phi(t)$ and AOA, $\alpha(t)$ are described in Eqs. (1) and (2).

$$\varphi(t) = \varphi_0[\cos(2\pi ft) - 1] \tag{1}$$

where $\varphi_0 = 0.602$.

During the reversal of the stroke, the change of the angle of attack with time is described as follows:

$$\alpha(t) = 0.5\dot{\alpha}_0\left[(t - t_r) - \frac{\Delta t_r}{2\pi}\sin(2\pi\frac{t - t_r}{\Delta t_r}) + \frac{\Delta t_r}{2}\right] t_r \le t \le t_r + \Delta t_r \tag{2}$$

where, $\dot{\alpha}_0$ =20000 degree/s, Δt_r is the duration of wing flip and t_r is the time of wing flip.

The profiles of flapping angle, angle of attack and their respective angular velocity in one cycle are shown in Figure 3.

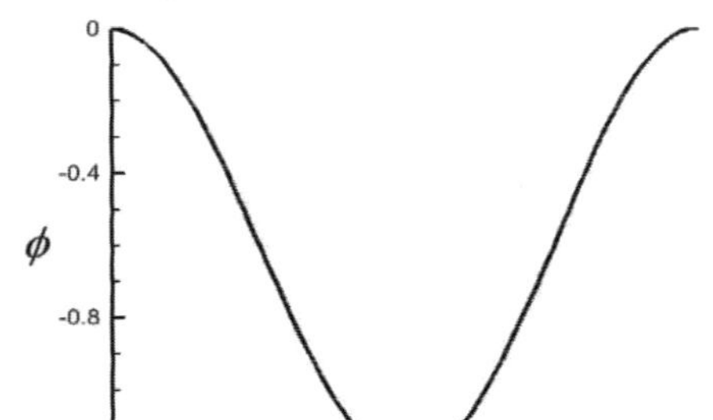

(a) Flapping angle

(b) Attack angle

Figure 3. The flapping angle and attack angle in one period

2.3 The model setup

The ANSYS CFX is used to solve the NS equations. In the model, air density $\rho = 1.185$ kg/m^3 and dynamic viscosity $\mu = 1.831\times10^{-5}\ kg\cdot m^{-1}\cdot s^{-1}$ are set in a still air. Transient analysis calculates 30 cycles to ensure the final solution in one period to attain convergence status. Time step is set to $5.56\times10^{-6}\ s$. The far field and moving boundary conditions are selected opening and non-slip wall boundary condition respectively.

2.4 Mesh convergence test

There are four meshes used to verify the convergence. The total mesh cells of the model are: 35566, 68497, 113614 and 233150 which can be found in Table 1.

Table 1. Mesh information

Mesh model	Wing surface mesh	Total mesh	Node point
Mesh1	1,024	35,566	6,566
Mesh2	1,024	68,497	12,058
Mesh3	2,786	113,614	19,995
Mesh4	6,084	233,150	40,746

Four monitoring points were picked up along the edge of the wing model shown in Figure 4. The instantaneous pressure values of those monitoring points are shown in Figure 5.

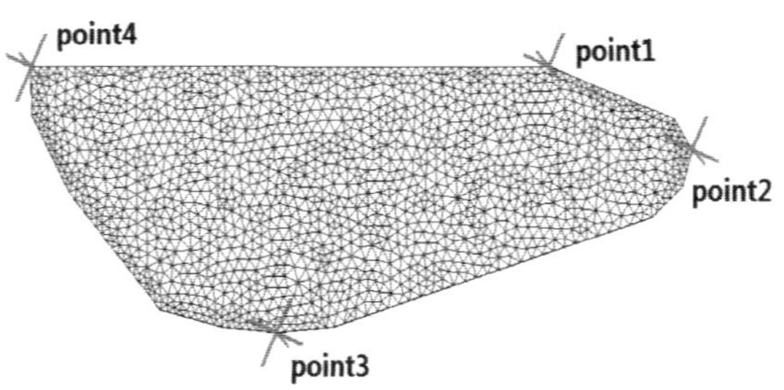

Figure 4. The position of the monitoring points on the wing model

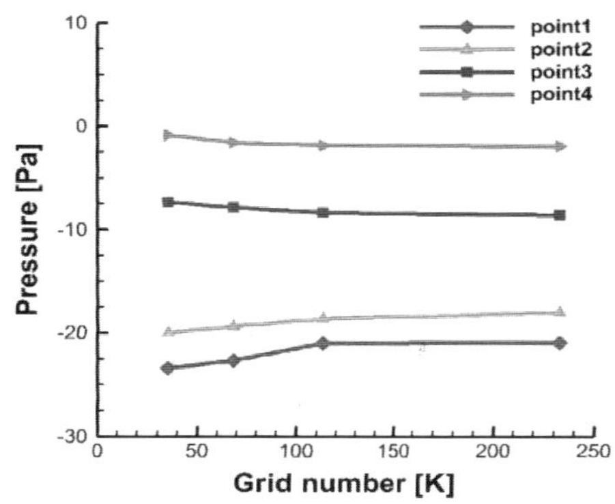

Figure 5. The instantaneous pressure values of the monitoring points under different grids

Figure 5 shows that with the increasing of mesh number, the instantaneous pressure values of the four monitoring points gradually trend to stable values. When the mesh amount increases to the third and the fourth series of meshes, the relative error of the instantaneous pressure values is below 3.24%. Therefore, the mesh could be considered reaching the convergence. Taking into account for the requirements of computational accuracy and time cost, the third set of grids is used as the computational mesh for subsequent studies.

2.5 Aerodynamic performance

In the study of insect flapping flight, the lift force coefficient, Cl and thrust coefficient, CT are most important to characterize the aerodynamic force. In the paper, the Cl denotes the aerodynamic component perpendicular to horizontal plane pointing to Y-Axis for lifting the weight of dragonfly while the CT represents the aerodynamic component orthogonal to the Cl direct to the Z-Axis. The Cl and CT are obtained by Eqs. (3) and (4) as followed.

$$CL = \frac{F_Y}{0.5\rho U^2 S_h} \tag{3}$$

$$CT = \frac{F_Z}{0.5\rho U^2 S_h} \tag{4}$$

where, $U = 4.0724$ *m/s*, is the dimensionless speed. S_h is the hindwing area, 5.152 cm^2.

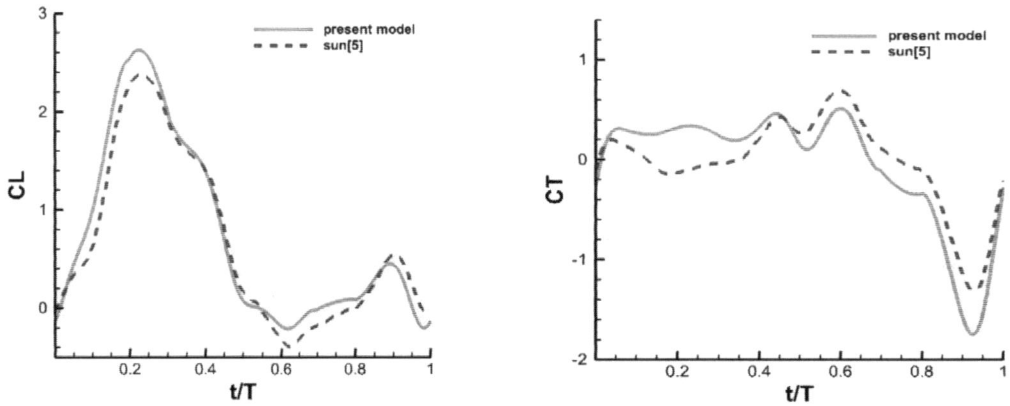

Figure 6. The profiles of CL and CT in one hovering motion cycle.

The Cl and CT are plot in Figure 6 and the results obtained by [5] are also plot for the comparison purpose. Figure 2.6 shows that the simulation results of the present model are approximately agreement with those in [5]. The period-averaged Cl of the present model during one period is 0.736 while the period-averaged Cl of reference [5] is 0.675. The relative error is about 9%.

3. The novel advanced model

In the paper, we present a novel advanced model in which the action of rapid pitch starts advanced before the traditional time moment, meaning the translation time of wing in downstroke becomes shorten and the translation time in upstroke becomes longer (seen in Figure 7). The advanced times in one flapping period, such as 2%, 4%, 6% and 8%, by comparing with the symmetry model, explored the effects on the aerodynamic characteristics of dragonfly.

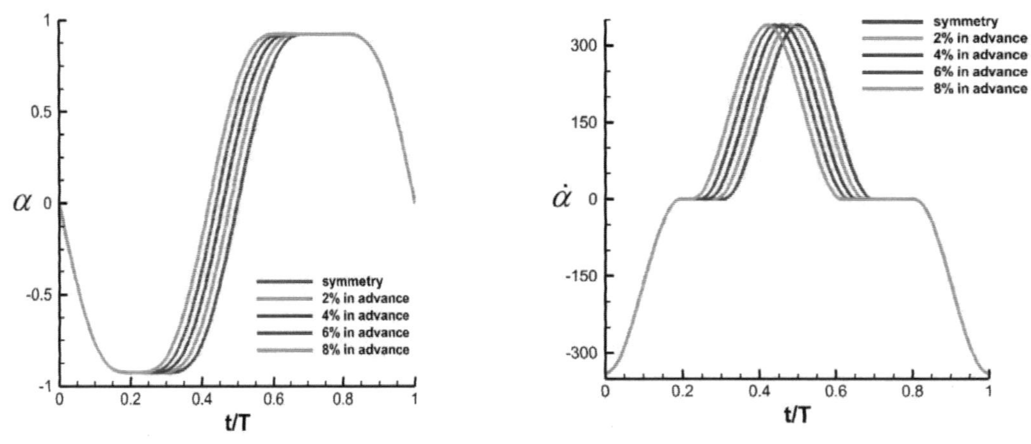

(a) attack angle (b) attack angle velocity

Figure 7. Angle of attack and angular velocity rates of four advanced rotations

3.1 Results of advanced models

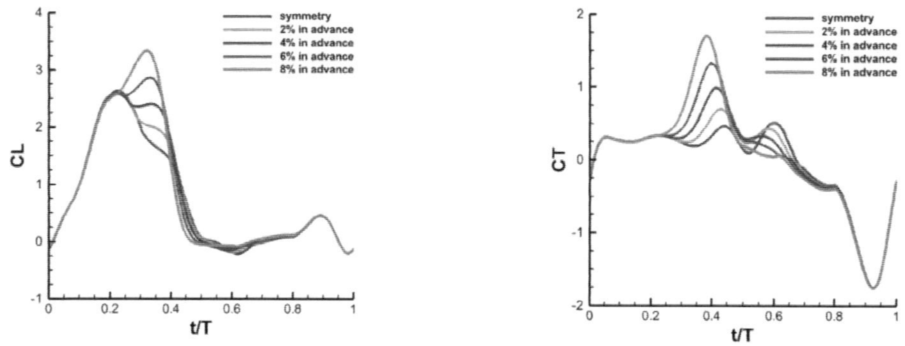

(a) the vertical force coefficients (b) the thrust force coefficients

Figure 8. Time courses of CL and CT for four advanced models

Figure 8(a) shows the profiles of CL for the four advanced models while Figure 8(b) shows the profiles of CT. The significant change happens during the time t/T from 0.22 to 0.41. CL has a dramatically increasing when the earlier advanced action happens. The period-averaged CL even increases 9% for 8% advanced model compared with no advanced mode. CT also increases during those time range which means resistance becomes larger in order to complete the advanced motion.

3.2 Energy consumption

Table 2. Time-averaged aerodynamic parameters for four advanced models

model	$\overline{CL}$	$\overline{CT}$	$\overline{CP}$	M
Symmetry	0.7374	-0.0593	0.196	1.328
2%	0.7642	-0.0368	0.214	1.242
4%	0.7946	-0.0124	0.242	1.128
6%	0.8240	0.0134	0.275	1.015
8%	0.8522	0.0401	0.312	0.916

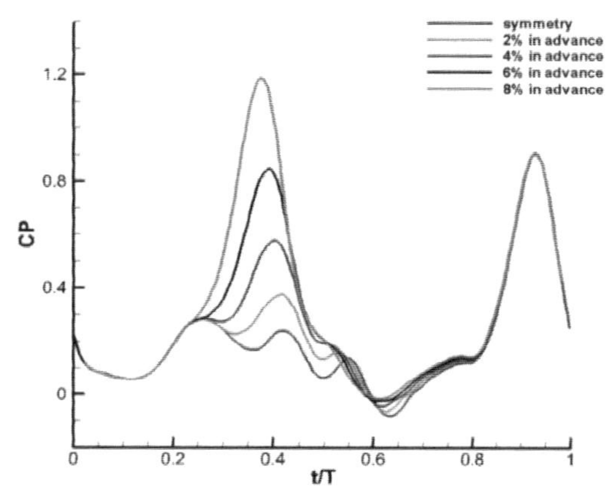

Figure 9. The profile of power coefficients Cps for four advanced models.

The aerodynamic parameters, including period-averaged CL, CT, power coefficient, CP, as well as figure of merit, M are calculated and summarized in table 2. And Figure 9 shows energy consumption of wing at one flapping period, which is calculated by the following formulation:

$$CP = \int_0^1 \oint_A (\vec{U}(t) \cdot \vec{n}(t))\, p(t) dA(t) d(t) \tag{5}$$

Moreover, the figure-of-merit M, is calculated by using the theory of Granlund, Ol, and Bernal [8] which is:

$$M = \frac{\overline{Cl}^{2/3}}{2P(2A_0 / c)^{1/2}} \tag{6}$$

Table 2 shows the time average CP increases with increasing the advanced time. The maximum value reaches to 0.3118 for 8% that means although the advanced model could improve lift force, the increasing of energy consumption has to be suffered.

4. Conclusion

In the paper, a novel numerical method is present to numerically investigate life force generation mechanisms of 3D dragonfly wing during hovering motions. A high lift force generated by implementing one specified advanced flapping action during hovering motion in which the translation time period of downstroke is shorten. The aerodynamic performance of different advanced models are numerically investigated compared with the values of the traditional symmetric model. The period-averaged CL increases 9% for 8% advanced model compared with no advanced mode. Although the advanced model could improve lift force, the increasing of energy consumption has to be suffered.

Acknowledgments

This research is supported by The National Natural Science Foundation, project No: 11672059. The financial support is gratefully acknowledged.

References

[1] Sane S P. The aerodynamics of insect flight[J]. J Exp Biol, 2003, 206(23):4191-208.

[2] Rozhdestvensky K V, Ryzhov V A. Aerohydrodynamics of flapping-wing propulsors[J]. Progress in Aerospace Sciences, 2003, 39(8):585-633.

[3] Lehmann F O. The mechanisms of lift enhancement in insect flight[J]. Die Naturwissenschaften, 2004, 91(3):101-22.

[4] Wang Z J. Dissecting insect flight[J]. Annual Review of Fluid Mechanics, 2005, 37(1):183-210.

[5] Sun M, Lan S L. A computational study of the aerodynamic forces and power requirements of dragonfly (Aeschna juncea) hovering[J]. J Exp Biol, 2004, 207:1887-1901.

[6] Norberg, R. A. The pterostigma of insect wings and inertial regulator of wing pitch[J]. Journal of Comparative Physiology, 1972, 81, 9-22.

[7] Norberg, R. A. Hovering flight of the dragonfly Aeschna juncea L., kinematics and aerodynamics[J]. In Swimming and Flying in Nature. 1975, pp. 763-781.

[8] Granlund, K., Ol, M., and Bernal, L. Quasi-Steady Response of Free-to-Pivot Flat Plates in Hover," Journal of Fluids and Structures, Vol. 40,July 2013, pp. 337–355.

IOP Conf. Series: Materials Science and Engineering **616** (2019) 012006 doi:10.1088/1757-899X/616/1/012006

Application of Arbitrary Vortex Method in Wind Tunnel Axial Fan Design

Xu Dachuan, Wang Haifeng, Qu Xiaoli, Yu Yongsheng, Shi Yu and Zhao Fang

Facility Design and Instrument Institute, China Aerodynamics Research and Development Center, No.6 southern second ring road, Mianyang 621000, China

xudc_ustc@hotmail.com

Abstract. The flow axial velocity is assumed equal everywhere in free vortex design method, and the pressure rise of flow pass through fan is assumed also equal at different radial height. But in arbitrary vortex design method, on the assumption that axial velocity along radial is variable, the flow rotational velocity along radial is not must an inverse distribution. Therefore, arbitrary vortex design method could have a more uniform axial velocity at the outlet of fan system by regulating the rotational velocity distribution. In the numerical simulation, the fan of a pilot wind tunnel is used as physical model. By comparison with the experimental data the reliability of the numerical simulation could be verified. Two fan physical models are established which are based on the designs of arbitrary vortex method and free vortex method. The result of numerical simulation indicated that compared to the free vortex method, the arbitrary vortex fan minished the separation on the tip and root of blade, obtained a more well uniform axial velocity distribution at the outlet of fan system. And the efficiency of blade is also increased by arbitrary vortex method.

1. Introduction

The energy which keeps air flow running in low speed wind tunnel is provided by fan. The energy loss of air flow in wind tunnel is in the shape of pressure drop. The function of fan is increasing the pressure of air flow. The power of large low speed wind tunnel was usually very high, which required the efficiency of fan was also very high; otherwise huge energy loss would be produced. And the low efficiency of fan indicated that the working condition of the fan was not good, which may generated big noise and the operation of wind tunnel would be affected. So as one of the core sections in wind tunnel, the performance of fan was very important. The fan design technique which could obtain high aero efficiency and low noise was one of the most important techniques in high quality wind tunnel design.

The most widely used fan design method was free vortex method which was based on 2D blade element theory in wind tunnel axial fan design. The assumptions of free vortex method were that the axial velocities of flow were equal everywhere and the pressure rise of flow passed through fan was also equal at different radial height. The calculation process of free vortex method was simple, and the method itself was verified by many engineering projects.

But in arbitrary vortex method, there was no such assumption. On the assumption that axial velocity along radial was variable, the flow rotational velocity along radial was not must an inverse distribution, and the pressure rise of air flow passed through the fan was not uniform. In arbitrary vortex design process the axial velocity distribution of incoming flow could be obtained by the measuring result of pilot facility or similar equipment. The rotational velocity distribution could be pilot calculated and optimized which depended on the experience of designers. When the axial velocity distribution of incoming flow was decided, different rotational velocity distributions corresponded to different axial velocity distributions of fan outlet. Therefore, there would always be a specific rotational velocity distribution which could meet the design requirement, when the uniformity of axial velocity distribution at fan outlet was one of the major objectives [1-2].

The blade aerodynamic load also would be change with different rotational velocity distribution. So some specific rotational velocity distribution could be selected to optimize blade aerodynamic load. In a pilot wind tunnel, for example, it was always required high aerodynamic efficiency of fan and low aero noise. Therefore, one of the basic design principles was avoiding blade stall, which was contradictory with raising efficiency. Arbitrary vortex method could adjust blade aerodynamic load according to the pilot calculation and performance of blade airfoil profile, so blade stall might be avoid meanwhile obtaining high efficiency. Arbitrary vortex method could predict blade load distribution along the blade radial direction more precise especially when the wind tunnel had numerous of operating conditions.

Compared to free vortex method, more complex calculation was a defect of arbitrary vortex method. If the uniformity of axial velocity distribution at fan outlet or the blade local stall was not the focus in wind tunnel design, free vortex method was used in most case. But when these were taken into account, arbitrary vortex method was recommended.

2. Calculation of arbitrary vortex method

Contrary to the free vortex design method arbitrary vortex method does not require radial equilibrium. Like the free vortex design method, it is assumed that the flow may be treated in concentric cylinders around the centre axis, whereby the contributions of the relatively small radial components may be neglected. It is further assumed that the flow leaves the stator without tangential components, hence should be essentially swirl free downstream of the fan system [3].

The figure 1 is the sketch map of fan system in wind tunnel. Position 1 is upstream of rotor, position 2 is between rotor and stator, position 3 is downstream of stator. Taking an infinitesimal circular passage, the radius is r, and dr is the width. According to the assumption of no radial flow, the air flow in infinitesimal circular passage does not mix with neighbouring flow.

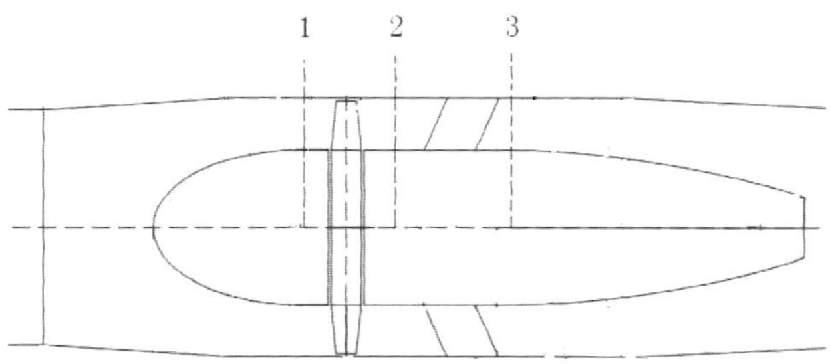

Figure 1. Sketch may of fan system

The total head equations in all three positions can then be written as follows:

$$H_1 = p_1 + \frac{1}{2}\rho u_1^2 \tag{1}$$

$$H_2 = p_2 + \frac{1}{2}\rho u_2^2 + \frac{1}{2}\rho(\omega_s r)^2 \tag{2}$$

$$H_3 = p_3 + \frac{1}{2}\rho u_3^2 \tag{3}$$

The affix s refers to the position between rotor and stator. The power applied to the shaft for that part of the flow may be expressed as (Q torque):

$$dQ = \rho u_1^2 2\pi r dr \omega_s r^2 \tag{4}$$

$$\Delta h_{th2} 2\pi r dr u_1 = \Omega dQ \tag{5}$$

Δh_{th2} is theoretical total pressure rise, and can be inferred as:

$$\Delta h_{th2} = H_2 - H_1 = \rho \Omega \omega_s r^2 \tag{6}$$

For the following calculations dimensionless quantities are introduced such as the local induced tangential rotation speed ε s, the local advance ratio λ and the local dimensionless pressure ratio k. Substitution in equation (6) gives

$$\varepsilon_s = \frac{\omega r}{u}, \quad \lambda = \frac{u}{\Omega r}, \quad k = \frac{\Delta h}{\frac{1}{2}\rho u^2}$$

$$k_{th2} = \frac{\Delta h_{th2}}{\frac{1}{2}\rho u_1^2} = \frac{2\varepsilon_{s2}}{\lambda_2} \tag{7}$$

This formula forms the basis for design of axial fans. For a free vortex design, for which $\omega_s r^2$ = constant. In case of an arbitrary vortex design it is customary to start with an estimated linear distribution of ε s with the radius r [4]. The value of ε s at the "mean" radius can be fixed with expression (7). In order to be able to calculate the mean velocity field in the fan plane, like in the case of a free vortex design, it is necessary to first calculate the velocity distribution downstream of the rotor. Equations (1), (2) and (6) contain 4 unknowns, however, so that for p2 an extra relation needs to be found. Wallis states that experience has taught that already close downstream of the rotor a new radial equilibrium develops so that the following expression may be applied [1].

$$\frac{dp_2}{dr} = \frac{\rho(\omega_s r)^2}{r} \tag{8}$$

To find the axial velocity distribution the following steps are now applied. Equations (1), (2) and (6) give:

$$\Omega \rho \omega_s r^2 = p_2 - p_1 + \frac{1}{2}\rho(u_2^2 - u_1^2) + \frac{1}{2}\rho(\omega_s r)^2 \tag{9}$$

After dividing these terms by the mean entrance velocity U and introduction of the mean dimensionless quantitie Fs λ_s and ε_s equation (25) may be rewritten as follows:

$$\left(\frac{u_2}{U}\right)^2 - \left(\frac{u_1}{U}\right)^2 = \frac{2\varepsilon_s}{\lambda_s} - \varepsilon_s^2 - \frac{p_2 - p_1}{\frac{1}{2}U^2} \tag{10}$$

To avoid a laborious trial and error calculation of p_2, Wallis states that in practice most velocity distributions may be approximated by a linear distribution with radius [1]. It is further recommended to select the radius for which the local axial velocity is equal to the mean velocity [5]. It can easily be shown that this radius is a function of the ratio x_b between the minimum (boss) and the maximum (house) radius of the annulus. The expression reads:

$$x_U = \frac{2(1 + x_b + x_b^2)}{3(1 + x_b)} \tag{11}$$

For this position p_2 may be calculated directly from equation (10) and forms thereby the integration limit for solving equation (8).

$$p_2 - p_{2xU} = \int_{xU}^{x} \frac{\rho(\omega_s r)^2 dx}{x} \tag{12}$$

Finally the axial velocity distribution after rotor can be inferred as follows:

$$\left(\frac{u_2}{U}\right)^2 = \left(\frac{u_1}{U}\right)^2 + \left|\frac{2\varepsilon_s}{\lambda_s} - (\varepsilon_s)^2\right| - \left|\frac{2\varepsilon_s}{\lambda_s} - (\varepsilon_s)^2\right|_{xU} - 2\int_{xU}^{x} \frac{(\varepsilon_s)^2 dx}{x} \tag{13}$$

Likewise, a similar expression can be derived for the velocity downstream of the stator vanes.

$$\left(\frac{u_3}{U}\right)^2 = \left(\frac{u_2}{U}\right)^2 + (\varepsilon_s)^2 - \left|(\varepsilon_s)^2\right|_{xU} + 2\int_{xU}^{x} \frac{(\varepsilon_s)^2 dx}{x} \tag{14}$$

3. Numerical simulation

Two fan physical models are established which are based on the designs of arbitrary vortex method and free vortex method. Fan A is a free vortex fan, fan B is an arbitrary vortex fan.

3.1. Physical models

Figure 2 shows the model of fan A. There are 10 blades and 7 stator vanes in fan A. The blade airfoil profile is Gö797. Installation angle at root is 44.2°. The chord length at root is 126mm, and 86mm at tip. The stator vane airfoil profile is NACA4415. Installation angle at root is 85.1°. The chord length at root is 66.5mm, and 110.9mm at tip.

Figure 3 shows the model of fan B. There are also 10 blades and 7 stator vanes in fan A. The blade airfoil profile is Gö797. Installation angle at root is 38.1°. The chord length at root is 126mm, and 86mm at tip. The stator vane airfoil profile is NACA4415. Installation angle at root is 81.3°. The chord length is 450mm from root to tip.

Figure 2. Model of fan A

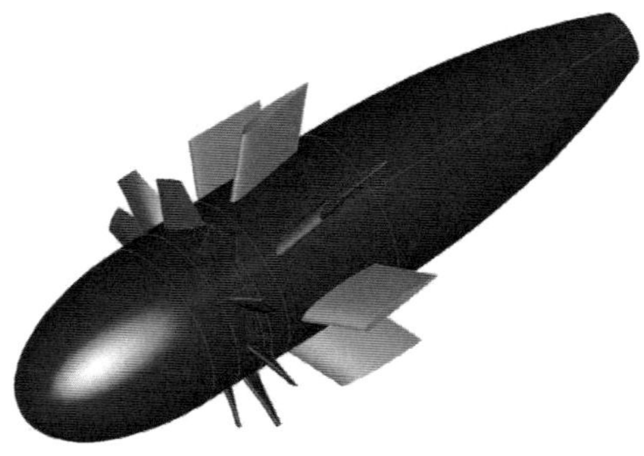

Figure 3. Model of fan B

The parameters of design point are listed in table 1.

Table 1. Parameters of design point

model	blade installation angle (deg.)	rotate speed (rps)	test section wind velocity (m/s)	theoretical pressure rise (Pa)
A	44.2	60	130	2900
B	38.1	60	130	2900

Figure 4 shows the mesh of models.

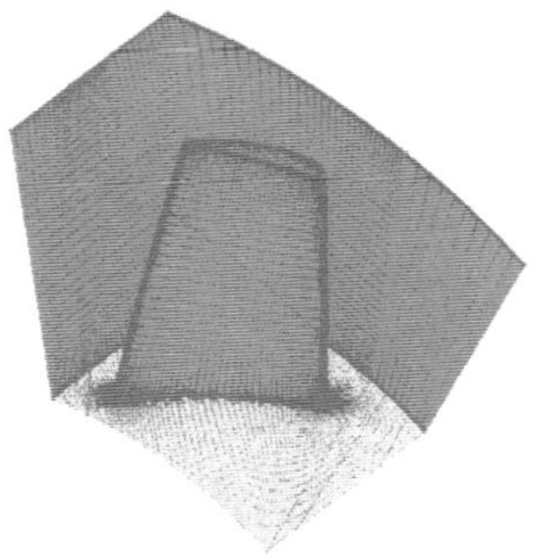

(a). mesh of fan A rotor

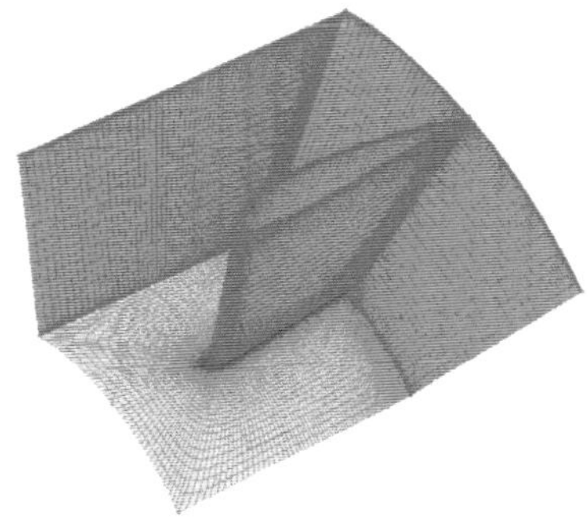

(b). mesh of fan A stator

(c). mesh of fan B rotor

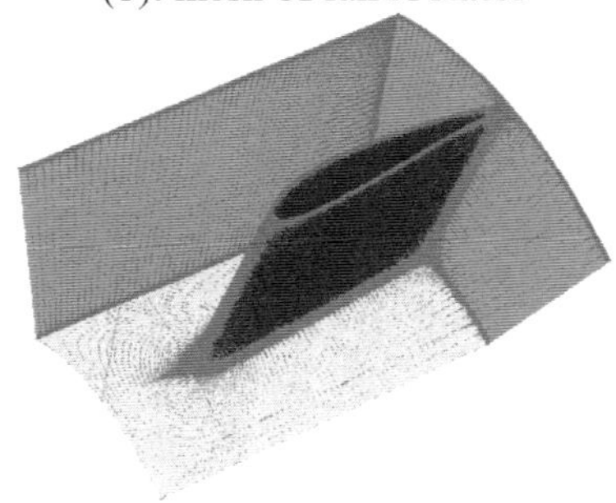

(d). mesh of fan B stator

Figure 4. Mesh of models

3.2. Result of numerical simulation

The axial velocity distribution of incoming flow is showed in figure 5, and the distribution at fan outlet is showed in figure 6. From these two figures, it can be indicated that the axial velocity distribution at arbitrary vortex fan outlet is more uniform than free vortex fan in case of the same velocity distribution of incoming flow. As mentioned above, it is because the aerodynamic load of blade is decreased at root and increased at tip. Meanwhile, these results also indicate that it is successful the more uniform axial velocity distribution at fan outlet is obtained by arbitrary vortex method.

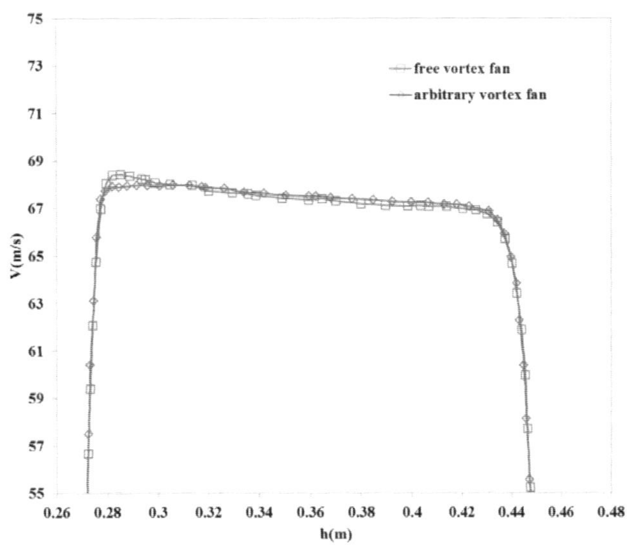

Figure 5. Axial velocity distribution of incoming flow

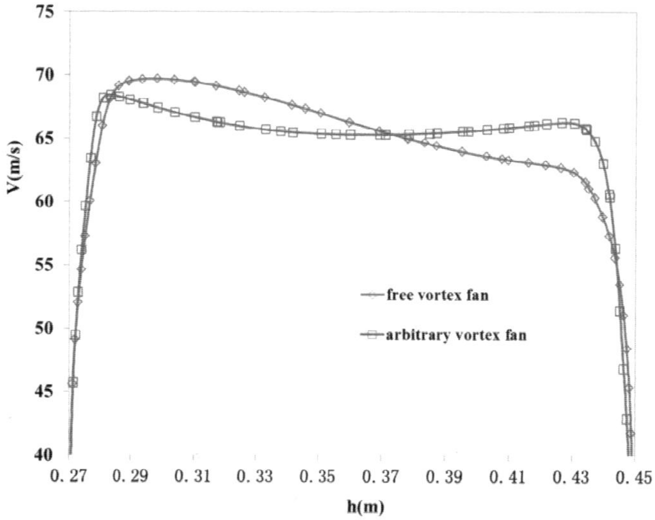

Figure 6. Axial velocity distribution at fan outlet

Figure 7 and figure 8 show oil streamline on blade upper surface of arbitrary vortex fan and free vortex fan respectively. In each figure, the left side is leading edge and the right side is trailing edge. The results indicate that there is no separation on most area of blade upper surface, not only arbitrary vortex fan but also free vortex fan, which prove these two fan design method are both successful. But at root and tip of trailing edge, separations are generated in both blades. The separation in arbitrary vortex fan is less than that in free vortex fan, which infers the performance of arbitrary vortex fan may be better. Because of the viscosity of air flow, there will be separation more and less in the boundary layer of trailing edge.

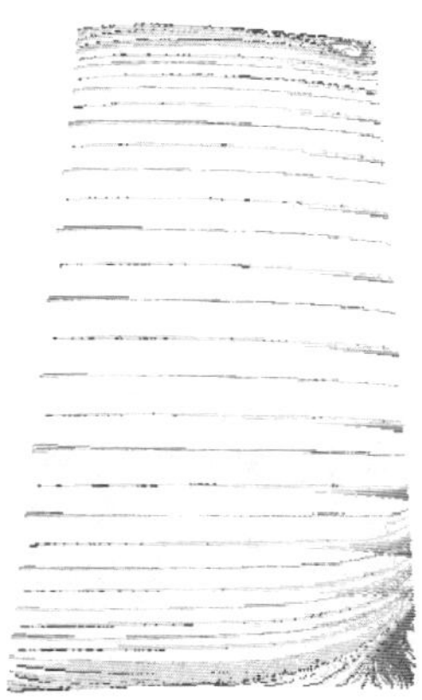

Figure 7. Oil streamline on blade upper surface of arbitrary vortex fan

Figure 8. Oil streamline on blade upper surface of free vortex fan

4. Conclusion

As the blade aerodynamic load can be adjusted in fan design process by arbitrary vortex method, there always will be a reasonable design. As mentioned above, the flow separation in arbitrary vortex fan is less than that in free vortex fan, and the axial velocity distribution at arbitrary vortex fan outlet is more uniform than free vortex fan in case of the same velocity distribution of incoming flow. The less

IOP Conf. Series: Materials Science and Engineering **616** (2019) 012006 doi:10.1088/1757-899X/616/1/012006

separation and more uniform velocity distribution are benefit for improving the performance diffuser which locate downstream of fan, and reducing the noise of fan.

References

[1] Wallis R *Axial Flow Fans, Design and Practice* (George Newnes Ltd, London, 1961)
[2] Suhas V *Numerical Heat Transfer and Fluid Flow* (Hemisphere Publish Corporation,1980)
[3] Wu C H 1952 *ASME.* **74** 363
[4] Kang S and Hirsch C 1996 *ASME Journal of Turbomachinery*, **118** 492
[5] Wu X and Chen M 1999 *PROGRESS IN NATURAL SCIENCE,* **9**

IOP Conf. Series: Materials Science and Engineering **616** (2019) 012007 doi:10.1088/1757-899X/616/1/012007

Comparative Analysis of the Flow Field in the Lower Bend at Different Rotation Angles

Zhaoliang Liu[1], Jinlong Wang[2] and Chengdong Yuan[2,*]

[1] Shandong university of science and technology, Shandong, China
[2]School of Weifang university, Shandong, China

*Corresponding author e-mail: wjl006908@163.com

Abstract. In order to obtain the influence of the distribution of the flow field in the lower bend at different rotation angles on the fluid flow, Fluent software was used to conduct numerical simulation of the flow field in the bend at different rotation angles, and the influence rule of the shape difference of the secondary flow in the lower bend at different rotation angles was simulated. The simulation results show that the 90 ° rotation Angle bend, its internal gas phase flow field better than other simulation Angle, secondary flow phenomena delay happened, reduce reflux area, syphon is helpful to improve the flow state; With the change of the rotation Angle, the tangential velocity of the outlet section of the bend increases continuously, which is conducive to the smooth gas-solid phase separation, but the pressure drop of the system also increases, resulting in the energy loss of the system. As a result, the Angle of 90 ° canary after rotation Angle for the simulation of the optimal point of view, for later engineering application provides data to support and help.

1. Introduction

In petroleum, chemical industry, machinery, power, water conservancy and other engineering applications, it is often designed to different rotation Angle of the bend flow problem. Taylor[1] made a great contribution in the early research period. He used LDV technology to make a detailed experimental study on the flow rate and pressure distribution of laminar and turbulent flows in horizontal pipe bends, and clearly proposed that there was a large pressure gradient and a secondary flow perpendicular to the flow direction in the pipe bend. Humphrey[2] also carried out experimental research on turbulent flows in different cross-sections in the horizontal bend, obtained flow field characteristics and pressure distribution data, and realized that the curvature of the bend had a certain impact on its internal turbulent flows. Hong Shang et al.[3] application of CTA hot wire anemometer and five hole probe to measure the circular cross section curvature at 0.87 level of 90 ° bend the inside of the turbulent flow, and the bent pipe under different Reynolds number is given different cross section parameters such as velocity, turbulent kinetic energy, total pressure distribution. In terms of theoretical research on curved pipes, Jue Ding et al.[4] studied the effects of different inlet directions on the structure and flow characteristics of the flow field in a horizontal bend by using the RNG - turbulence model. Hongming Fan et al.[5] used Largeeddysimu2lation (LES) method to simulate the fluid flow in a horizontal bend in three-dimensional space, which deepened the understanding and analysis of the flow in the bend. The basic equation of the large vortex model[6,7] is a numerical simulation and prediction method based on the cascade decay theory of statistics and vortices. However, the above research objects mainly focus on the horizontal bend. Although the influence of different flow directions on the flow field in the bend has been studied in the literature, the influence of the structure change of the rotation Angle of the bend on the flow field has not been taken into account.

Many bend in the engineering application is a rotation Angle, its structure and horizontal bend, there exist certain differences even in the same flow conditions, the internal flow field and horizontal bending also exist bigger difference, the exit flow field often exist velocity deflection and stress distortion, fluid energy loss not only, and can lead to erosion wear and bend particles in the wall of the deposition, affecting the normal operation of the device. Therefore, it is necessary to study the gas-phase flow field of such a bend. Fluent software was used to conduct numerical simulation of the flow field in the bend at different rotation angles, and the influence rules of secondary flow and jet flow in the bend at different rotation angles on the fluid flow in the pipe were mainly investigated, so as to provide references for engineering design.

2. Geometric model and boundary conditions

2.1. The geometric model

As shown in Figure 1, the elbow is divided into horizontal section, bending section and outlet section (its geometric size is shown in Figure 1).As the rotation Angle of the bend, represents the curvature of the bend, *c* is at the near center of the bend, *d* is at the far center of the bend, and the overall size of the bend is shown in the following table 1.

Table 1 geometric dimensions of elbow

size	a(mm)	b(mm)	θ (°)
Data	70	60	variable

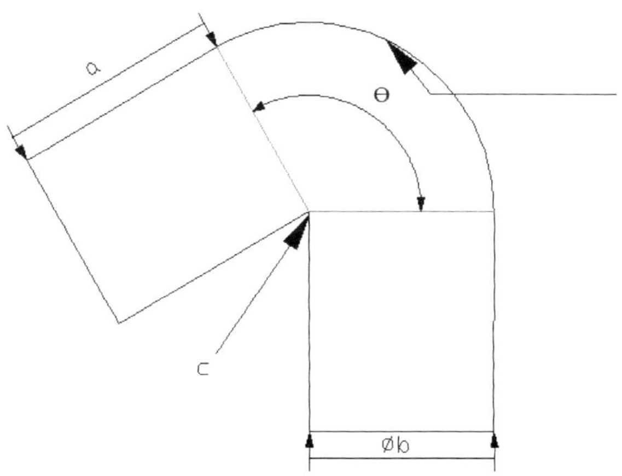

Figure 1. geometric sketch of the elbow

2.2. The boundary conditions

(1) Inlet boundary conditions for fluid to flow uniformly into the bend, its inlet velocity nu nu =35m/s, and inlet turbulence intensity I=4.21%.

(2) The exit conditions were treated according to the fully developed pipe flow conditions, that is, the axial gradient of all variables at the exit section was zero.

(3) Wall condition the wall is treated by non-slip boundary condition and the standard wall function method is used in the vicinity of the wall

3. Simulation results and analysis

3.1. Simulation and analysis of pressure distribution of lower elbow with different rotation angles

As shown in Figure 2. Syphon in order to study the different pressure distribution of flow field simulation scheme take 60° bend 120°, 90°bend but bend data simulation and analysis, through to the three different curved pipe pressure field simulation, analysis of the stress field in the bend structure distribution.

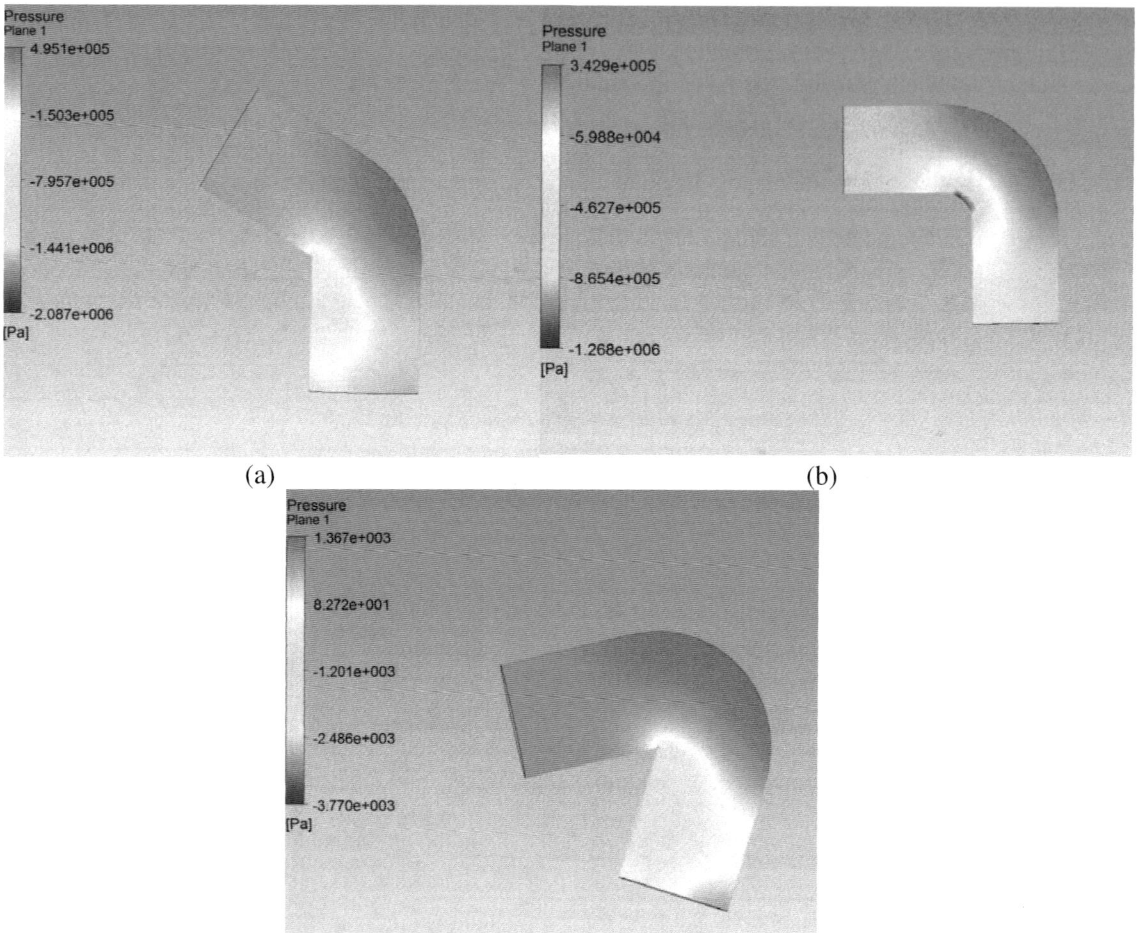

Figure 2. (a) and (b), (c), respectively, 60°, 90°and 120° bend stress field distribution

According to the simulation analysis results, it can be seen that under the three simulated environments, the pressure at the proximal center is lower than that at the distal center, which is consistent with the existing research[8].When the rotation Angle 60°,curved pipe pressure injection close at the heart of change, in the region of the exports, the pressure drop caused by the corner big, make close to the inside of the inlet pressure is greater than near the inner of the exit. Because of the difference of pressure distribution, there will be a secondary flow at the outlet of the elbow. Syphon when the rotation Angle 90°, near the heart without injection pressure changes, in export area and pressure drop phenomenon, no secondary flow phenomena happened in bend the exit; When the rotation Angle 120°, bending tube near the heart jet is 60°more obvious change, at the exit, a greater pressure drop area, secondary flow phenomenon is more obvious than when 60°.Comprehensive the above analysis, in the numerical simulation of stress field, is superior to 90°rotation Angle 60°and 120°.

3.2. Velocity distribution and simulation analysis of lower elbow with different rotation angles

As shown in Figure 3. Syphon to study the flow field velocity distribution, simulation scheme take 60°, 90° and 120° bend data simulation and analysis, through the simulation of velocity field in three different bend tube, analysis the distribution of velocity field within the bend structure.

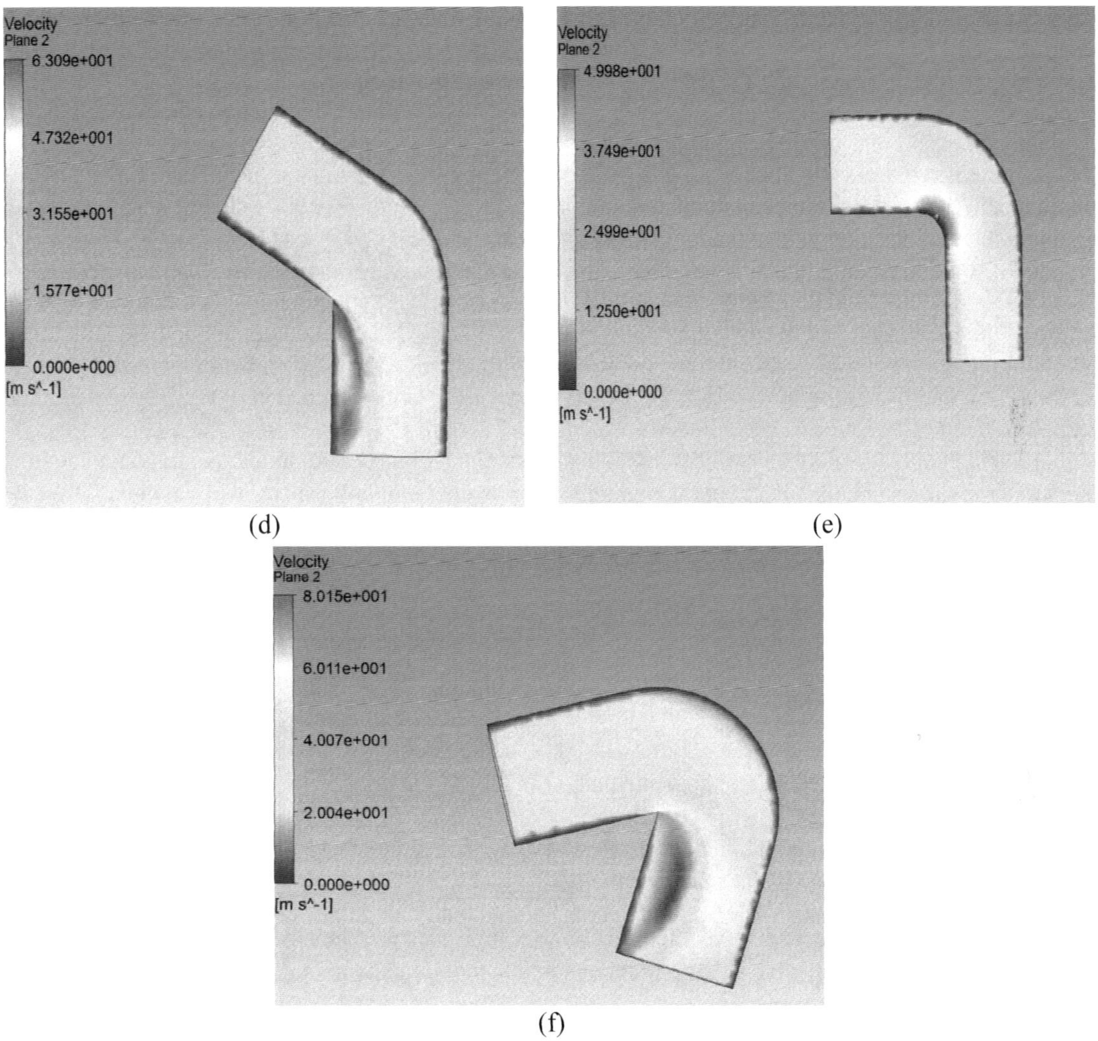

Figure 3. (d), (e), (f) respectively 60° , 90° and 120° bend velocity field distribution

According to the simulation analysis results, it can be seen that there are certain dead zones in the inner wall of the elbow under the three simulated environments. Syphon when the rotation Angle 60°, the central area of velocity field is uniform, but in recent heart point near the exit jet flow phenomenon exists, will affect the use of the bent pipe; Syphon when the rotation Angle 90°, the central area of velocity field is uniform, near the heart without bending tube jet flow phenomenon, near the heart flow velocity is higher than far heart; Syphon when the rotation Angle 120°, the central region is relatively homogeneous at the entrance, but close at the heart, jet flow phenomenon is more serious than the 60°,

IOP Conf. Series: Materials Science and Engineering **616** (2019) 012007 doi:10.1088/1757-899X/616/1/012007

so much so that the exit velocity of stratification. Serious impact on the use of elbow. Based on the analysis of the above, is superior to 90°rotation Angle 60°and 120°.

4. The distribution of flow velocity and pressure field is summarized

From the above analysis, it can be seen that the pressure field and velocity field under different rotation angles all present regular distribution. When the rotation Angle is too large or too small, jet flow, secondary flow and other phenomena appear, which is not conducive to the flow of fluid in the elbow and affects the use of the elbow. In the experiment selected three angles, the most suitable 90° bend. Reflected in the vicinity of 90°, the distribution of pressure field and velocity field showed a trend of kind of normal distribution. Therefore, in the actual production application, according to the actual needs, to choose the right Angle of rotation to produce the elbow.

5. Conclusion

(1) In the pressure field distribution at different rotation angles, the pressure at the proximal center is smaller than that at the distal center. Therefore, the strength at the distal center should be strengthened in practical use. Measures such as inner wall thickness at the distal center and so on can be adopted.

(2) Different rotation Angle of stress field distribution, too big or too small will produce secondary flow, affect the bent pipe is used, according to the simulation data, when take 90° as the best rotation Angle.

(3) In the velocity field distribution with different rotation angles, there is a certain dead zone in the inner wall, which is caused by a variety of factors. Measures such as increasing the inlet velocity and reducing the inlet diameter can be adopted to solve this problem.

(4) Different rotation Angle velocity field distribution, too big or too small will produce jet flow phenomenon, affect the use of the bent pipe, according to the simulation data, when take 90° as the best rotation Angle.

References

[1] Taylor A M K P, Whitelaw J H, Yianneskis M. Curved ducts with secondary motion: velocity measure of developing lam and turbulent flow. Journal of Fluids Eng.1982,104(2):350-359.

[2] Humphrey J A C. Turbulent flow in a square duct with strong curvature Fluid Mech.1981,103:443-463.

[3] Hong Shang,Qingli Kong,Shangjin Wang.90° bend tube with circular cross section of three-dimensional turbulent flow experimental research.journal of aerospace power.1994,9(3):263-266.

[4] Jue Ding,Peifen Weng.90° bend pipe flow of theoretical models and numerical studies of flow characteristics.Journal of computers.2004,21(3):314-321.

[5] Hongming Fan,Zhongyi He,Xiaohua Wang.Large eddy simulation of flow in curved pipe sections.Research and progress in hydrodynamics.2001,16(1):78-83.

[6] Zhaoshun Zhang,Guixiang Cui,Chunxiao Xu.Turbulence theory and simulation.Beijing tsinghua university press .2005:236.

[7] Griffiths W D ,Boysan F.Computational fluid dynamics(CFD) and empirical modeling of the performance of a number of cyclone samplers. Acrosol Science. 1996.27(2):281-304.

[8] Ming Li,Jing Gong.Mathematical model of severe slug flow in Marine riser system[J].Oil and gas storage and transportation.2013,32(5):462-46.

Study on Large-scale Dynamic Measurement of Lunar Rover with iGPS

Qiwei Chen, Laiying Tang, Zaihua Yang, Linzheng Wang, Wei Feng, Shengyu Zheng and Wangmin Yi

Beijing Institute of Spacecraft Environment Engineering, Beijing Engineering Research Center of the Intelligent Assembly Technology and Equipment for Aerospace Product, Beijing, China

wherewherechen@163.com

Abstract. In the motion simulation test, the real-time position and orientation of lunar rover need to be measured accurately in a wide area to test lunar rover's motion performance on simulated lunar soil. The iGPS system, whose accuracy is better than 1mm and sampling frequency is higher than 10 Hz, is applied to measure the position and orientation of lunar rover in the test. Large-scale dynamic measurement of lunar rover with iGPS is studied in this paper. This paper mainly discusses the layout methods of laser transmitters and receiving sensors, and analyzes the position and orientation measurement accuracy of different layout methods to achieve optimization layout. It can provide some reference for the research of large-scale dynamic measurement of other rigid objects.

1. Introduction

In order to ensure lunar rover's performance in the process of lunar exploration, a simulation test is carried out on the ground to test lunar rover's motion performance on simulated lunar soil[1]. The position and orientation of lunar rover need to be measured accurately, and then compared with the data of the lunar rover's guidance navigation and control system to test its mobile control performance. Meanwhile, the position and orientation of lunar rover need to be transmitted to the lunar rover in real time as data input to test its position and orientation determination performance[2]. Large-scale dynamic measurement of lunar rover with iGPS is studied in this paper. The optimization layout methods of laser transmitters and receiving sensors are mainly discussed.

2. Selection of Measuring Scheme

The dynamic measurement of lunar rover's position and orientation has the characteristics of large-scale, high-precision, real-time, complex motion state. Indoor position and orientation measurement methods mainly include total station system, laser tracker system, photogrammetry system and iGPS system[3].

The total station system requires the tracking target to remain stationary during measurement, and has delays. The laser tracker system can only measure one target ball once, and will not be able to continue tracking measurement if there is a break of light in the measurement process. The photogrammetry system requires a large number of cameras to achieve a wide range of measurement, and the static measurement accuracy is not high enough. The iGPS system is a factory-wide measuring, positioning and tracking system, which is characterized by large-scale, high-precision and high-frequency. Therefore the iGPS system is the most suitable measuring scheme for large-scale dynamic tracking measurement. The Metris iSpace iGPS system[4], whose accuracy is better than 1mm and sampling frequency is higher than 10 Hz, is applied to measure the position and orientation of lunar rover in the motion simulation test.

3. Design of measuring scheme

The iGPS system is mainly composed of laser transmitters and receiving sensors[5]. The transmitters activate a measurement field with laser signals as large as an entire room or facility. The receiving sensors' position is calculated by the laser signals received from transmitters. Before measurement, several laser transmitters should be arranged to cover the testing room with laser signals, and several receiving sensors should be installed on lunar rover to receive laser signals. When the lunar rover moves into the testing room, the receiving sensor on the lunar rover will receive the laser signal in real time to complete the measurement.

3.1. Layout of laser transmitters

The layout of the laser transmitter needs to be designed to make the laser signals cover the whole measurement room, and make the signal intensity in any area basically the same. The maximum measuring distance of laser transmitter is 50m, the maximum measuring azimuth is -145° ~145°, and the maximum measuring elevation is -30° ~30°, as shown in figure 1.

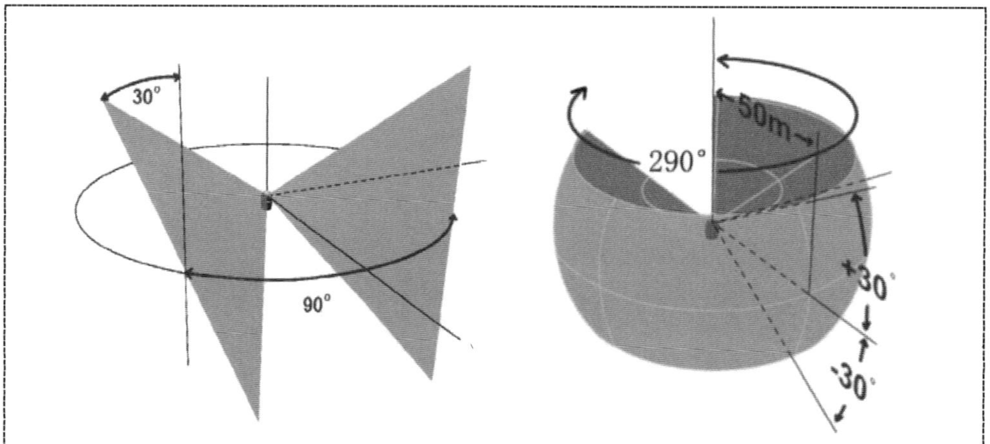

Figure 1. The measuring range of laser transmitter

At least two transmitters are needed to calculate the position of the receiving sensor, which can receive the different laser signals from these transmitters at the same time. Extra transmitters increase the positional accuracy of receiving sensors as better triangulation possible. The line-of-sight between laser transmitters and receiving sensors should be unblocked, otherwise partial or global occlusion of the signals will cause errors or drifts in data calculation.

The layout of the laser transmitters mainly involves its number, relative height and relative angle.

3.1.1. Number. Limited by the optimal measurement range and intersection angle, the number of laser transmitters is mainly determined by the size and shape of the measurement area.

The number of laser signals received at the same time will directly determine the positional accuracy of receiving sensors, as shown in figure 2. When only two laser transmitters cover a measuring point at the same time, the measurement accuracy can only reach 1.5mm (3σ). When three laser transmitters cover the measuring point at the same time, the measurement accuracy can reach 0.535mm (3σ). When four laser transmitters cover the measuring point at the same time, the measurement accuracy can reach 0.375mm (3σ). But when more than four laser transmitters cover the measuring points at the same time, the measuring accuracy is basically not increased. Therefore, in order to achieve high-precision measurement in the whole area, which is better than 0.375mm, any measuring point in the measurement area should receive laser signals from at least four laser transmitters at the same time.

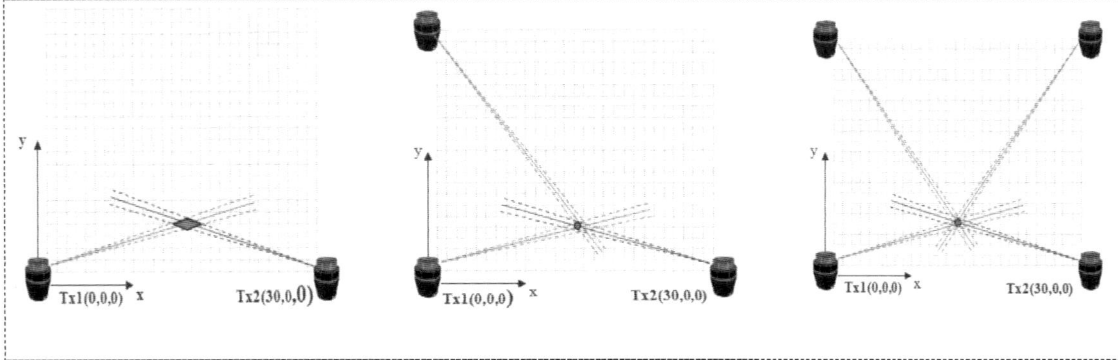

(a)Two laser transmitters (b)Three laser transmitters (c)Four laser transmitters

Figure 2. The measurement accuracy of iGPS with different numbers of laser transmitters

3.1.2. Height and angle. Mainly limited by the maximum measuring elevation of laser transmitters, -30° ~30°, the relative height and angle of laser transmitters should be arranged to cover the whole area with laser signals evenly in the vertical direction. Laser signals from at least two laser transmitters can be received at any measuring point within the measuring range.

3.2. Layout of receiving sensors

Within the measurement area, an unlimited number of receiving sensors can operate concurrently. To meet different measurement requirements, there are mainly four different structures of receiving sensors that can be chosen, which are Single Detector, Mini Vector Bar, Handheld Vector Bar and Handheld iProbe, as shown in figure 3. Single Detector and Mini Vector Bar are fixed measurement sensors, mainly suitable for dynamic measurement. Handheld Vector Bar and Handheld iProbe are mobile measurement sensors, mainly suitable for static measurement. Mini Vector Bar has twice as many sensors as Single Detector, so Mini Vector Bar is mainly used in dynamic measurement of large-scale and complex motions.

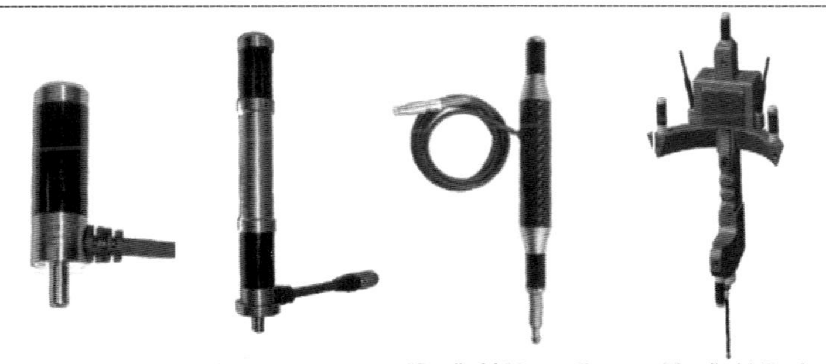

Single Detector Mini-Vector Bar Handheld Vector Bar Handheld iProbe

Figure 3. The four different structures of receiving sensors

Mini Vector Bar is a kind of omnidirectional receiving sensor. The maximum measuring azimuth of Mini Vector Bar is -180° ~180°, and the maximum measuring elevation is -45° ~45°. According to the real-time position of the receiving sensors, which are arranged on lunar rover, the real-time position and orientation of lunar rover can be calculated to track the local coordinate system of lunar rover. At least three non-collinear receiving sensors need to be arranged to calculate the orientation of lunar rover. To achieve the optimal measurement accuracy, the horizontal and vertical spacing of the receiving sensor should be as far as possible. Generally, the receiving sensor should be located on the outermost side of lunar rover. Furthermore, the line-of-sight between receiving sensors and laser transmitters should be unblocked as far as possible during the movement.

4. Measurement accuracy

4.1. Position measurement accuracy

According to the typical measurement accuracy of iGPS system[4], the dynamic position measurement accuracy of lunar rover is 0.375mm(3σ). The accuracy of the iGPS system is uniform across the entire measurement volume, regardless of the size of the metrology enabled area.

4.2. Orientation measurement accuracy

The orientation measurement accuracy of the lunar rover is mainly determined by position measurement accuracy and the spacing of receiving sensors.The receiving sensor is isotropic, so the measurement model of the receiving sensor can be simplified to two-dimensional, as shown in figure 4.

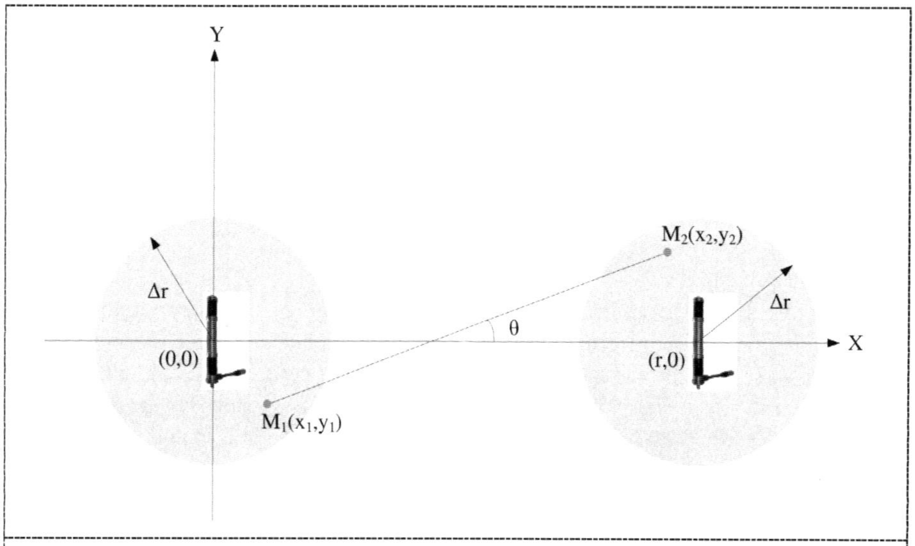

Figure 4. The simplified two-dimensional measurement model

Suppose the distance between sensors is r, the position measuring error is Δr and the angle between sensors is θ.

$$\theta = \arctan((y_2 - y_1)(x_2 - x_1)^{-1}) \tag{1}$$

$$\left.\begin{aligned} x_1^2 + y_1^2 &= \Delta r^2 \\ (x_2 - r)^2 + y_2^2 &= \Delta r^2 \end{aligned}\right\} \tag{2}$$

Then

$$\theta = \arctan(((\Delta r^2 - (x_2 - r)^2)^{1/2} + (\Delta r^2 - x_1{}^2)^{1/2}))(x_2 - x_1)^{-1}) \tag{3}$$

$$\theta_{max} = \arctan(2\Delta r (r^2 - 4\Delta r^2)^{-1/2}) \tag{4}$$

Therefore, if the orientation measurement accuracy is required to be better than 0.1°, the distance between sensors should be more than 429.72mm. If the orientation measurement accuracy is required to be better than 0.01°, the distance between sensors should be more than 4297.2mm. The relationship between orientation measurement accuracy and receiving sensor spacing is shown in figure 5.

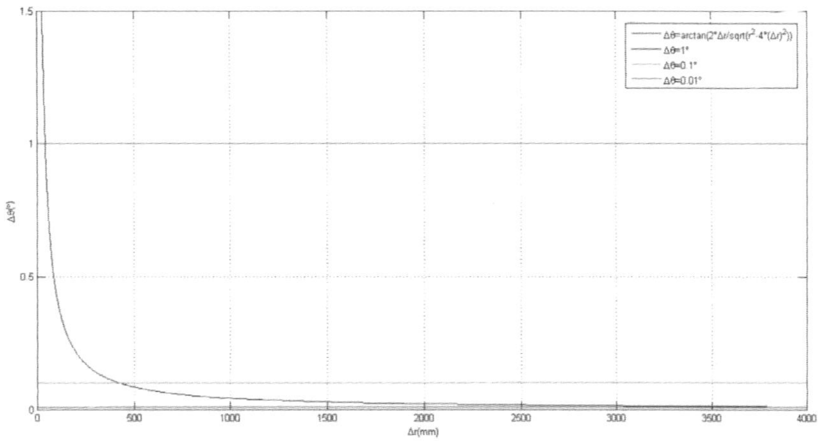

Figure 5. The relationship between orientation measurement accuracy and distance of receiving sensors

4.3. Environmental factors

The main environmental factors affecting the measurement accuracy of iGPS system are vibration, infrared light interference, temperature and air pressure. Vibration will directly affect the calibrated location of the laser transmitters and the parameters of the rotating head, and cause displacement and drift of the laser transmitters. The influence of infrared light on measurement system will become significant with the increase of measurement distance. Temperature and pressure will affect the optical system[3].

5. Application

The site topography of the motion simulation test is shown in figure 6. The left half of the test site is a rectangular experimental area, 15m*30m. The right half is a semi-circular test area with a diameter of 30m. There are three slopes in the test site, with a maximum slope of 20° and a maximum height of 2m. Four laser transmitter stations are installed outside the test site. These stations have different installation heights, ranging from 1.5m to 3.5m, and installation angles, ranging from -30° to 0°.

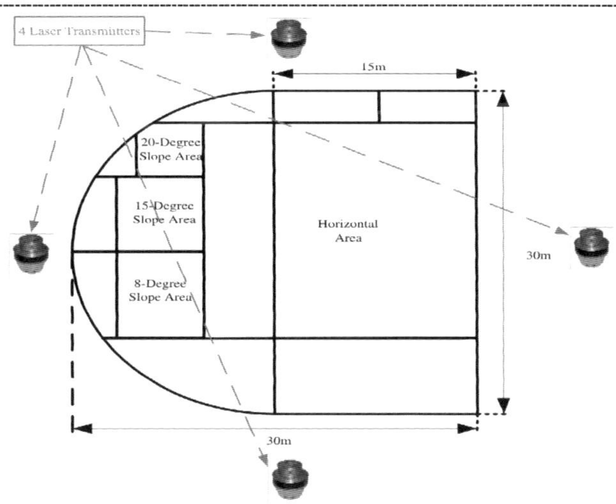

Figure 6. The layout of laser transmitters in the motion simulation test

IOP Conf. Series: Materials Science and Engineering **616** (2019) 012008 doi:10.1088/1757-899X/616/1/012008

The simplified configuration of lunar rover is shown in figure 7. The overall dimension of lunar rover is about 1.5mm*1m*1m. Four Mini Vector Bars are installed on lunar rover as dynamic tracking unit. The redundancy design of eight receiving sensors is applied to ensure the stability and reliability of the dynamic measurement. Each spacing between the Mini Vector Bars is over 0.5m to ensure the accuracy of orientation measurement, better than 0.1°.

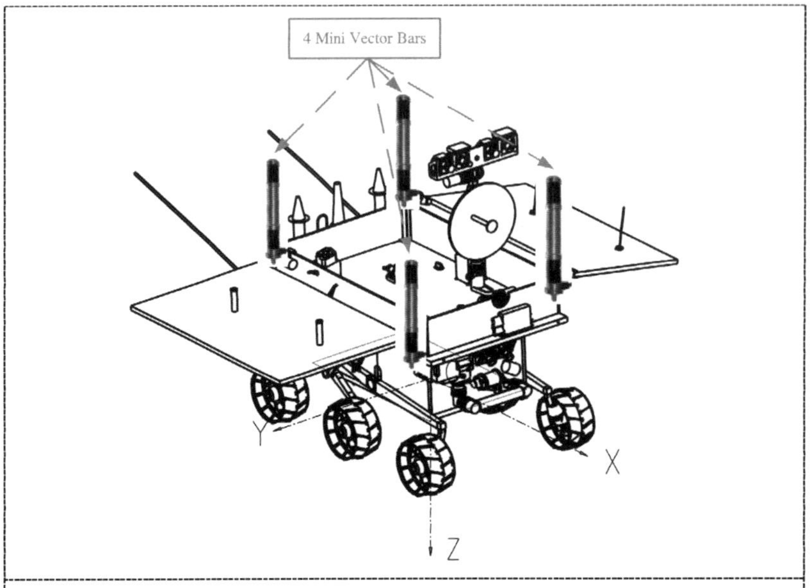

Figure 7. The layout of receiving sensors on lunar rover

6. Summary

iGPS system has great application prospects in the field of large-scale, high-precision, high-frequency dynamic measurement. In this paper, the dynamic measurement of lunar rover with iGPS is studied. The layout methods of laser transmitters and receiving sensors are emphatically discussed. The measurement accuracy of different layout methods is analyzed. It can provide some reference for the research of large-scale dynamic measurement of other rigid objects.

Reference

[1] Jia Yang, Chen Jianxin, Shen Zhenrong. Lunar Rover [J]. Aeronautical Manufacturing Technology, 2007(12): 44-47.

[2] Yang Zaihua, Chen Qiwei, Tang Laiying, Yi Wangmin, Sun Gang. Technology of lunar rover position and orientation measurement in test[J]. Journal of Harbin Institute of Technology. 2013(45), 124-128.

[3] Li guangyun, Li zongchun,. The principles and applications of industrial measuring systems[M]. Beijing: SinoMaps, 2011

[4] Metris Canada, Inc. ISPACE USER MANUAL AND STARTUP GUIDE[M]. Waterloo, Ontario, Canada, 2009.

[5] Kang Haidong, Fan Baixing, Li Zongchun, et al. iGPS measurement principle and precision analysis[J]. Bulletin of Surveying and Mapping, 2012(3): 12-15.

IOP Conf. Series: Materials Science and Engineering **616** (2019) 012009 doi:10.1088/1757-899X/616/1/012009

Analysis of Thermochemical Nonequilibrium Ablation Flow for a Hypersonic Hemispherical Nose

Chun Shao, Yingchun Chen, Miao Zhang and Jun Shu

Shanghai Aircraft Design and Research Institute, Commercial Aircraft Corporation of China Ltd, Shanghai, 201210, China

Corresponding author's e-mail address: shaochun@comac.cc

Abstract. The thermal protection materials enveloping hypersonic vehicles ablate. This paper develops a numerical methodology of thermochemical nonequilibrium flow and structure temperature field to study the key flow and structure parameters of a hypersonic hemispherical nose under the condition of carbon-based thermal protection material ablation. The multi-species chemical reaction model of flow field and the corresponding numerical algorithm are constructed. Oxidation and sublimation ablation process are considered in ablation model coupled with flow field solver by gas-solid interaction method. The computational results of a hemispherical nose indicate that carbon-based thermal protection material ablation has significant effects on thermochemical nonequilibrium flow. Details of the ablation flow characteristics are reported to highlight the influences of ablation on microcosmic reaction species and macroscopic flow parameters including shock wave position, pressure and temperature.

1. Introduction

Serious aerodynamic heating will lead to the ablation of carbon-based thermal protection material on the surface of hypersonic vehicles the aircraft, which affects the characteristics of thermochemical nonequilibrium flow field. The ablation mechanism of thermal shielding materials and the characteristics of thermochemical nonequilibrium ablation flow field of hypersonic vehicle have been studied extensively. Based on JANAF sublimation carbon group model[1], two kinds of surface ablation models named after chemical equilibrium model[2] and finite rate model[3, 4] are developed. The collision integration model and the rate coefficient of forward and reverse reaction are measured to simulate the ablation flow field of the hypersonic vehicle using the carbon-phenolic thermal protection materials[5]. Considering the pyrolysis effect of the resin inside the carbon-phenolic ablation material at a lower temperature, a more accurate ablation reaction model is developed[6]. Coupled with the relationship of mass and energy balance on the ablation surface, the numerical results based on NS equations are in good agreement with experiment values[7]. Multi-component chemical reaction model is used to calculate polypropylene ablation flow field of blunt cone re-entry and the influence of ablation products on electron number density in wake area is detailed studied[8,9]. The prediction of continuous ablation and heat transfer in the whole re-entry process is realized by aerodynamic engineering estimation method and nonlinear heat conduction equation[10].

The objective of this paper is to develop appropriate simulation methods for capturing the key parameters of the thermochemical nonequilibrium ablation flow and reveal the mechanism by which the ablation influences the chemical component characteristics of a hypersonic hemispherical nose.

The two-way coupled effect of flow field and structure temperature field is considered and the ablation flow field is calculated by NS equations to guarantee precision.

2. Thermochemical nonequilibrium reaction model

According to the flight conditions of hypersonic vehicle and the characteristics of carbon-based material, following species may appear in the flow field: O, N, O_2, N_2, NO, $NO+$, CO, C_1, C_2, C_3, CN, CO_2 and e. In consideration of chemical nonequilibrium, forward reaction rate k_f and backward rate k_b are defined by Arrhenius form:

$$k_f = A_f T^{B_f} e^{\left(-\frac{E_f}{T_r}\right)} \tag{1}$$

$$k_b = A_b T^{B_b} e^{\left(-\frac{E_b}{T_r}\right)} \tag{2}$$

The values of A_f, B_f, E_f and A_b, B_b, E_b are also shown in Gupta's air reaction model[11] and Blottner's ablation products reaction model[12]. The governing temperatures of reactions are determined by the type of reactions in Park's model[13]. The equations of state and energy conservation for mixed gas are given as:

$$P = \hat{R}\left(\rho T \sum_{i=1}^{ns-1} \frac{c_i}{M_i} + \rho T_v \frac{c_e}{M_e}\right) \tag{3}$$

$$E = E_{tr} + E_e + E_v + E_0 + \frac{1}{2}\left(u^2 + v^2 + w^2\right) \tag{4}$$

According to Wilke's semi-empirical formula, the viscosity coefficient, vibration coefficient and electronic thermal conductivity of mixed gas are given as:

$$\mu = \sum_{s=1}^{ns} \frac{X_s \mu_s}{\Phi_s}, \quad k_{tr} = \sum_{s=1}^{ns} \frac{X_s k_{tr,s}}{\Phi_s}, \quad k_v = \sum_{M} \frac{X_s k_{v,s}}{\Phi_s}, \quad k_e = \frac{X_e k_{e,e}}{\Phi_s} \tag{5}$$

The diffusion coefficients D_s of component s and electron are written as:

$$\rho D_s = \frac{(1-c_s)\mu}{(1-X_s)S_c} \tag{6}$$

$$\rho D_e = M_e \sum_{I} \frac{c_s}{M_s} \rho D_s / \sum_{I} c_s \tag{7}$$

3. Carbon-based material ablation model

The surface ablations account for both oxidation and sublimation process for different surface temperatures. The ablation reaction rate is affected by two factors: one is the rate at which oxygen diffuses to the wall or the material sublimation product diffuses to the air, the other is the rate at which the material reacts with the air.

The oxidation process is:

1) $2C + O_2 \rightarrow 2CO$
2) $C + O \rightarrow CO$

The chemical reaction using JANAF sublimation model is:

1) $jC \rightarrow C_j \left(j = 1,2,3\right)$
2) $C + N \rightarrow CN$
3) $2C + N_2 \rightarrow 2CN$
4) $C + O \rightarrow CO$

5) $2C + O_2 \rightarrow 2CO$

The mass ablation rate $\dot{m}_{-\infty}$ of carbon-based material contains the mass loss rate $\dot{m}_C$ caused by ablation and the mass loss rate $\dot{m}_t$ caused by mechanical denudation. The ablation degradation rate and mechanical denudation factor are presented respectively as:

$$v_{-\infty} = \dot{m}_{-\infty} / \rho \tag{8}$$

$$f_r = \dot{m}_{-\infty} / \dot{m}_C \tag{9}$$

where

$$f_r = \begin{cases} 1.0 & P_e < 20.26\times10^5 Pa \\ 1.2 & P_e \geq 20.26\times10^5 Pa \end{cases} \tag{10}$$

4. Numerical Algorithm

The flow field governing equations are calculated by Advection Upstream Splitting Method by Pressure-based Weight function[14] (AUSMPW+) while the diffusion terms are central difference. The implicit Lower-Upper Symmetric Gauss-Seidel Relaxation[15] (LU-SGS) is employed as time marching algorithm. Besides, the structure temperature field is solved by heat balance method and the time marching algorithm is explicit Runge-Kutta method[16].

The coupled solving procedure of ablation flow field and structure temperature field is presented as:

1. Calculate the thermochemical non-equilibrium flow field and obtain heat flux, temperature, pressure and mass fraction of each chemical reaction element at fluid-solid interface.
2. Use the results in step 1 as boundary conditions to calculate structure heat transfer and ablation reaction.
3. After structure temperature field marches 10 s, the ablation products and interface temperature are applied as the boundary conditions to calculate the chemical non-equilibrium ablation flow field at that moment.
4. The characteristics of flow and temperature fields at different time are obtained by repeated iteration of interface information.

5. Results and discussion

The flight conditions are given in Table 1 and the hemispherical nose geometry with the radius of 0.5 *m* are shown in Figure 1. The computational grid contains two regions: flow field in blue and structure temperature field in red. As a compromise between accuracy and computational cost, the grid for flow field comprises 3.5 million cells and the uniform grid for structure temperature field is 0.3 million.

Table 1. Freestream conditions for the nose cone.

Altitude, *km*	Velocity, *m/s*	Mach No.	Angle of attack, *degree*	Temperature, *K*	Time, *min*
70	6000	20.2	0	220	20

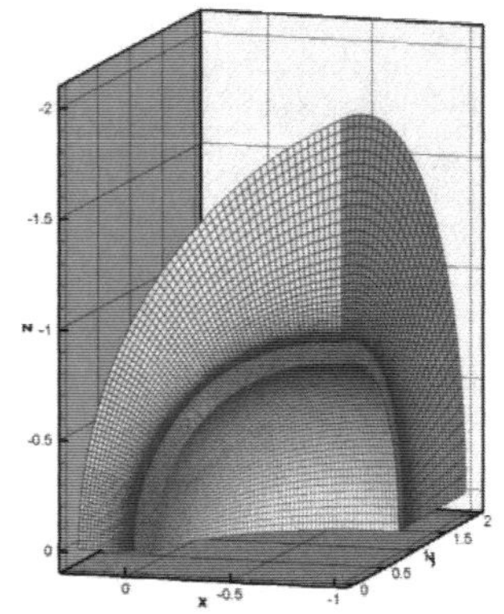

Figure 1. Grids of flow field and structure thermal field.

The flow field without ablation is firstly studied under non-catalytic wall condition and the species mass fraction distribution is given in Figure 2. O_2 almost fully dissociates behind the shock and part of it forms *NO* with N_2. Parts of N_2 dissociate as flow passes through the shock and composite again near the surface. The associative-ionization reaction forms *NO+* and e and the mass fractions of them are much less than other species.

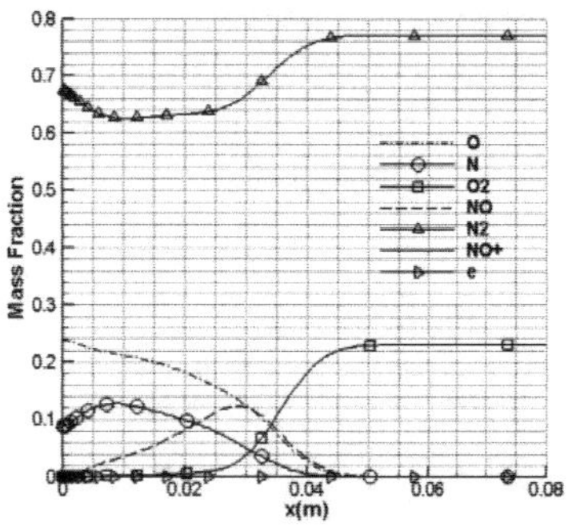

Figure 2. Species mass fraction along the stagnation line without ablation.

After the consideration of aerodynamic heating to the structure, the stagnation temperature rises from 220 *K* to approximately 2600 *K* as shown in Figure 3. The ablation of thermal protection material occurs during the heating process and the ablation degradation rate of ablation wall is given in Figure 4. Within the first four minutes, the ablation degradation rate is basically zero and controlled by weak oxidation reactions. During the fourth to tenth minutes, the ablation degradation rate near the stagnation point increased rapidly, indicating that the strong oxidation reactions occur. In the last ten minutes, the ablation degradation rate increases slowly and the ablation process is controlled by sublimation reactions.

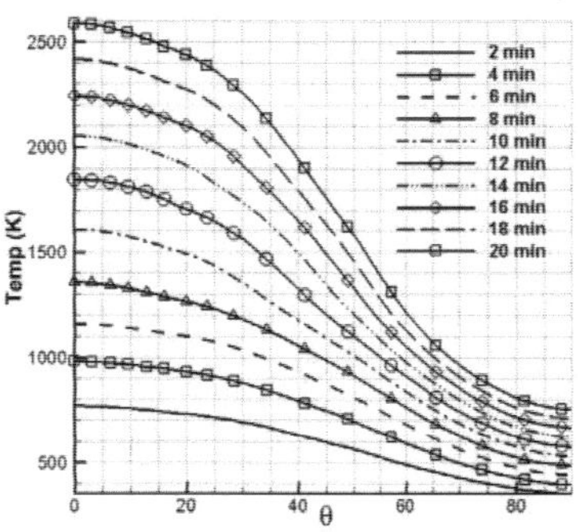

Figure 3. Temperature distribution of nose wall along meridian at different time.

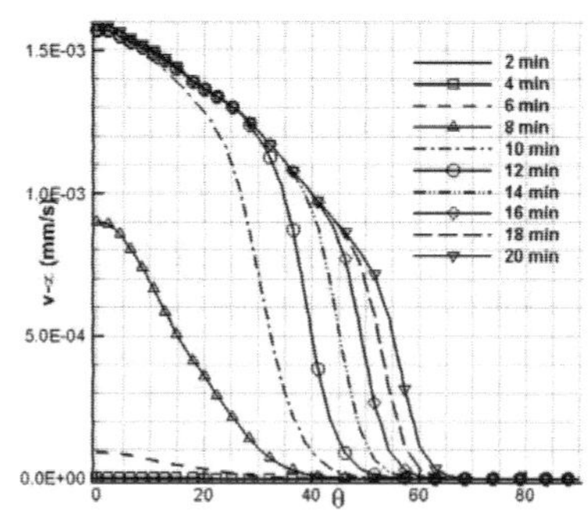

Figure 4. Ablation degradation rate of nose wall along meridian at different time.

The ablation plasma flow field at 20 *min* is further analyzed and the species mass fraction distribution along the stagnation line is given in Figure 5 and Figure 6. Note that nearly all *O* are used to form *CO* rather than *NO* and *NO+*. Meanwhile, parts of *N* form *CN* with *C*. The mass fraction of *CO* near the stagnation point is much larger than other ablation products. The mass fraction distribution of the other air species changed little compared with the non-ablation state.

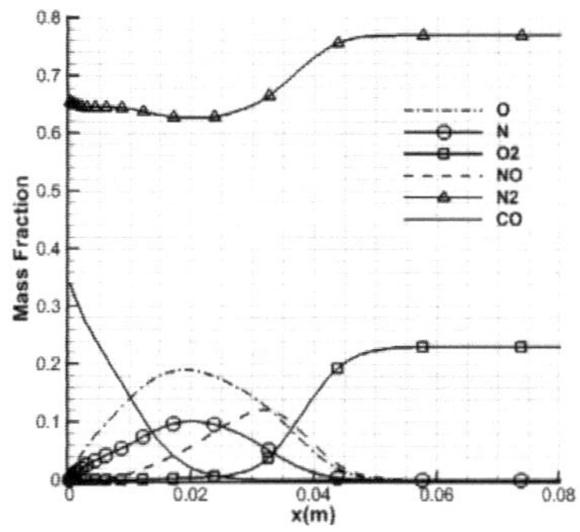

Figure 5. Mass fraction of major species along stagnation line at 20th minute.

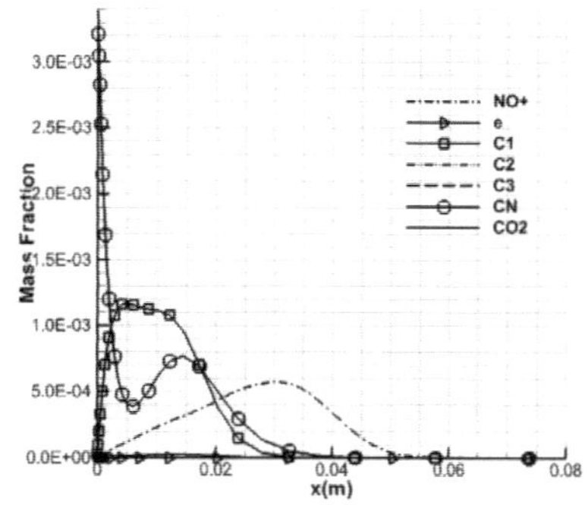

Figure 6. Mass fraction of minor species along stagnation line at 20th minute.

The pressure and temperature distributions along stagnation point line of initial time versus 20 min are shown in Figure 7 and Figure 8. Ablation effect makes the shock detachment distance increase and affects the pressure and temperature distribution of flow field. Meanwhile, the peak temperature near stagnation point in ablation flow field is lower than that in the non-ablation state, due to the absorption of heat by chemical reactions between ablation products and air after shock wave.

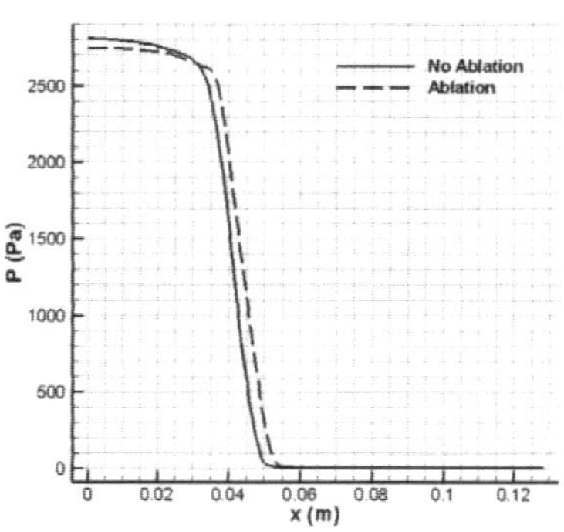

Figure 7. Comparison of pressure distribution along stagnation line for initial time and 20th minute.

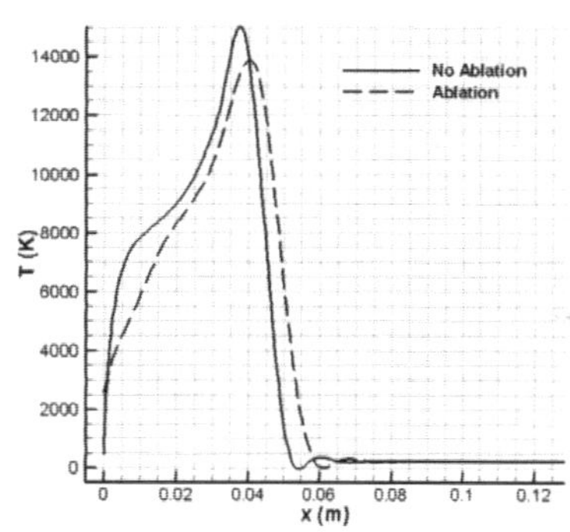

Figure 8. Comparison of temperature distribution along stagnation line for initial time and 20th minute.

6. Conclusions

In this paper, a numerical approach was developed to study the thermochemical nonequilibrium flow field of a hypersonic hemispherical nose with carbon-based thermal protection material ablation. The ablation flow field was simulated by using a full N-S approach coupled with the heat conduction in structure temperature field. The ablation reaction model was developed based on material oxidation and sublimation process.

Under the condition tested, the ablation of thermal protection material was obvious, resulting in the change of chemical species in flow field. In oxidation process, the ablation reactions consumed much of *O* in nose regions and the primary product was *CO*. With the increase of surface temperature, sublimation products appeared in the flow field. It was noted that ablation had significant effects on thermochemical nonequilibrium flow. The ablation completely consumed *O* surrounding the stagnation point and changed species distribution. Besides, carbon-based material ablation influenced macroscopic flow parameters including shock wave position, pressure and temperature.

Acknowledgments
This research was sponsored by the Shanghai Sailing Program (grant number 19YF1459800)

References

[1] Stull, D.R. and H. Prophet, JANAF thermochemical tables. 1971, DTIC Document.
[2] Powars, C.A. and R.M. Kendall, User's Manual Aerotherm Chemical Equilibrium (ACE) Computer Program. Aerotherm Corp., 1969.
[3] Park, C. and H.-K. Ahn, Stagnation-point heat transfer rates for Pioneer-Venus probes. *Journal of thermophysics and heat transfer*, 1999, 13(1): 33-41.
[4] S.V. Zhluktov, T. Abe, Viscous shock-layer simulation of airflow past ablating blunt body with carbon surface, *Journal of thermophysics and heat transfer,* 1999, 13(1): 50–59.
[5] Park C, Jaffe RL, Partridge H. Chemical-Kinetic Parameters of Hyperbolic Earth Entry. *Journal of Thermophysics and Heat Transfer*, 2001, 15(1): 76-90.
[6] Martin A, Boyd ID. Assessment of Carbon-phenolic-in-air Chemistry Models for Atmospheric Reentry. *10th AIAA/ASME joint thermophysics and heat transfer conference*, 2010, 4656.
[7] Bianchi D, Nasuti F. Navier–Stokes Simulations of Hypersonic Flows with Coupled Graphite Ablation. *Journal of Spacecraft and Rockets*, 2010, 47(4): 554-562.
[8] Cheng Xiaoli, Dong Yonghui, Li Tinglin. Computational and Experimental Studies of Ablation Effect on Electronic Characteristics in Vehicle Wake. *ACTA AERONAUTICA ET ASTRONAUTICA SINICA*, 2007, 28(4): 796.
[9] He Kaifeng, Gao Tiesuo, Jiang Tao. Numerical study for the effects of ablation on electron number densities of flow over reentry vehicles. *ACTA AERODYNAMICA SINICA*, 2009, 27(1): 57-61.
[10] Wang Jun, Pei Hailong, Wang Naizhou. Research on Ablation for Crew Return Vehicle Based on Re-entry Trajectory and Aerodynamic Heating Environment. *ACTA AERONAUTICA ET ASTRONAUTICA SINICA*, 2014, 35(1): 66-71.
[11] Gupta, R.N. A review of reaction rates and thermodynamic and transport properties for an 11-species air model for chemical and thermal nonequilibrium calculations to 30000 K. *NASA STI/Recon Technical Report N*, 1990, 90: 27064.
[12] Blottner, F. Prediction of electron density in the boundary layer on entry vehicles with ablation. *NASA Special Publication*, 1971. 252: 219.
[13] Park, C. Assessment of two-temperature kinetic model for ionizing air. *Journal of Thermophysics and Heat Transfer*, 1989. 3(3): 233-244.
[14] Kim, K.H., C. Kim, and O. H. Rho, Methods for the accurate computations of hypersonic flows: I. AUSMPW+ scheme. *Journal of Computational Physics*, 2001. 174(1): 38-80.
[15] Yoon, S. and A. Jameson, Lower-upper symmetric-Gauss-Seidel method for the Euler and Navier-Stokes equations. *AIAA Journal*, 1988. 26(9): 1025-1026.
[16] Pareschi L, Russo G. Implicit–explicit Runge–Kutta schemes and applications to hyperbolic systems with relaxation. *Journal of Scientific computing*, 2005, 25(1): 129-155.

Design of an End-effector for Spacecraft Automatic Assembly Robot

Bin Zhang[1,2*], Chengli Zhang[1,2], Fugui Xie[3], Wangmin Yi[1,2,3], Shaohua Meng[2] and Wei Wang[1,2]

[1] Beijing Institute of Spacecraft Environment Engineering,Beijing,China, 100094
[2] Assembly Processing Technology Center for Aerospace Product of CASC, Beijing, China, 100094
[3] Tsinghua University, Beijing, China, 100084

*E-mail: zhangbin_bisee@126.com

Abstract: With the increasing of large-weight equipment on spacecraft, the traditional assembly method using hoisting manual guidance has some shortcomings, such as poor flexibility, high risk of bumping, low assembly efficiency, etc. [1]. In recent years, robot assembly system has been widely used in spacecraft assembly process. End-effector is an important part of automatic assembly robot. In this paper, an end-effector for batch equipment installation with fast connection and fast release function was designed. Through simulation analysis and experimental verification, it proves that the end-effector has good performance and can meet the application requirements. It helps to raise efficiency and reduce cost for spacecraft assembly.

1. Introduction

The traditional spacecraft assembly operations are usually based on manual operation. For the heavy equipment which cannot be afforded by hand, we usually use crane to us to install it. With the continuous improvement of spacecraft integration, more and more heavy equipment are employed on spacecraft. For the large-weight equipment outside the spacecraft cabin, the crane and other tools can be used to help operators to complete the installation. For the large-weight equipment inside the cabin or in some special locations, the space is so narrow that it is difficult to use lifting tools. The feasibility of operation is low and the risk of collision is high. In recent years, a robot-based assembly system for large-scale equipment has been developed. Robots are used to grasp the installed objects and transport them to the assembly location accurately, which can make the assembly process of large-weight equipment safer and more efficient [2-4]. End-effector is a device that directly contacts and operates with the equipment and tools. It is one of the key components of automatic assembly system, which directly affects the effect and efficiency of robot assembly work[5-6]. Spacecraft products show the characteristics of small batch and diversification, and the shape and size are not uniform. For the connecting devices used for the clamping of medium-sized equipment in spacecraft (the weight range is 20kg-150kg), they were designed separately for each equipment in the past. The connecting devices are not universal, the assembly and disassembly work is complex, the utilization rate is low, and the cost is high. Therefore, it is necessary to develop an end-effector with fast loading and unloading function.

2. Design Constraint

2.1 Requirement analysis and design scheme

There are 7 types of equipment with mass above 30 kg on a spacecraft, as shown in Table 1. These seven types of equipment need to be installed by robot assembly system. Different kinds of equipment have different interface sizes and requirements. According to past experience, seven types of end-effectors need to be designed to connect the robot assembly system and the equipment to be installed. The design work is heavy and the resources are expensive.

Table 1. List of equipment above 30kg on a spacecraft

Equipment Type	Interface Size(L×B)	Interface Bolts	Horizontal Projection Size	Weight
A1	377mm×148mm	4-M6（10mm）	659mm×410mm	30kg
A2	430mm×226mm	4-M10（10mm）	696mm×480mm	65kg
A3	490mm×260mm	4-M10（10mm）	832.3mm×583mm	100kg
A4	630mm×300mm	4-M10（12mm）	974.3×700mm	129kg
A5	329mm×258mm	4-M6（18mm）	406.5mm×362mm	32kg
A6	180mm×170mm	4-M10（15mm）	426mm×333mm	65kg
A7	170mm×175mm	4-M5（8mm）	378mm×366mm	40kg

At the same time, the installation process is complex. The installation steps with traditional end-effector are as follows: 1) Connect End-effector 1 and the Robot; 2) Connect end-effector 1 and Equipment A1; 3) Grab the Equipment A1to the Spacecraft and install Equipment A1; 4) Remove End-effector 1from Equipment A1; 5) Remove End-effector 1from the Robot. Then repeat steps 1)-5) to install equipment 2. Repeat the steps until all the equipment are installed.

For the convenient consideration, a kind of fast connection and fast release end-effector is designed. The end-effector is divided into two parts: robot connecting part and equipment connecting part. The robot connecting part is connected to the six-dimensional force sensor on the end of the robot. The equipment connection part is connected to the equipment to be installed. And the two parts are connected by a fast connection and fast release connector. The assembly system is shown by Figure1.

For the installation requirements of various equipment with different interface sizes, the equipment connecting parts with different interface sizes are designed. These connecting parts have the same interface with the robot connecting parts, and the interface is a fast connection and fast release interface with simple operation, which is convenient for switching between different interface devices and supports a batch of equipment installation.

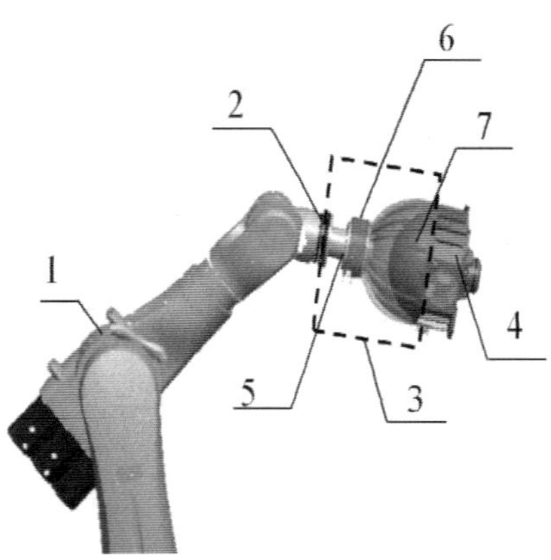

1. Robot;
2. Six-dimensional force sensor;
3. End-effector;
4. Equipment;
5. Robot connecting part;
6. Fast connection and fast release connector;
7. Equipment connecting part.

Figure 1. Robot assembly system for spacecraft large-weight equipment installation

IOP Conf. Series: Materials Science and Engineering **616** (2019) 012010 doi:10.1088/1757-899X/616/1/012010

2.2 Work process

The connecting parts of the end effector robot are installed on the robot. When the robot is initially installed, the locking actuator is in the initial position. The connecting parts of equipment are installed separately on each equipment to be installed, and then the robot reaches the position of the equipment to be installed, and docks the robot connecting part with the docking slot on the equipment connecting part, triggers the docking switch after docking, lights up the docking indicator on the controller, and drives the actuator to turn forward and extend to start the locking action through the locking button on the controller. After the lock is completed, the indicator light on the controller is turned on, and the connection between the robot and the equipment to be installed is completed. After the robot installs the equipment 1 to the installation position, and completes the installation operation, it drives the actuator to reverse and retract to the initial position through the release button on the controller. The lock indicator light on the controller is turned off and the release indicator light is turned on. At this time, the robot reaches the position of the equipment2 to be installed, repeats the above actions, and completes the robot connecting part with the end-effector on the robot. The parts are connected with the connecting parts of the equipment on the equipment 2 to be installed. Repeat the above steps to complete the installation of all the equipment. The locking and loosening operation is simple and fast, which is convenient for batch installation of interface equipment of different sizes and improves the spacecraft assembly efficiency.

2.3 Interface Conditions

1) According to the Table1, the maximum weight of the equipment is 129kg, and for the consideration of some expansion, the maximum load of the end-effector is designed to be 150kg. The safety coefficient is 3.

2) For the safety consideration, the end-effector should have enough stiffness. If under the worst condition, the maximum deformation at full load should no more than 1 mm.

The six-dimensional force sensor employed in this paper is Omega190, which can load 150 kg weight. So the interface of robot connecting part which will be docked to the six-dimensional force sensor must be designed to be matched to the interface of Omega190.

3. System Design

3.1 Structure Design

The installation of end-effector and robot can be divided into coaxial type and suspension type[7].Coaxial means that the axis of the end effector is installed parallel to the axis of the flange of the robot. Suspension means that the axis of the end effector is installed vertically with the axis of the flange of the robot. In this paper, coaxial axis coincidence mode is chosen. The end-effector axis can coincide with the geometric axis of the equipment in order to avoid the vertical installation space of the equipment in the cabin.

The structure of the fast connection and fast release end-effector is shown by Figure 2. It is consisted by supporting structure, fast connector, connection and release actuator, equipment connecting structure, and force sensor.act..Fast connector is the part for fast connection and release. The connection and release actuator is the key executive part for connection and locking.

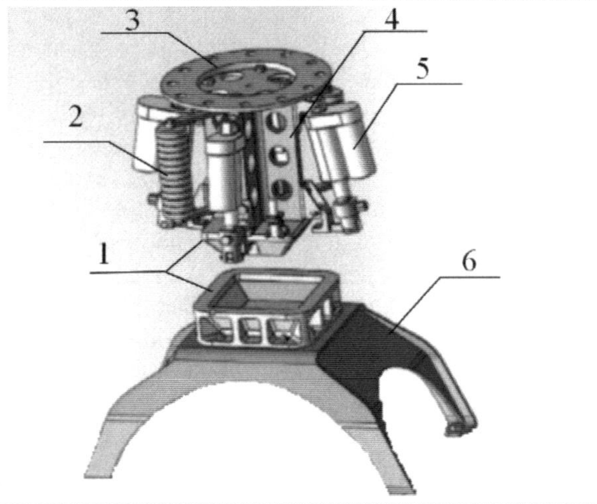

1.Fast connector;
2. Handler;
3. Force sensor interface;
4. Supporting structure;
5.Connection and release actuator;
6.Equipment connecting structure.

Figure 2. Framework of fast connection and fast release end-effector

3.2 Control system

The main components of the control system include control cabinet, PLC, sensor signal processor, relay, switch lamp and other control components, as well as proximity switch, force sensor and actuator sensing and executing devices. PLC, sensor signal processor, relay and switch lamp are installed on the control cabinet, and proximity switch, force sensor and actuator are installed on the terminal of the robot. Connecting with control cabinet through 24V DC specification cable in joint.

4. Simulation analysis

The material of the main structure is 30CrMnSi, whose yield point is 800MPa, and the tensile strength is1100MPa.

To simplify the fixed support scheme mentioned above, the finite element analysis model was obtained. Loading simulation calculation is carried out in three directions: X, Y and Z, with the maximum mass of the load is 150 kg. Y direction is the direction of vertical equipment connection surface, and the weight is distributed to four bolt connection holes on average. The strength of end-effector is analysed and the deformation is analysed.

4.1 Strength analysis result

The stress is simulated. The result is shown by Figure 3, and the maximum stress of X, Y, and Z direction are shown in table 2. From the result of the simulation, we can see the maximum stress is 195.16MPa on Z direction, and the minimum safety coefficient is 4.1, which is larger than 3. So strength of the main structure can satisfied the requirement.

Table 2. The results of strength analysis

Serial number	Simulated working condition	Maximum stress (MPa)	Safety coefficient
1	X	189.77	4.2
2	Y	42.03	19
3	Z	195.16	4.1

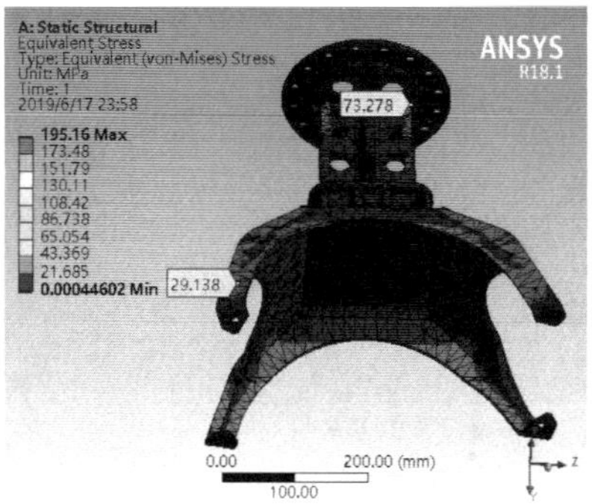

Figure 3. Strength analysis result of Z direction

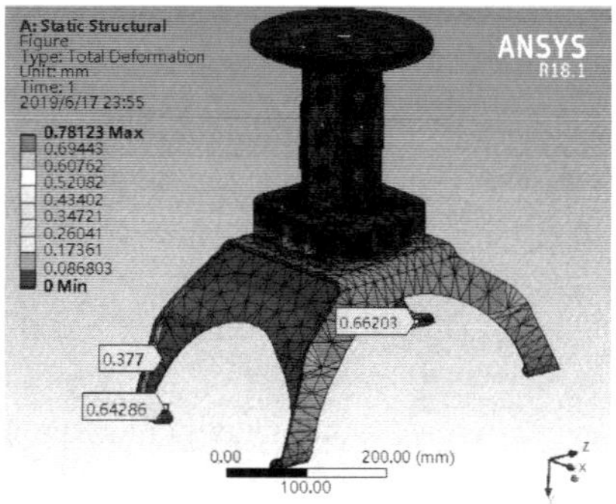

Figure 4. Deformation analysis result of Z direction

4.2 Deformation analysis result

The deformation is simulated. The result is shown by Figure 4, and the maximum deformation of X, Y, and Z direction are shown in table 3. From the result of e simulation, we can see the maximum deformation is 0.78mm on Z direction, which is shorter than1mm. So deformation of the main structure can satisfied the requirement.

Table 3. The results of deformation analysis

Serial number	Simulated working condition	Maximum deformation (mm)
1	X	0.40
2	Y	0.38
3	Z	0.78

5. Test verification

In order to verify the functions of the end-effector, a testing system was built, as shown by Figure 5, including KUKA KR210 Robot, Omega190 six-dimensional force sensor, end-effector design in this paper, and weight block which is simulating the equipment.

First function check was carried out, such as fast connection function check, connection locking function check, connection releasing function check, and etc., it is verified in good conditions. Then, loading test is carried out on the end-effector to simulate the actual installation of the equipment. Each connection of the end-effector and the equipment is loaded 100kg in directions of X, Y and Z, respectively. Deformation of the end-effector is measured. Trough measurement, the maximum deformation is 0.2mm. It can meet the requirements of the design.

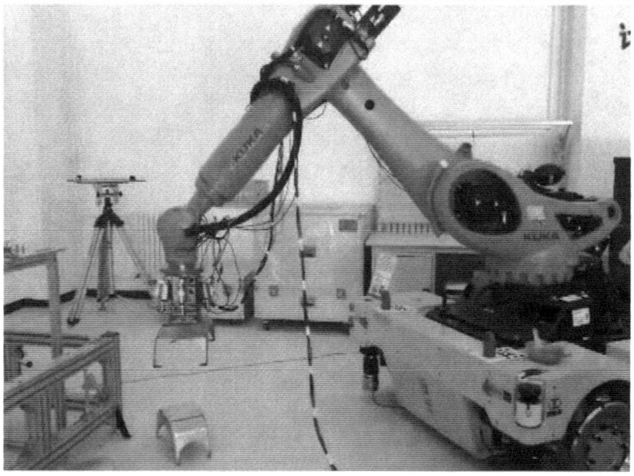

Figure 5. Testing system for end-effector based on robot

6. Conclusion

In this paper, an end-effector used for spacecraft assembly robot with fast connection and fast release function was developed in order to meet automatic installation requirements for several large-weight equipment. After design, the End-effector was manufactured, assembled and tested. It is verified in good functions, and can provide a strong basis for following research.

Reference

[1] Hu R Q, Zhang L J, Meng S H, Dong Q, Long C Y, Robotic Assembly Technology for Heavy Comoponent of Spacecraft Basedon Compliance Control, *JOURNAL OF MECHANICAL ENGINEERING*, 2018 54(11) :85-93

[2] Yi W M, Long C Y, Hu R Q, Design and Application of Spacecraft Assembly and Testing Robot System, Aerospace China, 2019 (2):30-33

[3] Wang J, Liu H. Design of Drilling End Effector for Aircraft Automatic Assembly, Journal of Nanjing University of Aeronautic and Astronautics, 2012 4(44):20-22

[4] Olsson T, Haage M, Kihlman H, Johansson R, Nilsson K, Robertsson A, BjÄorkman M, Isaksson R, Ossbahr G, Brog°ardh T. Cost-eficient drilling using industrial robots with high-bandwidth force feedback. Robotics and Computer-Integrated Manufacturing, 2010, 26(1): 24-38

[5] Kubela T, Pochyly A, Singule V, Flekal L. Force-torque control methodology for industrial robots applied on finishing operations. Mechatronics: Recent Technological and Scien-tific Advances. Berlin: Springer-Verlag, 2012. 429-437

[6] DeVlieg R, Feikert E (2008) One-up assembly with robots. In: SAE 2008 Aerospace Manufacturing and Automated Fastening Conference & Exhibition, North Charleston, SC, USA.SAE Technical Papers 2008,44(01):297~302.

[7] Cheng D X. Mechanical Design Manual (Volume 4) [M]. 4th Edition Beijing: Chemical Industry Press, 2002.

Research on Test Technology of Ionizing Radiation Effect of Spacecraft Material

Li Man[a], Liu Yuming, Zhao Chunqing, Zhang Kai, Liu Xiangpeng and Xu Yun

Beijing Institute of Spacecraft Environment Engineering, Beijing, China,100094

[a] li-man8319@163.com

Abstract. ^{60}Co γ-ray source and 45keV electron source are used as radiation sources respectively to study the ionizing radiation effect of spacecraft material heat-shrinkable sleeve, and compare the radiation effect similarities and differences between the two radiation sources. The results show that in the damage mechanism, both γ-rays and electrons cause ionization and excitation of electrons outside the nucleus in the material, producing ionization effects, and resulting in degradation and cross-linking of the molecular chains of the material. However, in terms of macroscopic mechanical property degradation, γ-rays can cause complete loss of mechanical properties of material, while 45keV electrons only cause partial damage to the mechanical properties of the material. In addition, at the same dose, γ-rays can damage the mechanical properties of the material to a greater extent than 45keV electrons. This paper explains the phenomenon and puts forward the applicable conditions of the two test technologies.
Keywords: Spacecraft material, ionizing radiation effect, heat-shrinkable sleeve, damage mechanism

1. Introduction

There are lots of electrons in the earth orbit of the spacecraft, which mainly come from the earth's radiation belt. These space electrons are irradiated on the spacecraft, which will produce electron irradiation effects on materials such as thermal control materials, optical materials, and insulating materials used by the spacecraft, thus affecting the working life and reliability of the spacecraft [1, 2]. Therefore, it is necessary to use the ground test technology to study the damage effect of the space electron radiation environment on materials.

According to the theory of atomic physics, electrons, because of their light weight and fast motion, ionize or excite the atoms of materials when they interact with them. Ionization loss is a major way of electron loss in the material, which in turn causes the ionizing radiation effect of the material being irradiated. The electron radiation environment in space mainly comes from the radiation belt, with several keV to 100 keV electrons as the main source, and the electron flux above 1MeV is relatively small. The damage effect of the induced material is mainly caused by the ionizing radiation effect. In the ground test research work, since γ-rays can also induce the ionizing radiation effect of materials, therefore, in addition to the use of electron sources for the research on ionizing radiation effects, γ-rays are often used as radiation sources for tests [3, 4]. There are two types of ground test technologies: γ-ray test method and electron irradiation test method. However, in the two test methods, whether the ionization damage caused by different irradiation sources (electrons, γ-rays) is consistent, especially whether the damage is consistent under the same total dose. There is little research in this area and no specific conclusions have been reached.

In this paper, the polyethylene heat-shrinkable sleeve is taken as the research object, and the electron source and ^{60}Co γ-ray source are used to study the effects of ionizing radiation induced by electrons and γ-rays respectively. The influence of test technologies on the ground by using different simulation sources on the test results is analyzed.

2. Test Design

The electron radiation environment of the geosynchronous orbit is selected for the space electron radiation environment, and the AE8 model is selected for the earth radiation belt model selects. Figure 1 shows the relationship between the ionizing radiation dose and the shielding thickness of the space electron radiation environment of the satellite in orbit for 15 years. The figure is based on the analysis results of the solid sphere shielding model.

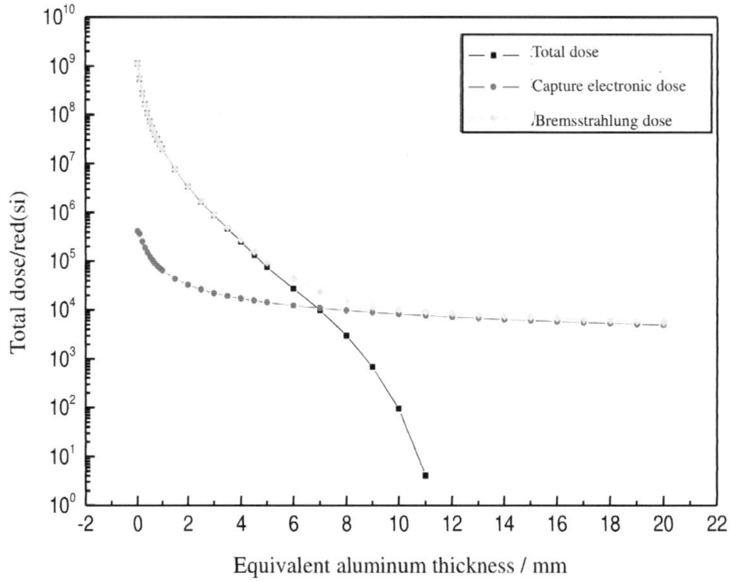

Figure 1. The dose deepness curve of space irradiation

Extrapolated from the calculated data, the total dose accepted by the exposed material in the 15-year GEO orbital environment without shielding is 2×10^9 rad (Si).

The test sample is a transparent heat-shrinkable sleeve for satellite. As one of the important protective materials of spacecraft cables, this material has been widely used in spaceflight.

In Table 1, the elongation at break is the mechanical property parameter of the high-polyethylene heat-shrinkable sleeve. Elongation at break refers to the ratio of the displacement value of the sample to the original length at the time of breaking, and the result is expressed as a percentage (%). Surface morphology, X-ray photoelectron spectroscopy and infrared spectroscopy are used to analyze the mechanism of material ionization damage.

Table1. Parameters of transparent heat-shrinkable sleeve and test instruments

parameter	test instrument
elongation at break	CMT4104 Eletronic Universal Testing Machine
surface morphology	JEM7401 scanning electron microscope
X-ray photoelectron spectroscopy	AXIS Ultra
infrared spectroscopy	-

The dose rate of the ^{60}Co γ-ray source irradiation test is 100 rad (Si) / s, and the total dose is 8.64×10^7 rad (Si). An electron energy of 45 keV and a total fluence of 2.0×10^{16}e/cm^2 are selected for the electron irradiation test which is equivalent to a 15-year GEO orbital electron radiation environment [5].

3. Test Results and Analysis

3.1 Consistency Analysis of Damage Mechanism

Since the space is an electronic radiation environment, when using γ-rays to study the ionizing radiation effect of spacecraft materials, first, it must be ensured that the damage mechanism of ionizing radiation effect on materials is the same. On this basis, it is possible to analyze whether the macroscopic performance degradation of materials is consistent.

Figure 2 is the surface morphology of the heat-shrinkable sleeve after different irradiation tests. As can be seen from the figure, the surface of the unirradiated heat-shrinkable sleeve is relatively flat with some wrinkles and particulate matters (Figure 2a). After γ-ray irradiation, the surface of the sample did not change significantly, and the surface of the sample was slightly wrinkled or slightly peeled, but in general, the morphology did not change significantly (Figure 2b). After the electron irradiation, the surface is rougher than the original sample, and there is a bulge similar to the particles, and the surface is uneven (Figure 2c).

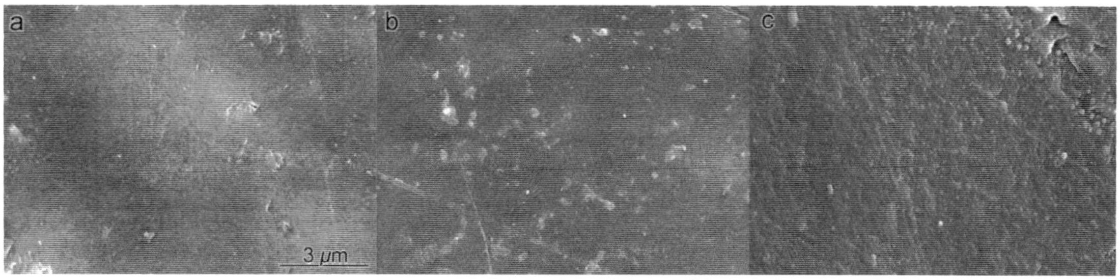

Figure 2. The SEM images of heat-shrinkable sleeves before and after the irradiated.
(a) Before irradiation. (b) Irradiated by γ ray. (c) Irradiated by electron beam

Table 2 shows the element content (XPS analysis) of the heat-shrinkable sleeve after different irradiation tests. The high-polyethylene heat-shrinkable sleeve is prepared by the silane cross-linking method, grafting reaction between polyethylene and organosilane is used to obtain cross-linked silane grafted polyethylene, and then the heat-shrinkable sleeve is prepared by subsequent catalytic hydrolysis condensation reaction. The main component is C_2H_4, and it also contains other materials such as silicon and oxygen.

Table 2. Elements contents variation of heat-shrinkable sleeves

Experimental condition	elements contents（%）						
	Na 1s	F 1s	O 1s	N 1s	C 1s	S 2p	Si 2p
sample	0.2	-	3.02	-	96.28	0.05	0.45
60Co γ ray.	0.2	-	7.74	0.25	91.37	0.06	0.37
45keV electron	0.18	1.08	15.19	0.44	79.54	-	3.57

The results showed that the oxygen content on the surface of the samples increased after the two irradiation tests. This is because under the action of irradiation, the molecules of the high-pressure polyethylene heat-shrinkable sleeve are ionized, the molecular bonds are broken, and free radicals are formed. Oxygen molecules combine with free radicals to form C-O bonds, which cause an increase in the oxygen content of the sample after irradiation. This shows that the formation of free radicals caused by ionization damage is an important mechanism for the damage of heat-shrinkable sleeve. It can also be seen from the results that the silicon content of the sample irradiated by electrons increased greatly. Combined with SEM observation results, electrons are likely to cause sputtering on the surface of the material, causing a small amount of polyethylene molecules to be completely destroyed, leaving silicon elements, which causes the surface silicon content to rise and the surface to have irregularities.

Figure 3 is the infrared spectrogram (FTIR) of the heat-shrinkable sleeve after different irradiation tests. Polyethylene is mainly C_2H_4 structure. From the infrared spectrum, it can be seen that there are 2928cm^{-1}CH_2 anti-symmetric stretching vibration, 2850 cm^{-1}CH_2 symmetric stretching vibration, 1472 cm^{-1}CH_2 plane bending vibration, 731cm^{-1} oscillating vibrations in the surface of the carbon chain and other main absorption peaks, and C-O vibration absorption peak near 1178 cm^{-1}. The infrared spectrum of the heat-shrinkable sleeve after irradiation is compared with the original, the original characteristic peak becomes stronger, and the heat-shrinkable sleeve after irradiation increases the absorption peak of 1715 cm^{-1}, corresponding to the C=O absorption peak, and the C=O absorption peak of the electron irradiated sample is significantly larger than that of the γ-ray irradiated sample. The absorption peak of C-O has also increased. It shows that after γ-ray and electron irradiation, the surface of the heat-shrinkable sleeve shows degradation and cross-linking [6][7].

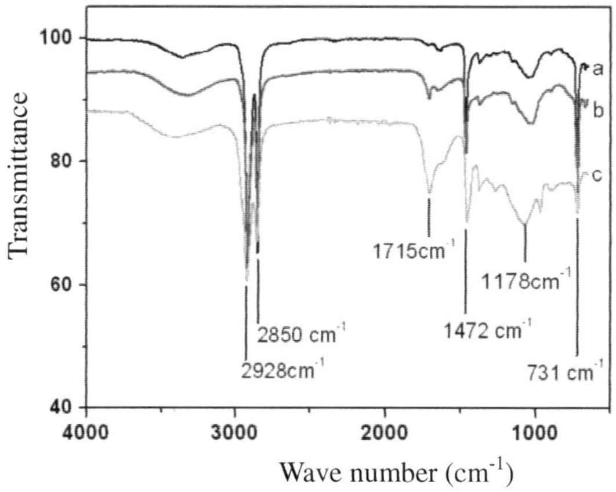

Figure 3. Analysis of infrared spectrogram before and after the irradiated.
(a) Before irradiation. (b) Irradiated by γ ray. (c) Irradiated by electron beam

Analysis of the heat-shrinkable sleeve by different means shows that the chemical bonds that hold the atoms together of the polyethylene heat shrinkable sleeve change under the irradiation environment. This phenomenon can occur in both macromolecular main chains and branched chains. As a result, active free radicals, ionization or excitation are generated in the macromolecules and chemical reactions are initiated. Among them, the most basic reactions affecting the properties of the polyethylene heat-shrinkable sleeve are the cross-linking reaction and the degradation reaction. Cross-linking refers to the re-formation of new chemical bonds between molecules, and degradation refers to the breakage of polymer molecular chains. In fact, degradation reactions and cross-linking reactions often occur simultaneously. For the consistency of the damage mechanism of electron and γ-ray irradiation on polyethylene heat-shrinkable sleeve, it can be seen that the changes of surface morphology between electron and γ-ray is different, and γ-ray almost did not cause the changes of surface morphology of heat-shrinkable sleeve, while electron irradiation caused bulges and particles on the surface. However, from the perspective of molecular chain degradation and cross-linking, the two irradiation methods are consistent. Since the surface morphology may be related to the sputtering effect of electrons and does not have a dominant influence on the overall mechanical properties of the sample, it can be considered that the mechanism of micro damage caused by electron irradiation and γ-ray irradiation on heat-shrinkable sleeve is basically the same.

3.2 Property Damage Consistency Analysis

The mechanical properties of the samples after different irradiation tests are shown in Figure 4. It can be seen from Figure 4 (a) that the elongation at break of polyethylene heat-shrinkable sleeve increases first and then decreases under γ-ray irradiation. This may be related to the change of the leading role of polyethylene degradation and cross-linking in the irradiation process. At the beginning of the radiation, the received radiation dose is very low, cross-linking plays a major role, and the properties of the polyethylene heat-shrinkable sleeve increase. But when the irradiation continues, the degradation begins to dominate, the properties of the polyethylene heat-shrinkable sleeve begin to decline. Since the total dose of the γ-ray irradiation test is only 8.64×10^7 rad (Si), which is the dose received by the exposed material in the GEO orbit environment of about 9 months. When the radiation dose reaches the 15-year GEO orbital radiation dose, the polyethylene heat-shrinkable sleeve will be completely cracked and without mechanical properties [8]. Therefore, the mechanical properties of the polyethylene heat-shrinkable sleeve will eventually be completely lost under γ-ray irradiation.

Figure 4(b) shows the property change of the heat-shrinkable sleeve under 45 keV electron irradiation. It can be seen from the figure that under the electron irradiation, the properties of heat-shrinkable sleeve does not first rise and then decline, but shows a rapid decline and then tends to be stable. This is because the dose deposited at the electron minimum fluence point in the material is much higher than the radiation dose range when the degradation is dominant, so there is no increase in properties. In addition, after the electron irradiation reached a certain fluence, the properties of the polyethylene heat-shrinkable sleeve showed a stable trend, indicating that the electron irradiation could not completely destroy the polyethylene. The reason is that 45 keV electrons can only be deposited in materials of several micrometers thick, while the polyethylene heat-shrinkable sleeve has a certain thickness, and electrons cannot act on the entire material. Since only a very thin layer of material loses its properties, the mechanical properties of the polyethylene heat-shrinkable sleeve are not completely lost due to low energy electron irradiation.

In addition, the total electron fluence is 2.0×10^{16}e/cm^2, which is equivalent to the 15-year GEO orbital electron radiation environment. It shows that under the same irradiation dose, the polyethylene heat-shrinkable sleeve is more seriously damaged by γ-ray irradiation, while under electron irradiation, the degree of damage is small.

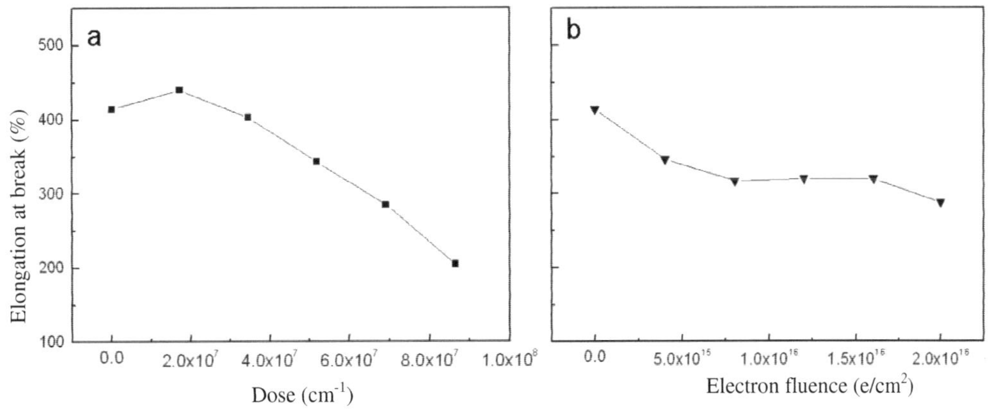

Figure 4. The changes of elongation at break of the polyethylene heat-shrinkable sleeve in irradiations. (a) Irradiated by γ ray. (b) Irradiated by electron beam

3.3 Analysis

The space electron radiation environment mainly induces ionizing radiation effects in materials. When the ionizing radiation effect on the polyethylene heat-shrinkable sleeve is carried out on the ground, from the damage mechanism of the material, both γ-ray sources and electron sources will cause degradation and cross-linking of polyethylene. Although there is a difference between electron irradiation and γ-ray irradiation in terms of surface properties, for bulk materials, this difference does not affect the relationship between macroscopic property change and damage mechanism. Therefore, from the damage mechanism, the ionizing radiation effect of γ-ray source and electron source to the materials researched is consistent.

From the macroscopic mechanical properties of the polyethylene heat-shrinkable sleeve, the degree of damage of γ-ray source and electron source to the mechanical properties of the heat-shrinkable sleeve is very different. First, when the γ-ray source is used for irradiation, due to the large penetration depth of γ-rays, the whole heat-shrinkable sleeve is affected by the rays, and the mechanical properties decrease until they are completely lost. When the 45keV electron source is used for irradiation, due to the small penetration depth of electrons, only the superficial heat-shrinkable sleeve is subject to electronic action. The mechanical properties of the heat-shrinkable sleeve decrease and then tend to be stable. In the end, the mechanical properties will be stable at a value and will not be completely lost. Secondly, at the same dose, γ-ray irradiation damages the mechanical properties of the heat-shrinkable sleeve more, which may also be related to the depth of γ-ray action. Only the surface properties of heat-shrinkable sleeve change under electron irradiation, so the overall property degradation is small. Through damage mechanism analysis and property damage analysis, it can be known that γ-ray can be used to study the ionizing radiation effect of materials in the space electron radiation environment, but it cannot be used to evaluate the change of the property damage degree of the material in the space electron radiation environment with the in-orbit time (the in-orbit time determines the radiation dose received by the material under the premise that the orbit is known). When evaluating the property damage of materials, it is necessary to select appropriate electron sources according to the dose depth distribution of spatial electron radiation deposited in the materials and carry out related evaluation work.

4. Conclusions

This paper compares two test technologies commonly used to study the ionizing radiation effects of space electron radiation environment on spacecraft materials, namely the test technologies using γ-ray sources and electron sources as simulated radiation sources, respectively. The results show that both γ-rays and electrons can produce ionizing radiation effects on the material, resulting in degradation and cross-linking of the polyethylene heat-shrinkable sleeve. This is the micro-damage mechanism that causes the degradation of the macroscopic mechanical properties of the heat-shrinkable sleeve. In this point, the two test methods are consistent.

From the perspective of macroscopic mechanical property degradation of the heat-shrinkable sleeve, the γ-rays and electrons have different penetration depths, so there is a big difference in the damage of macroscopic mechanical properties of the material. γ-rays can cause overall damage to the material, while electrons can only cause damage to the material within the depth of electron penetration. And γ-rays damage the material more than electrons at the same dose. In this point, the two test methods are inconsistent.

In conclusion, when studying the ionizing radiation effect of spacecraft materials, γ-rays or electrons can be used as radiation sources to carry out tests; When assessing the degradation of the in-orbit performance of spacecraft materials, the electron source should be selected as the radiation source, and detailed test design and test research should be carried out according to the space electron radiation environment.

Reference

[1] Wang Li, Xing Yan. Materials in space applications and the support technology[J]. Spacecraft Environment Engineering, 2010, 27(1): 35-40

[2] XUE Yu-xiong,YANG Sheng-sheng, BA De-dong, AN Heng, LIU Qing, SHI Hong, CAO Zhou. Analyze of spacecraft system failures and anomalies attributed to the natural space radiation environment [J], Vacuum and Cryogenics, 2012, 18(2): 63-70

IOP Conf. Series: Materials Science and Engineering **616** (2019) 012011 doi:10.1088/1757-899X/616/1/012011

[3] WANG Xue-qin, ZHANG Chun-xi, JIN Jing, SONG Ningfang, YUAN Huizheng. Radiation-induced attenuation effect on special optical fibers applied in space [J]. Infrared and Laser Engineering, 2011, 40(12): 2516-2520

[4] TIAN Hai, FENG Zhan-zu, YANG Sheng-sheng. Evaluation of charged particle radiation effect on spacecraft optical lens[J]. Journal of Northwest Normal University, 2014, 50(5):33-36

[5] long-term degradation of solar absorptance properties of geostationary satellite thermal control coatings[J]. Chinese Space Science and Technology. 2005, 25 (3): 34-40]

[6] Sun jiansheng, et al. Study of the Irradiation Ageing Test of the Materials for Class 1E Cables for Nuclear Power Plant[J]. ELECTRIC WIRE & CABLE, 2010, 1:26-28

[7] Mansour S A. Effects of γ-irradiation on the Mechanical and Relaxation Behavior of High and Low Density Polyethylene [J]. Egypt.J.sol, 2001, 24(1):89-100.

[8] Liu yuming, et al. Radiation effects on the insulation properties of heat-shrinkable sleeves [J]. Spaceraft Environment Engineering. 2010, 27(1):59-62.

IOP Conf. Series: Materials Science and Engineering **616** (2019) 012012 doi:10.1088/1757-899X/616/1/012012

Analysis on Synchronizing and Gear Shifting Force of Commercial Vehicle Powertrain

Zhengfeng Yan[1], Peisheng Zhao[1] and Yajun Wu [2*]

[1.] School of Mechanical Engineering, Hefei University of Technology, Hefei, 230009, China;
[2.] Research and Development Center, China First Automobile Group Co., LTD, Changchun, 130011, China

[*]Corresponding author: wuyajun@faw.com.cn

Abstract. Based on a commercial vehicle transmission as the research object, the paper analyses the influence for commercial vehicle powertrain synchronization and shift performance by the rotational inertia of the clutch driven plate assembly and transmission input speed. According to the synchronization and shifting mechanism of the transmission, transmission shift dynamics model was established. Analyzing the causes of shift impact and the critical shifting force for mis-engagement put forward the shift control strategy to avoid shift impact. We can draw a conclusion that optimizing the clutch driven plate inertia parameter during clutch design can effectively improve the shift performance of the transmission.

1. Introduction

For traditional manual transmission(MT), the driver completes the shift of the target shift position through the shift fork. With the application of hydraulic and electronic control systems in the transmission, automatic mechanical transmission (AMT) has been developed, which has the advantages of simple operation, low cost and reliable operation. In recent years, AMT has been widely used in heavy-duty commercial vehicle.

At present, the proportion of assembling AMT on heavy commercial vehicles in Europe and the United States is gradually increasing. It is predicted that in the next few years, nearly 50% of MT in Europe will be replaced by AMT [1]. The AMT transmission has high reliability, but the shifting performance of the transmission still needs to be studied in depth. For example, during the shifting process, due to the fact that the current gear speed and the target gear speed are not consistent, this will causes mis-engagement problem. Such an operation would increase shift impact and severely damage the synchronizer, affecting the life of the transmission. Literature [2] used ADAMS to simulate and analyse the various stages of the shifting process, and the relationship between shifting force and shifting displacement is obtained, which provides a basis for the transmission shift control. Literature [3] established the clutch dynamics model, analysed and optimized the stiffness and damping of the clutch by ADAMS software to improve the performance of the clutch. Literature [4] analysed the influence of the clutch's output shaft inertia and stiffness on the performance of the wet clutch by bench test and proposed an effective scheme for optimizing the shift performance. Literature [5] studied all factors affecting the gear shifting force by fault tree analysis and found the fundamental reason of the problem is that the capacity of synchronizer torque is not enough and the force of self-locking spring is too big. To solve the problem of shift impact and synchronizer abnormal abrasion in electronic automatic mechanical transmission (AMT) of hybrid electric vehicle (HEV), the mechanism of shift impact and the causes were analysed and experimented[6]. The existing research results analysed the shifting characteristics of the clutch, self-locking spring and transmission shaft, and there is no analysis of the clutch driven plate inertia. In current paper analysed the influence of clutch driven plate inertia on shifting performance by modelling and analysing the shifting process. Furthermore, using the transmission test bench verified the proposed shift strategy.

2. Analysis of shift process

The equivalent concentrated inertia method is used to simplify the vehicle powertrain system into a linear multi-degree-of-freedom elastic concentration inertia system, and the relevant components exist in the form of concentrated inertia [7].

2.1. Shift performance evaluation index

Shifting force. The main components affecting the shifting force during the shifting process are the fork shaft, the shift fork, the locating sleeve and the synchronizer, and the friction of the surface of each component when in contact [8].Refer to literature [9], the shifting force can be obtain,

$$F = \frac{J\Delta\omega\sin\theta}{iR\mu\eta\Delta t} \tag{1}$$

Where J is the synchronization inertia; $\Delta\omega$ is the rotational speed difference; Δt is the synchronization time; R is the effective radius of the synchronization ring cone; μ is the friction coefficient; i is the leverage ratio; η is the operating system efficiency; θ is the synchronization ring cone angle.

Shift impact. The degree of impact is generally used to evaluate shift performance. The mathematical expression of the impact j is:

$$j = \frac{da}{dt} = \frac{d^2v}{dt^2} \tag{2}$$

Where j is the vehicle impact; a is the vehicle acceleration; v is the vehicle speed; t is the time. In general, the upper limit $j_{\max}$ of the German standard shift impact index is $10\,m\cdot s^{-3}$.

2.2. Critical shifting force for mis-engagement

At the beginning of the synchronous shifting phase of the mechanical synchronizing state, under normal circumstances, the synchronization ring rotates half a tooth width angle under the action of the friction torque, and the engaging sleeve teeth are brought into contact with the synchronization state; If the engaging sleeve moves too fast, it may move the distance d before the synchronization ring moves half a tooth width angle, and will directly mesh with the synchronization ring miss the synchronizing phase. At this time, the engaging sleeve and the ring gear still have a large rotational speed difference. When they are mesh, a serious toothing phenomenon will occur, which is called mis-engagement toothing [10].

According to literature [10] the critical value of the shifting force:

$$F_{shift}' = f(F_0) \tag{3}$$

When, $F_{shift} \leq F_{shift}'$,mis-engagement phenomenon does not occur.

3. Theoretical shifting force calculation

The transmission scheme is shown in figure 1. Refer to literature [9], the input part should be converted to the output shaft during the calculation of the shifting force.

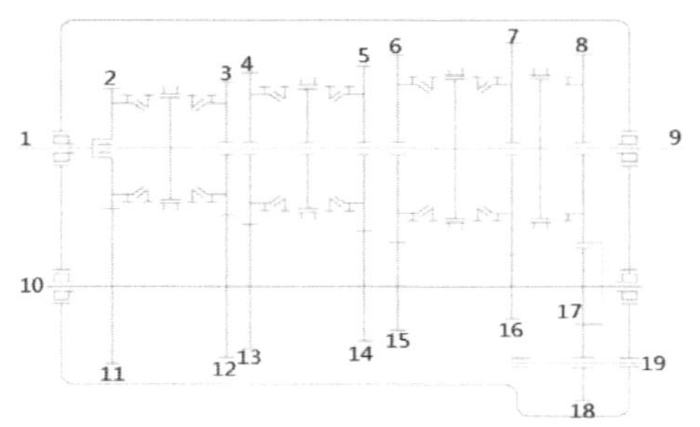

Figure 1. Transmission schematic.

3.1. Theoretical shifting force calculation

The parameters of the selected test transmission samples are listed in tables 1 to table 3.

Table 1. Moment of inertia of each axis $kg \cdot m^2$

	Input shaft	intermediate shaft	Output shaft	clutch driven plate
moment of inertia	4.315	12.05	1.939	37.308

Table 2. Intermediate axis parameters

intermediate shaft	1st	2nd	3rd	4th	5th	6th	Rear
Teeth	11	20	27	35	40	43	11
moment of inertia($kg \cdot m^2$)	0	0	0	0	9.9	13.8	0

Table 3. Output axis parameters

Output shaft	1st	2nd	3rd	4th	5th	6th	Rear
Teeth	43	47	38	30		21	40
moment of inertia($kg \cdot m^2$)	26.8	15.8	10.1	5.6	9.8	2.9	15.8

The theoretical shifting force of the 2 → 1 gears with different speeds and inertias calculated by equation (1) is listed in table 4.

Table 4. Calculation of theoretical shifting force/ N

Gear position	moment of inertia /($kg \cdot m^2$)	Effective radius of the ring cone / m	Engine speed / $r \cdot \min^{-1}$		
			1 500	2 000	2 500
2nd -1st gear [a]	2.793 409 265	0.178 2	59.61	79.48	104.34
	2.796 119 265		62.63	82.51	109.38
	2.797 175 265		64.65	84.54	114.42

[a] Note: The synchronization time is 0.6s.

The calculation and analysis of the shifting force can be concluded that the shifting force increases with the increase of the engine speed under the same gear, and the shifting force increases as the moment of clutch driven plate inertia increases. It can be seen from the calculation of the theoretical shifting force that the clutch driven plate inertia has an influence on the shifting force.

4. Simulation analyses

4.1. Shifting force simulation analysis

Built shifting process model on the Matlab/Simulink platform and simulated the shift process. Taking 2→1 gears as an example. Using the shifting Simulink model simulate the 2 → 1 gears shifting process, the synchronizer is first engaged with the 2nd gear; at the 5th second, given the shift signal, and the synchronizer disengaged with the 2nd gear and then engaged with 1st gear. In this process, the positional relationship between the engaging sleeve and the locating pin is shown in figure 2.

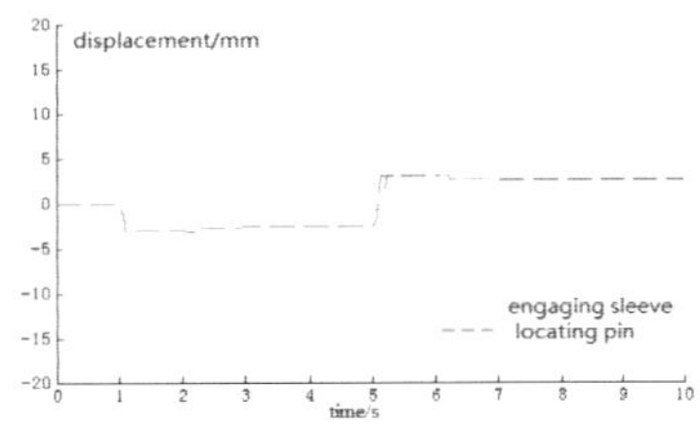

Figure 2. positional relationship between the engaging sleeve and the locating pin

The equivalent moment of inertia J_{rd} of the input shaft was changed to 2.793 409 265, 2.796 119 265 $kg \cdot m^2$, respectively. The simulation results of the shifting force are shown in figure 3. It can be conclude from the simulation results that the shifting force increases with the increase of the equivalent moment of inertia J_{rd} at the same engine speed.

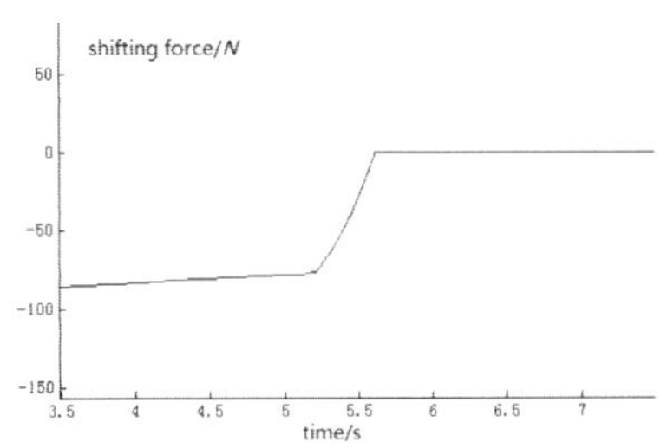

(a) J_{rd} =793 409 265 $kg \cdot m^2$

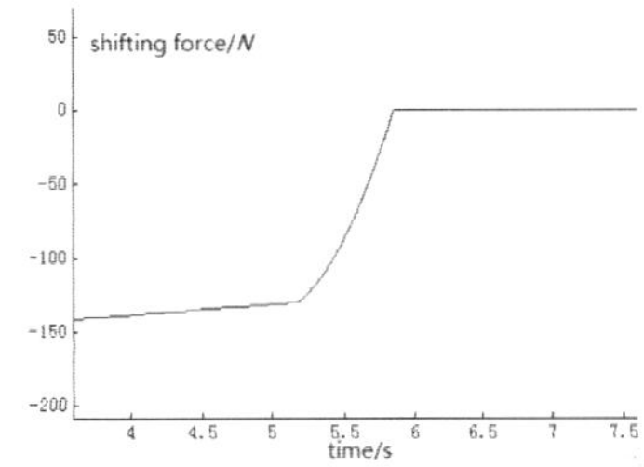

(b) J_{rd} =2.796 119 265 $kg \cdot m^2$

Figure 3. Shifting force under different inertia (N)

The engine speed was changed to 1 500, 2 000 $r \cdot min^{-1}$ for simulation. The simulation results are shown in figure 4. It can be conclude from the simulation results that the shifting force increases with the increase of the engine speed under the same equivalent moment of inertia J_{rd} .

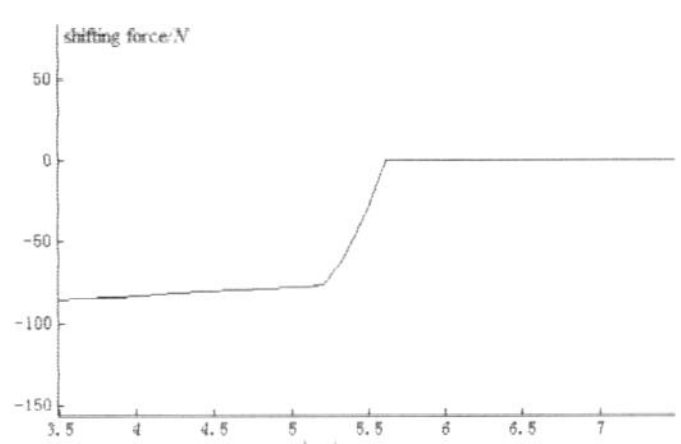

(a)engine speed n =1 500 r / min

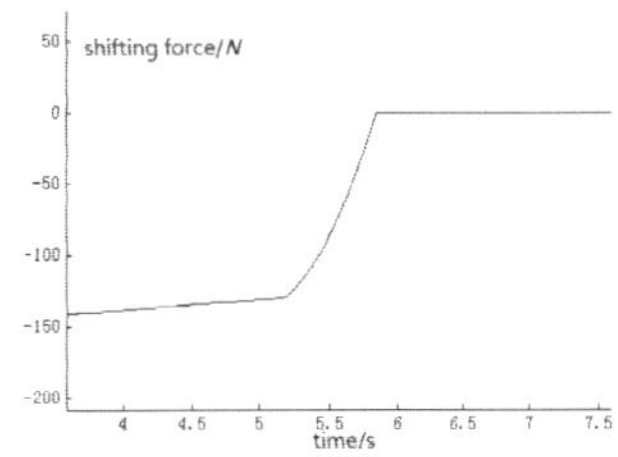

(b)engine speed n =2 000 r / min

Figure 4. Shifting force (N) at different engine speeds

4.2. Shift impact simulation

The shift impact simulation was performed at the same engine speed, equivalent moment of inertia J_{rd} of the input shaft was changed to 2.793 409 265, 2.796 119 265 kg · m^2, respectively, and the simulation results are shown in the figure 5.

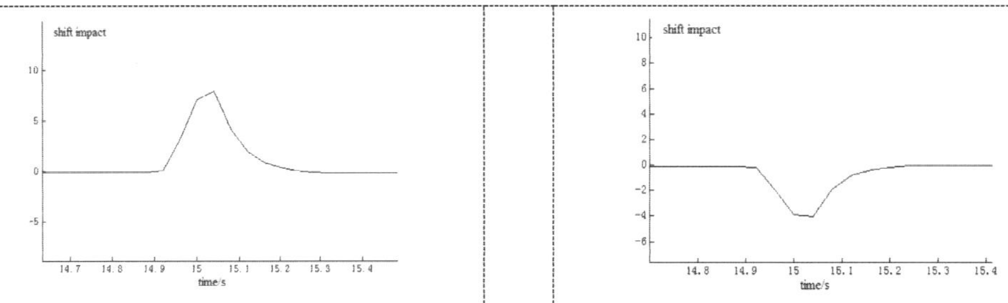

(a) J_{rd} =2.793 409 265 $kg \cdot m^2$ (b) J_{rd} =2.796 119 265 $kg \cdot m^2$

Figure 5. Shifting impact under different inertia

It can be seen from the simulation results of the shifting impact of figure 5 that the maximum shift impact is within the range of the upper limit $j_{\max} = 10m \cdot s^{-3}$ of the German standard. The increase of the clutch driven plate inertia will increase the shifting impact and affect the shifting performance. Therefore, in the design of the clutch, the influence of the clutch driven plate inertia on the manipulability of the transmission should be fully considered. At the same time, the shift control strategy should be optimized to avoid shifting impact.

4.3. Studies on Avoiding Shift Impact Control Strategy

Avoiding mis-engagement phenomenon can reduce the shift impact. From the previous analysis, the critical shifting force F_{shift} to avoid mis-engagement teething is:

$$F_{shift} \le F_{shift}'$$

For AMT, the shifting force can be controlled by controlling the cylinder chamber pressure of the shift actuator cylinder at different stages [11]. A flow chart of the control strategy for avoiding shift impact is shown in figure 6.

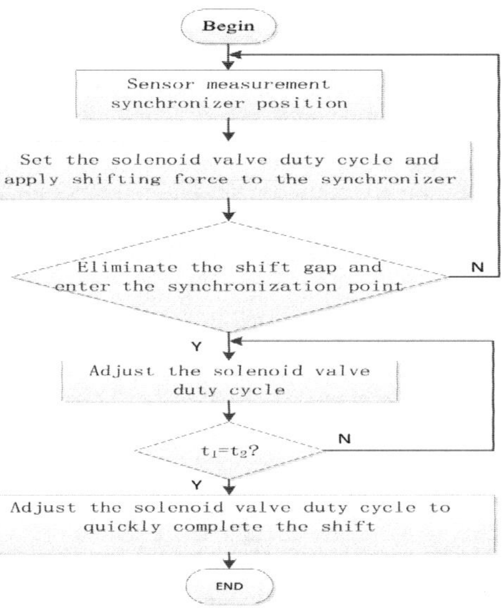

Figure 6. Shift synchronization process control flow chart

First, the position of the shift actuator is detected by the shift stroke displacement sensor. Then, the shifting actuator eliminates the shifting clearance. After the shifting clearance is eliminated, the synchronizer starts to enter the synchronization phase, and the electronic control unit controls the pressure of the shifting actuator cylinder to increase the force acting on the engaging sleeve, so that the synchronizer quickly enters the synchronization phase. In this process, the electronic control unit continuously adjusts the pressure of the cylinder according to the difference between the actuator position, the current gear position and the target gear speed. When detecting that the shift shaft speed is tighter than the target speed, the cylinder will adjusted again and quickly complete the shift process [12-13].

5. Bench test verification

5.1. Transmission test bench

For the above shift control strategy, an experiment was conducted on a test bench as shown in figure 7. The test bench can simulate the process of transferring the whole vehicle power from the clutch driven plate to the wheel during the shifting process. The main components are an electrical system, a pneumatic shifting actuator, a data acquisition system, and a simulated road resistance mechanism. The drag motor is connected with the inertia flywheel to simulate the inertia of the vehicle during driving; the transmission is connected to the clutch driven plate at the input end to simulate the inertia of the input end of the vehicle when shifting. During the test, the data acquisition system collects the sensor's signal in real time, and controls the movement of the shifting mechanism, and records the shifting data [14-15].

Figure 7. Transmission test bench

5.2. Bench test results

Take the test data of 2→1 gears for analysis. The data for the bench test are listed in table 5.

Table 5. Shifting force bench test results (N)

	Equivalent moment of inertia J_{rd} /($kg \cdot m^2$)	Engine speed/($r \cdot \min^{-1}$)		
		1 500	2 000	2 500
	2.793 409 265	64.12	90.60	108.26
2nd→1st [a]	2.796 119 265	68.28	91.47	112.91
	2.797 175 265	73.14	96.76	114.79

[a] Note: The synchronization time is 0.6s.

5.3. Analysis of test results

Take the test simulation and test data of $2 \rightarrow 1$ gears for analysis. Through the bench test, the transmission shifting force data under different conditions is obtained. According to the experimental data, the relation curve between the actual shift force and synchronization time was obtained. As shown in figure8 and figure9.

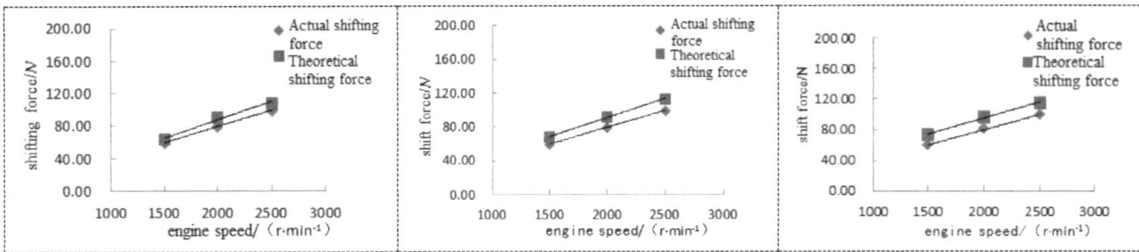

(a) J_{rd} =2.793 409 265 $kg \cdot m^2$ (b) J_{rd} =2.796 119 265 $kg \cdot m^2$ (c) J_{rd} =2.797 175 265 $kg \cdot m^2$

Figure 8. Actual shifting force and theoretical shifting force curve under different inertia

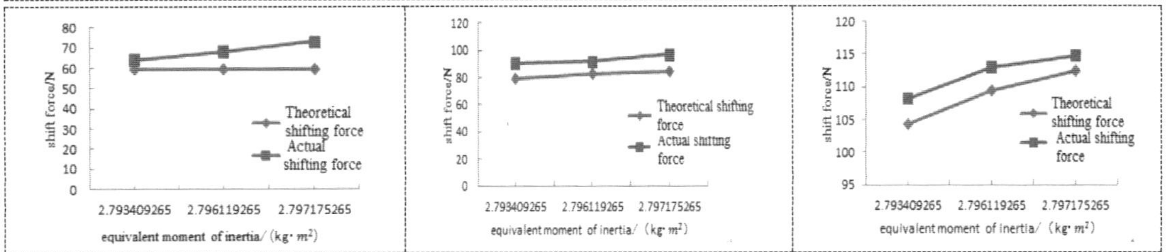

(a) n =1 500 $r/\min$ (b) n =2 000 $r/\min$ (c) n =2 500 $r/\min$

Figure 9. Actual shifting force and theoretical shifting force curve at different engine speeds

It can include from the shifting force curves of figure 8 and figure 9 that the change trend of the shifting force increases with the increase of the inertia of the clutch driven plate assembly at the same input shaft speed during the $2 \rightarrow 1$ gears; Under the same equivalent moment of inertia J_{rd}, as the input shaft speed increases, the change trend of the shifting force increases with the increase of the input speed. Comparing the test value and theoretical value and simulation value of the shifting force, it can be seen that the changing trend of the shifting force is basically the same, the average error is about 8%, and the overall error is basically within 10%. The error is mainly caused by the acquisition of the shift process data and the accuracy of the test bench.

6. Conclude

Through research, we can draw the following conclusions:

(1) By establishing the power transfer model of the shifting process and analyzing the factors affecting the shifting force simplified the theoretical calculation formula of the shifting force.

(2) Analyzed the shifting process of the synchronizer and the mechanism of the shift impact, and obtain the optimal shifting force that achieves the shifting requirement without generating the toothing phenomenon.

(3) The dynamic model of the shifting process is established, and the model is simulated by Matlab/Simulink. The simulation results show that the inertia of the clutch driven plate assembly affect the shift performance.

(4)The test results show that the moment of inertia of the clutch driven plate assembly affect the shifting force of the different gears.

From the above analysis, it is shown that the moment of inertia of the clutch driven plate assembly affect the shift performance. Therefore, optimizing the clutch driven plate inertia parameter during clutch design can improve the shifting performance of the transmission.

References

[1] Lu X T and Hou G Z 2004 Introduction of the AMT control system structure and main foreign AMT products **5** *Automobile Technology* 19-22

[2] Chen Y X, Zang M Y and Chen Y 2012 Shift force analysis of manual transmission based on virtual prototyping technology **23(8)** *China Mechanical Engineering* 996-1000

[3] Yuan D and Pan Y X 2009 Optimal design for stiffness and damping of automobile friction clutch *International Conference on Natural Computation. IEEE* 369-373.

[4] Fatima N, Marklund P and Larsson R 2013 Influence of clutch output shaft inertia and stiffness on the performance of the wet clutch **56(2)** *Tribology Transactions* 310-319.

[5] Yue L J,Si Z M and Chen J 2012 Analysis of causes of large gear shifting force on a light bus and the improvement **35(11)** *Journal of Hefei University of Technology(Natural Science)* 1465-68

[6] Liu C W and Liu K L 2011 Research and control on AMT shift torque of hybrid electric bus **25(1)** *Journal of HuBei Automotive Industries Institute* 22-24

[7] Wang X F 2010 Automotive chassis design *Beijing: Tsinghua University Press* 127-133

[8] Cheng X X,Chen H X and Tian G Y 2014 Investigation on influence of shift force on shift impact of integrated motor-transmission system **(4)** *Automotive Technology* 1-5

[9] Wang Y ,Xi J Q and Chen H Y 2009 A study on the mechanism and countermeasures for shift-impact in AMT **31(3)** *Automotive Engineering* 253-257.

[10] Tian J Y 2009 Construction principle and design method of vehicle automatic transmission *Beijing: Peking University Press* 154-186

[11] Liu Y G 2010 Study on integrated control of passenger vehicles equipped with Dual Clutch Transmissions PhD thesis *College of Mechanical Engineering of Chongqing University, Chongqing, China*

[12] Zhu X F 2014 Research on shift control strategy of Dual Clutch Transmission M.A. thesis *Xi'an University of Science and Technology, China*

[13] Hoshino H 1999 Analysis on Synchronization Mechanism of Transmission *International Congress & Exposition.*

[14] Kim J, Sung D and Seok C 2002 Development of shift feeling simulator for a manual transmission *International Body Engineering Conference & Exhibition and Automotive & Transportation Technology Congress* 502–513.

[15] Hu J J ,Li G H and Wu G Q 2008 Accurate calculation of clutch torque transmission during vehicle starting **30(12)** *Automotive engineering* 1083-86

IOP Conf. Series: Materials Science and Engineering **616** (2019) 012013 doi:10.1088/1757-899X/616/1/012013

Parameter Effects of the Hydrostatic Propulsion Drive System on the Operation Accuracy of a Tamping Machine

Wenlin Wang[1,2*], Xiong Cai[2] and Zirong Zhou[1]

[1]School of Mechanical Engineering, Dongguan University of Technology, Dongguan 523808, China.
[2]College of Mechanical and Vehicle Engineering, Hunan University, Changsha 410082, China

[*]Corresponding e-mail: pianowwl@163.com

Abstract. A dynamic mathematical model of the hydrostatic propulsion drive system of a tamping machine is established in this study, and a simulation model is built in the AMEsim software environment using the mathematical model. Parameter effects of the hydrostatic propulsion drive system on the operation accuracy of the tamping machine are analysed. Simulation results show that the effective moment of inertia of the powertrain and the oil volumetric elastic modulus have obvious influences on the operation accuracy of the tamping machine, the effect of the total volume of the high-pressure circuit of the hydraulic system on the displacement accuracy is not obvious, but it has some influence on the velocity amplitude. The models established and results obtained in this work are of significance and instructive for design optimization and maintenance of tamping machines.

1. Introduction

Fluid power drives are widely used in modern industrial apparatus [1] and railway construction and maintenance machineries [2,3]. In a railway tamping machine [3,4], the hydrostatic propulsion drive system [5,6] plays an important role in the machine's operation accuracy, which includes the velocity stability and displacement accuracy.

Yan proposed a hydraulic drive system to obtain [7] both high speed for mobility and low speed for tamping, Wang et al. studied different acceleration modes [8] for a hydraulic engineering vehicle by using a hydraulic proportional pump and a hydraulic variable motor. Shi investigated the problems of out of step [9] in velocity and traction force of a tamping machine, Hu et al. improved the pressure stability [10] of the hydraulic clamping system of a tamping machine, by adding a resistance orifice in the circuit, and Zeng et al. [11] and Zhai et al. [12] respectively investigated the adjustment and maintenance technology for the hydraulic system of their tamping machines.

A dynamic mathematical model of the hydrostatic propulsion drive system of a tamping machine is established in this study, and a simulation model is built in the AMEsim software environment based on the mathematical model. Parameter effects of the hydrostatic propulsion drive system on the operation accuracy of the tamping machine are analysed, simulation results show that the effective moment of inertia of the powertrain, the total volume of the high-pressure circuit of the hydraulic system and the oil volumetric elastic modulus have obvious influences on the operation accuracy of the tamping machine. The models established and results obtained in this work will be instructive for tamping machine design and maintenance.

2. Mathematical model of the hydrostatic propulsion drive system

Figure 1(a) shows that the tamping machine is driven by two hydraulic motors, and Figure 1(b) illustrates that the hydrostatic propulsion drive system of the tamping machine is a typical closed-type hydraulic system employing a two-way variable displacement pump and a two-way constant displacement motor. Varying the displacement of the hydraulic pump will lead to running speed change of the tamping machine, and alternating the rotation direction of the hydraulic pump will lead to moving direction alternation of the tamping machine.

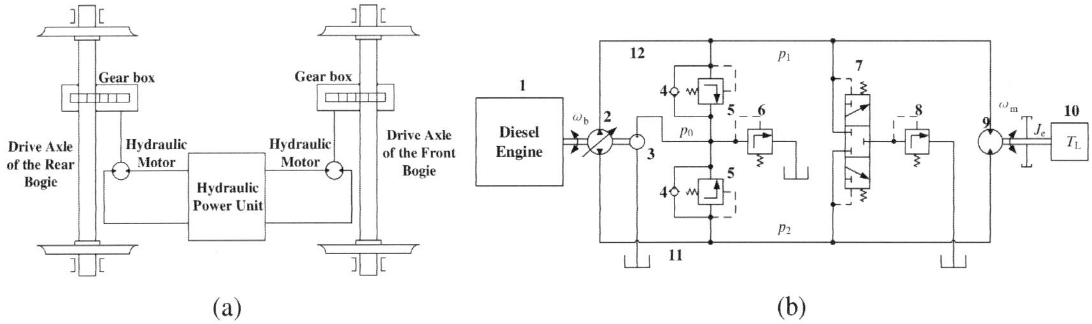

Figure 1. Mechanism (a) and hydraulic power circuit (b) of the hydrostatic propulsion drive system of a tamping machine (1. diesel engine, 2. variable pump, 3. fluid compensation pump, 4. check valve, 5. high-pressure relief valve, 6. fluid compensation relief valve, 7. flush valve, 8. back-pressure valve, 9. hydraulic motor, 10. load, 11. reservoir).

Referring to Figure 1(b), the flow continuity equation of the variable pump is formulated by

$$Q_b = K_b\omega_b\varphi_b - C_{ib}\left(p_1 - p_0\right) - C_{eb}p_1 \tag{1}$$

where Q_b, K_b, ω_b, ϕ_b, C_{ib} and C_{eb} are the output flow, the gradient of displacement change, the rotation speed, the swashplate angle, the inner leakage coefficient and the outer leakage coefficient of the pump, respectively; p_1 and p_2 are pressures of the high-pressure chamber and the back-pressure chamber of the pump. Therefore, taking the Laplace Transform of Equation (1) to obtain

$$Q_b(s) = K_b\omega_b\varphi_b(s) - C_b p(s) \tag{2}$$

The flow continuity equation of the high-pressure circuit of the hydrostatic propulsion drive system can be expressed by

$$Q_b - C_{im}\left(p_1 - p_0\right) - C_{em}p_1 = D_m\omega_m + \frac{V_0}{\beta_e}\frac{dp_1}{dt} \tag{3}$$

where C_{im}, C_{em}, D_m and ω_m are the inner leakage coefficient, the outer leakage coefficient, the displacement and the shaft rotation speed of the hydraulic motor, respectively; V_0 is the total volume of the high-pressure circuit of the hydraulic system, and β_e is volumetric elastic modulus of the oil. Similarly, taking the Laplace Transform of Equation (3) to obtain

$$Q_b(s) - C_m p(s) = D_m\omega_m(s) + \frac{V_0}{\beta_e}sp(s) \tag{4}$$

where C_m is the total leakage coefficient of the hydraulic motor, and $C_m=C_{im}+C_{em}$.

Thus, combining Equations (2) and (4) will have

$$K_b\omega_b\varphi_b(s) = D_m\omega_m(s) + \left(C_t + \frac{V_0}{\beta_e}s\right)p(s) \tag{5}$$

where C_t is the total leakage coefficient of the hydraulic system, $C_t=C_b+C_m$ and $C_b=C_{ib}+C_{eb}$.

The torque equilibrium equation of the hydraulic motor and load can be described as

$$D_{\mathrm{m}}\left(p_1 - p_0\right) = J_{\mathrm{e}}\frac{\mathrm{d}\omega_{\mathrm{m}}}{\mathrm{d}t} + B\omega_{\mathrm{m}} + T_{\mathrm{L}} \tag{6}$$

where J_e is the effective moment of inertia of the powertrain, B is the viscous damping coefficient, and T_L is the load torque. If neglecting the air drag resistance, the load torque can be formulated by

$$T_{\mathrm{L}} = r\sum F = r\left[mg\cos\beta(f_0 + kv) + mg\tan\beta + m\delta\frac{dv}{dt}\right] \tag{7}$$

where r, $\sum F$ and m are the wheel radius, the total resistance force and the mass of the tamping machine, (f_0+kv) is the coefficient of rolling friction of the wheels, β is the slope angle, and δ is an effective coefficient of the rotation mass of the tamping machine.

Thus, considering the vehicle speed $v=r\omega_m$, so combining Equations (6) and (7), and taking the Laplace Transform to obtain

$$D_{\mathrm{m}}p(s) = \frac{J_{\mathrm{e}} + B}{r}v(s) + (m\delta rs + mg\cos\beta rk)v(s) \tag{8}$$

Thus, if in terms of the swashplate angle of the pump φ_b as the input signal and vehicle speed v as the output signal, combing Equations (5) and (8) to obtain the transfer function of the hydraulic system, as

$$G(s) = \frac{v(s)}{\varphi_{\mathrm{b}}(s)} = \frac{rK_{\mathrm{b}}\omega_{\mathrm{b}}D_{\mathrm{m}}}{\dfrac{V_0\left(J_{\mathrm{e}} + m\delta r^2\right)}{\beta_{\mathrm{e}}}s^2 + \left[C_{\mathrm{t}}\left(J_{\mathrm{e}} + m\delta r^2\right) + \dfrac{V_0\left(B + mgr^2k\cos\beta\right)}{\beta_{\mathrm{e}}}\right]s + C_{\mathrm{t}}\left(B + mgr^2k\cos\beta\right) + D_{\mathrm{m}}^2} \tag{9}$$

In addition, if in terms of the load torque T_L as the input signal and vehicle speed v as the output signal, the transfer function of the hydraulic system can be described as

$$G_{\mathrm{L}}(s) = \frac{v(s)}{T_{\mathrm{L}}(s)} = \frac{1}{m\delta rs + mg\cos\beta rk} \tag{10}$$

Thus, based on the above mathematical modelling of the hydrostatic propulsion drive system, the dynamic control block diagram of the hydraulic circuit can be obtained and demonstrated by Figure. 2.

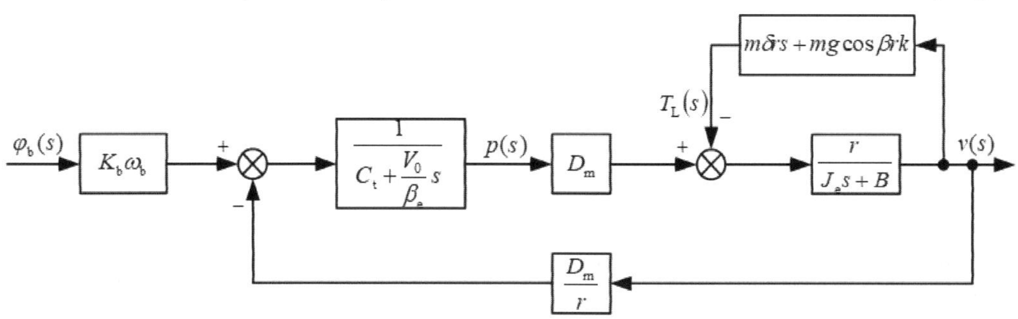

Figure 2. Dynamic control block diagram of the hydrostatic propulsion drive system.

3. Simulation model

Referring to Figure. 1(a) and the mathematical model established in Section 2, a simulation model of the hydrostatic propulsion drive system of the tamping machine is built in the AMEsim software environment, and the simulation model is illustrated by Figure. 3.

Simulation of the parameter effects of the hydrostatic propulsion drive system on the operation accuracy of the tamping machine is performed, the parameters and values used in the simulation are summarized in Table 1.

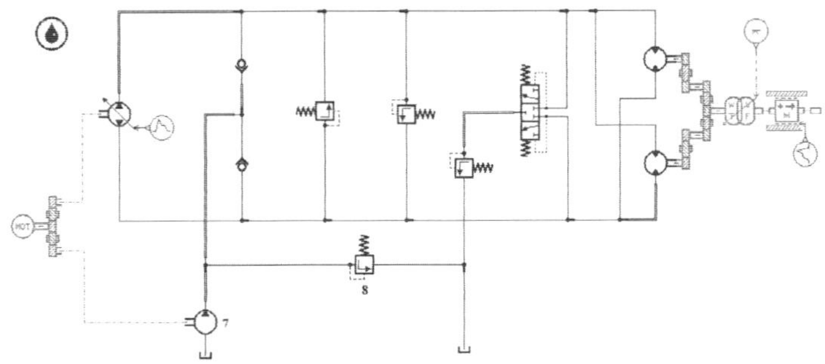

Figure 3. Simulation model of the hydrostatic propulsion drive system in the AMEsim environment.

Table 1. Parameters and values used in the simulation

Parameter	Value	Parameter	Value
Rotation speed of engine (r/min)	2300	Gear ratio of gear box	4.73
Maximum displacement of variable pump (mL/r)	125	Full laden mass of tamping machine (Kg)	60000
Displacement of fluid compensation pump (mL/r)	28.3	Wheel radius (m)	0.42
Set pressure of main relief valve (MPa)	40	Rotation damping coefficient	0.039
Set pressure of fluid compensation relief valve (MPa)	2.5	Braking damping coefficient	0.05
Relief pressure of flush valve (MPa)	2	Volumetric efficiency of hydraulic pump and motor	0.95
Displacement of hydraulic motor (mL/r)	500	Mechanical efficiency of wheel gear box	0.92

4. Parameter effects analysis

Figure 4 and Table 2 demonstrate the effect of the effective moment of inertia of the powertrain J_e on the operation accuracy of the tamping machine, and shows that J_e has an obvious influence on displacement accuracy of the tamping machine. For instance, if with a bigger J_e=12 kgm^2, the tamping machine will have a relative error of 3.78% on the displacement of the 1st cycle and a relative error of 5.84% on the velocity amplitude of the 1st cycle.

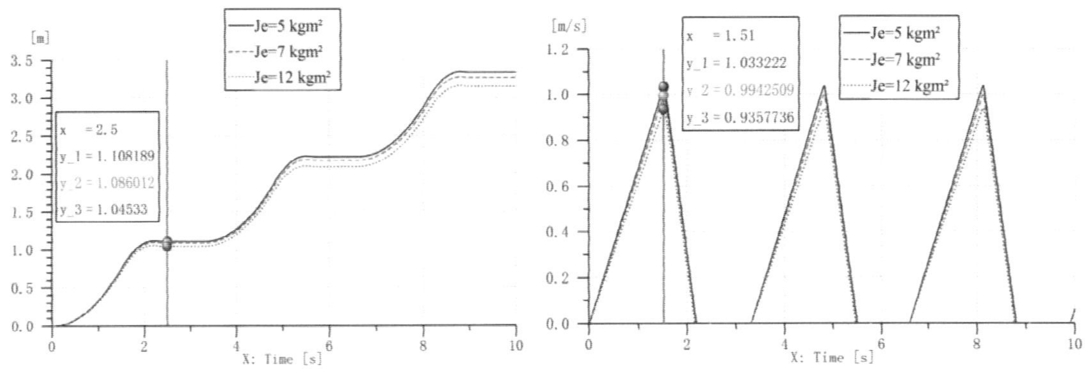

(a) (b)

Figure 4. Effects of effective moment of inertia of the powertrain J_e on (a) the displacement accuracy and (b) the velocity stability of the tamping machine.

IOP Conf. Series: Materials Science and Engineering **616** (2019) 012013 doi:10.1088/1757-899X/616/1/012013

Table 2. The Effect of J_e on operation accuracy of the tamping machine

Operation Accuracy Indices	J_e=7 kgm^2 (Standard)	J_e=5 kgm^2	J_e=12 kgm^2
Displacement of the 1st cycle (m)	1.086	1.108 (Relative error 2.03%)	1.045 (Relative error 3.78%)
Displacement of the 2nd cycle (m)	2.174	2.218 (Relative error 2.02%)	2.092 (Relative error 3.78%)
Displacement of the 3rd cycle (m)	3.262	3.328 (Relative error 2.02%)	3.140 (Relative error 3.74%)
Velocity amplitude of the 1st cycle (m/s)	0.994	1.033 (Relative error 3.92%)	0.936 (Relative error 5.84%)

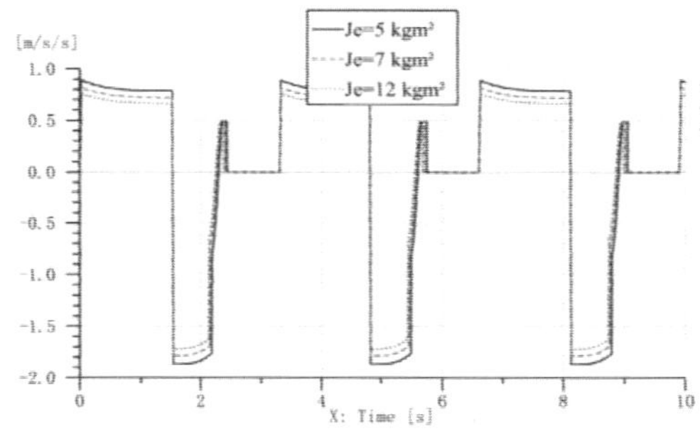

Figure 5. Effects of J_e on acceleration performance of the tamping machine.

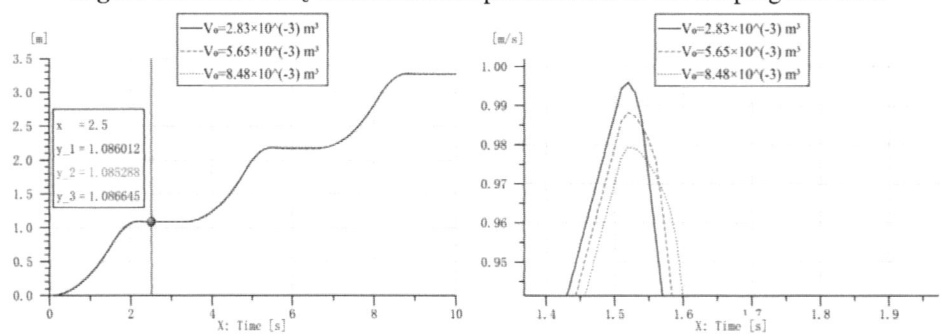

(a) (b)

Figure 6. Effects of the total volume of the high-pressure circuit of the hydraulic system V_0 on (a) the displacement accuracy and (b) the velocity stability of the tamping machine.

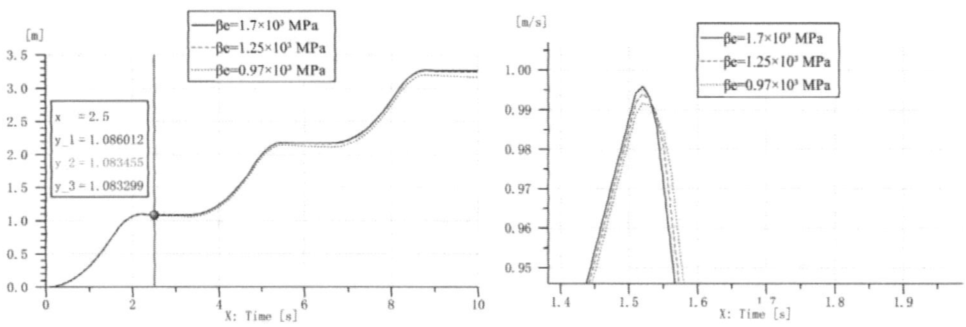

(a) (b)

Figure 7. Effects of the oil volumetric elastic modulus β_e on (a) the displacement accuracy and (b) the velocity stability of the tamping machine.

Figure 5 also shows that the acceleration performance of the tamping machine will be obviously affected by the effective moment of inertia of the powertrain.

Figure 6 shows that the effect of the total volume of the high-pressure circuit of the hydraulic system V_0 on the displacement accuracy is not obvious, but V_0 has some influence on the velocity amplitude. However, Figure 7 shows that the oil volumetric elastic modulus β_e has obvious effects on both of the displacement accuracy and velocity amplitude.

Generally speaking, a smaller J_e and V_0 are expected in the engineering development of a tamping machine, while β_e is easily affected by the entrapped air ratio in the oil, so a bigger β_e is expected in the operation and maintenance of the tamping machine.

5. Conclusions

- The effective moment of inertia of the powertrain J_e has an obvious influence on operation accuracy of the tamping machine, a smaller J_e will improve the response and operation accuracy of the tamping machine, and can be pursued by means of lightweight designs.
- The effect of the total volume of the high-pressure circuit of the hydraulic system V_0 on the displacement accuracy is not obvious, but V_0 has some influence on the velocity amplitude. A smaller V_0 will improve the response and operation accuracy of the tamping machine.
- The oil volumetric elastic modulus β_e has obvious effects on both of the displacement accuracy and velocity amplitude. β_e is easily affected by the entrapped air ratio in the oil, so a bigger β_e is expected in the operation and maintenance of the tamping machine.
- The mathematical model established and results obtained in this work will be instructive for further tamping machine design optimization and maintenance.

Acknowledgements
The authors thank financial support from the National Natural Science Foundation of China (NSFC) under Grant No. 11572123, the Hunan Provincial Natural Science Foundation under Grant No. 2017JJ4015 and the Research Fund for High-level Talent of Dongguan University of Technology under Project No. GC200906-30.

References

[1] Yang H, Sun W and Xu B 2007 New Investigation in energy regeneration of hydraulic elevators *IEEE/ASME Transactions on Mechatronics* **12**(5) 519–526

[2] Kache M 2014 Investigation an all-hydraulic hybrid system for diesel-hydraulic rail cars *European Transport Research Review* (6):181–189

[3] Zhou Y Z 2011 *Research on the fluid power drive system of a continuous tamping machine* (Master's Degree Thesis, Changsha: Central South University, China)

[4] Wang X 2008 *Research on the hydrostatic transmission running system of 08-32 tamping machine* (Master's Degree Thesis, Changsha: Central South University, China)

[5] Dasgupta K 2000 Analysis of a hydrostatic transmission system using low speed high torque motor *Mechanism and Machine Theory* **35**(10) 1481–1499

[6] Ho T H and Ahn K K 2010 Modeling and simulation of hydrostatic transmission system with energy regeneration using hydraulic accumulator *J. Mechanical Science and Technology* **24**(5):1163–1175

[7] Yan C G 2010 The analysis of 08-16 tamping machine's hydraulic drive system in the high-speed running *Equipment Manufacturing Technology* (10) 28–29

[8] Wang X, Hu J K and Long P 2008 Research on the acceleration model of hydraulic driving vehicle *Mechanical research and application* **21**(4) 28–29, 31

[9] Shi T L 2013 Simulation and improvement of the hydraulic driving system of tamper machine based on AMESim *J. Zhengzhou University (Engineering Science)* **34**(5) 89–92

[10] Hu Y, Hu J K and Fang J K 2014 Research on pressure stability of clamping hydraulic system in railway tamping device *J. Railway Science and Engineering* **11**(3) 146–150

[11] Zeng Y, Chen J B and Hu J K 2006 Adjustment of hydraulic system for railway subgrade tamping operation *J. of Construction machinery and maintenance* (6) 156–157

[12] Zhai S C and Huang Z J 2008 Tamping equipment's hydraulic system of 08-32 tamping machine and it's use and service *Hydraulics Pneumatics & Seals* **28**(3) 55–57

IOP Conf. Series: Materials Science and Engineering **616** (2019) 012014 doi:10.1088/1757-899X/616/1/012014

High-Speed Performance of a Tamping Machine with Closed-Type Hydrostatic Propulsion Drive

Zirong Zhou[1], Xiong Cai[2] and Wenlin Wang[1,2*]

[1]School of Mechanical Engineering, Dongguan University of Technology, Dongguan 523808, China.
[2]College of Mechanical and Vehicle Engineering, Hunan University, Changsha 410082, China

[*]Corresponding e-mail: pianowwl@163.com

Abstract. It is significant to understand and optimize the parameters of the hydraulic system to pursue both better low-speed operation and high-speed mobility performance of a tamping machine. A mathematical model formulating the high-speed performance of a tamping machine with a closed-type hydrostatic propulsion drive system is established in this study, a simulation model is built in the AMEsim software environment based on the mathematical model. Simulation results show that the tamping machine with a closed-type hydrostatic propulsion drive system has got a 35-100 km/h scope for high-speed mobility, and the acceleration performance is also satisfactory. The mathematical model established and results obtained in this work will be instructive for further parameter sensitivity analysis and tamping machine design optimizations.

1. Introduction

Hydraulic tamping machines [1–3] are crucial and efficient apparatus in modern railway construction and maintenance [4,5]. In a railway tamping machine, the hydrostatic propulsion drive system [3,6] has two working modes, i.e., one low-speed mode for operation, and another high-speed mode for mobility, thus, it is significant to understand and optimize the parameters of the hydraulic system to obtain both better low-speed and high-speed performance of a tamping machine.

Wang et al. investigated different acceleration modes [7] for a fluid-power-driven engineering vehicle by using a hydraulic proportional pump and a hydraulic variable motor, Long [8] studied the acceleration performance of a military utility vehicle with hydrostatic propulsion drive, Yan proposed a hydraulic drive system to obtain [9] both high speed for mobility and low speed for tamping. Shi investigated the problems of out of step [10] in velocity and traction force of a tamping machine, Hu et al. improved the pressure stability [11] of the hydraulic clamping system of a tamping machine, by adding a resistance orifice in the circuit, and Ji et al. [12] fault diagnosis technology for the hydraulic system of their tamping machine.

A mathematical model formulating the high-speed performance of a tamping machine with closed-type hydrostatic propulsion drive system is established in this study, and a simulation model is built in the AMEsim software environment based on the mathematical model. Simulation results show that the tamping machine has got a maximum speed of 100 km/h for fast mobility, the acceleration performance at various speeds are feasible for operations with high efficiency. The model established and results obtained in this work will be instructive for further parameter sensitivity analysis and tamping machine design optimizations.

2. Mathematical model of the high-speed hydrostatic propulsion drive

Figure 1 demonstrates that, in high-speed mode, the tamping machine is driven by three variable displacement hydraulic motors. Figure 2 illustrates the concrete high-speed hydrostatic propulsion drive system of the tamping machine, the hydrostatic propulsion drive system is a typical closed-type hydraulic system employing a two-way variable displacement hydraulic pumps and a two-way variable displacement hydraulic motors. The speed and direction of the tamping machine can be adjusted and alternated by volumetric speed control of the closed-type hydraulic system.

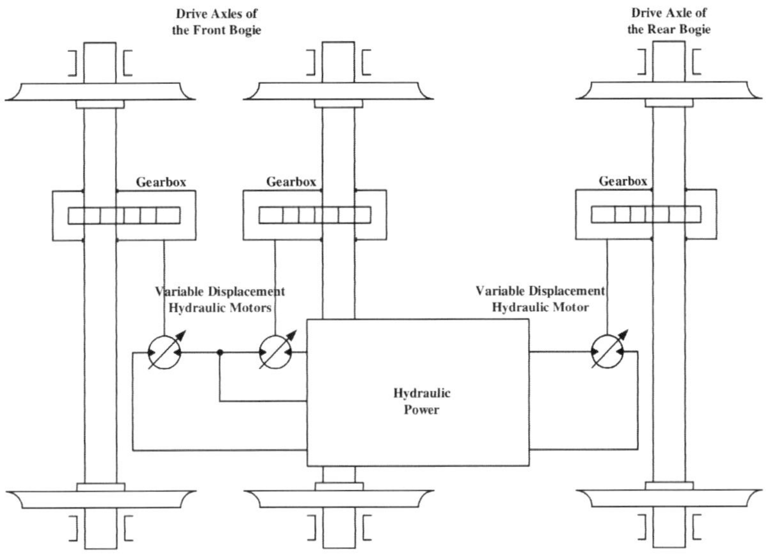

Figure 1. Mechanism of the high-speed hydrostatic propulsion drive system of a tamping machine.

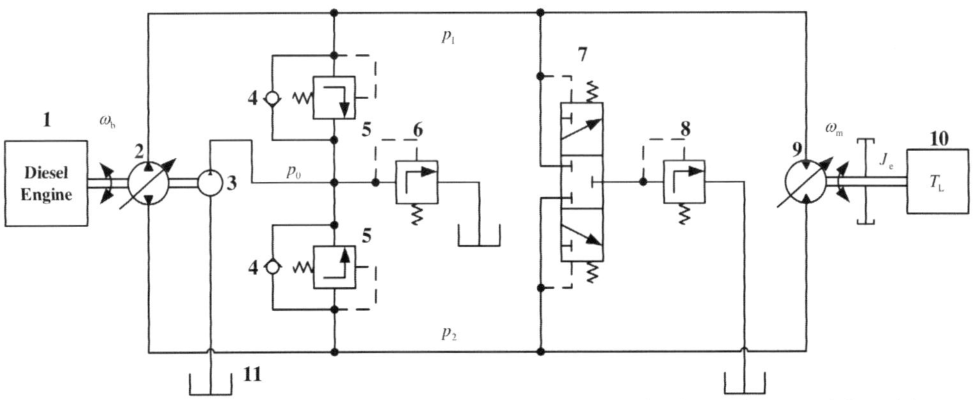

Figure 2. Closed-type hydraulic power circuit of the high-speed hydrostatic propulsion drive system of a tamping machine (1. diesel engine, 2. variable pump, 3. fluid compensation pump, 4. check valve, 5. high-pressure relief valve, 6. fluid compensation relief valve, 7. flush valve, 8. back-pressure valve, 9. hydraulic motor, 10. load, 11. reservoir).

Referring to Figure 2, the output flow of the hydraulic pump Q_b can be formulated by

$$Q_b = x_b D_b \omega_b - C_{ib}(p_1 - p_0) - C_{eb} p_1 \tag{1}$$

where x_b, D_b, ω_b, C_{ib}, C_{eb}, p_1 and p_0 are the adjustment coefficient, the maximum displacement, the rotation speed, the inner leakage coefficient, the outer leakage coefficient, high pressure and back pressure of the pump, respectively. Thus, taking the Laplace Transform of Equation (1) to obtain

$$Q_{\mathrm{b}}(s)=x_{\mathrm{b}}(s)D_{\mathrm{b}}\omega_{\mathrm{b}}-C_{\mathrm{b}}p(s) \tag{2}$$

The flow continuity equation of the high-pressure circuit of the hydrostatic propulsion drive system can be formulated by

$$Q_{\mathrm{b}}=C_{\mathrm{im}}(p_1-p_0)+C_{\mathrm{em}}p_1+x_{\mathrm{m}}D_{\mathrm{m}}\omega_{\mathrm{m}}+\frac{V_0}{\beta_{\mathrm{e}}}\frac{dp_1}{dt} \tag{3}$$

where C_{im}, C_{em}, x_{m}, D_{m} and ω_{m} are the inner leakage coefficient, the outer leakage coefficient, the adjustment coefficient, the maximum displacement and the rotation speed of the hydraulic motor, respectively; V_0 is the total volume of the high-pressure circuit of the hydraulic system, and β_{e} is volumetric elastic modulus of the oil. Similarly, taking the Laplace Transform of Equation (3) to obtain

$$Q_{\mathrm{b}}(s)=C_{\mathrm{m}}p(s)+x_{\mathrm{m}}(s)D_{\mathrm{m}}\omega_{\mathrm{m0}}+D_{\mathrm{m}}\omega_{\mathrm{m}}(s)+\frac{V_0}{\beta_{\mathrm{e}}}sp(s) \tag{4}$$

where C_{m} is the total leakage coefficient of the hydraulic motor, and $C_{\mathrm{m}}=C_{\mathrm{im}}+C_{\mathrm{em}}$, ω_{m0} is the initial rotation speed of the hydraulic motor.

Because there exists

$$v(s)=\omega_{\mathrm{m}}(s)r \tag{5}$$

where v and r are the vehicle speed and the wheel radius, respectively.

Thus, combining Equations (2), (4) and (5) will have

$$x_{\mathrm{b}}(s)D_{\mathrm{b}}\omega_{\mathrm{b}}-x_{\mathrm{m}}(s)D_{\mathrm{m}}\frac{v_0}{r}=\left(C_{\mathrm{t}}+\frac{V_0}{\beta_{\mathrm{e}}}s\right)p(s)+D_{\mathrm{m}}\frac{v(s)}{r} \tag{6}$$

where v_0 is the initial vehicle speed, C_{t} is the total leakage coefficient of the hydraulic system, and $C_{\mathrm{t}}=C_{\mathrm{b}}+C_{\mathrm{m}}$, $C_{\mathrm{b}}=C_{\mathrm{ib}}+C_{\mathrm{eb}}$.

In addition, the torque equilibrium equation of the hydraulic motor and load can be described as

$$x_{\mathrm{m}}D_{\mathrm{m}}(p_1-p_0)=J_{\mathrm{e}}\frac{d\omega_{\mathrm{m}}}{dt}+B\omega_{\mathrm{m}}+T_{\mathrm{L}} \tag{7}$$

where J_{e} and B are the effective moment of inertia and viscous damping coefficient of the powertrain, T_{L} is the load torque on the wheel, and can be formulated by [13]

$$T_{\mathrm{L}}=r\sum F=r\left[mg\cos\beta(f_0+kv)+mg\tan\beta+\frac{C_{\mathrm{D}}Av^2}{21.15}+m\delta\frac{dv}{dt}\right] \tag{8}$$

where $\sum F$, m, A and C_{D} are the total resistance force, the mass, the windward area and the drag coefficient of wind of the tamping machine, (f_0+kv) is the coefficient of rolling friction of the wheels, β is the slope angle, and δ is an effective coefficient of the rotation mass of the tamping machine.

Thus, combining Equations (7) and (8), and taking the Laplace Transform to obtain

$$x_{\mathrm{m}}(s)D_{\mathrm{m}}p_{10}+D_{\mathrm{m}}p(s)=\frac{J_{\mathrm{e}}s+B}{r}v(s)+\left(m\delta rs+mgrk\cos\beta+\frac{C_{\mathrm{D}}Ar}{21.15}\right)v(s) \tag{9}$$

Based on the above mathematical modelling of the high-speed hydrostatic propulsion drive system, the dynamic control block diagram of the hydraulic circuit can be obtained and demonstrated by Figure 3.

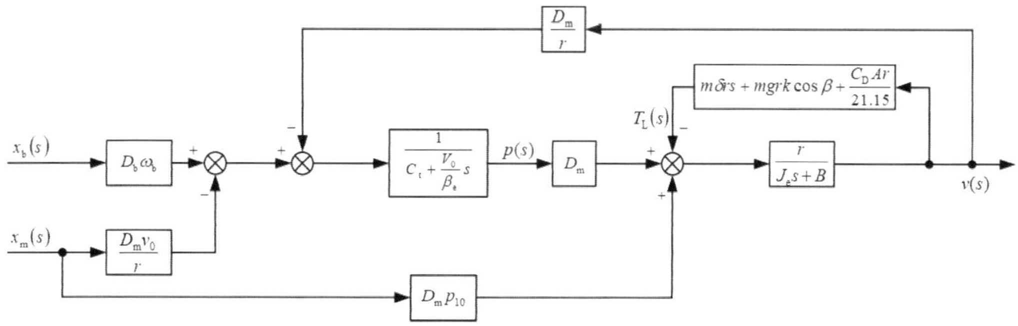

Figure 3. Dynamic control block diagram of the high-speed hydrostatic propulsion drive system of the tamping machine.

Thus, the transfer function of the high-speed hydrostatic propulsion drive system of the tamping machine can be formulated by

$$v(s)=\frac{D_b D_m \omega_b x_b(s)-\left[\frac{D_m^2}{r}v_0-\left(C_t+\frac{V_0}{\beta_e}s\right)D_m p_{10}\right]x_m(s)}{as^2+bs+c} \tag{10}$$

where the coefficients are described as

$$\begin{aligned} a&=\frac{V_0\left(J_e+m\delta r^2\right)}{r\beta_e}, \\ b&=\frac{V_0}{\beta_e}\left(\frac{B}{r}+mgrk\cos\beta+\frac{rC_D A}{21.15}\right)+C_t\left(J_e+m\delta r\right), \\ c&=\frac{D_m^2+C_t B}{r}+mgrkC_t\cos\beta+\frac{rC_D AC_t}{21.15}. \end{aligned} \tag{11}$$

3. Simulation and result analysis

As shown by Figure 4, a simulation model of the high-speed hydrostatic propulsion drive system of the tamping machine is built in the AMEsim software environment, and simulations are carried out.

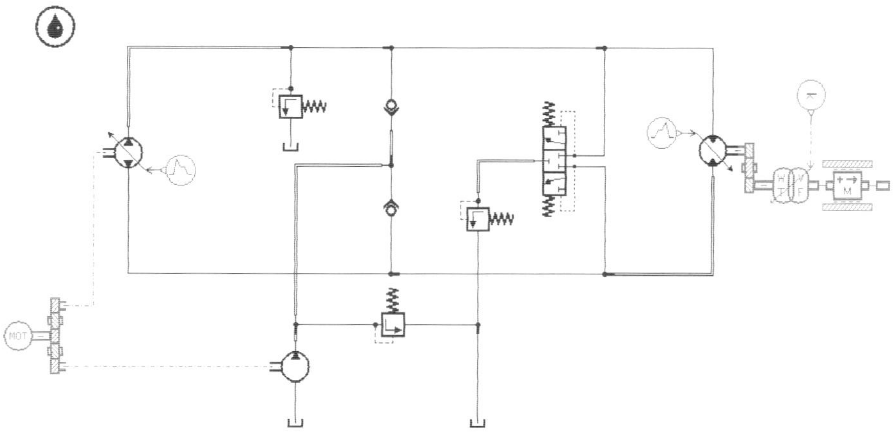

Figure 4. Simulation model of the high-speed hydrostatic propulsion drive system of the tamping machine.

Figure 5 demonstrates the profiles of the pressure and the output flow of the hydraulic motor during acceleration of the tamping machine, i.e., during volumetric speed control of the closed-type hydraulic system. Figure 5 shows that the hydraulic motor will get a stable pressure of 40 MPa in 20 seconds and a maximum and stable output flow of 668.1 L/min in about 50 seconds, as the result, Figure 6(a) shows the obtained acceleration speed of the tamping machine and calculated that the tamping machine has got a maximum speed of 100 km/h for mobility.

Figure 6(b) compares the acceleration performance of the tamping machine at various operation speeds, and shows that the acceleration periods are 32 seconds, 65 seconds and 143 seconds or so when the vehicle is accelerated to 50 km/h, 75 km/h and 100 km/h, respectively.

Thus, the tamping machine with a closed-type hydrostatic propulsion drive system has got a 35-100 km/h scope for high-speed mobility, and the acceleration performance is also satisfactory.

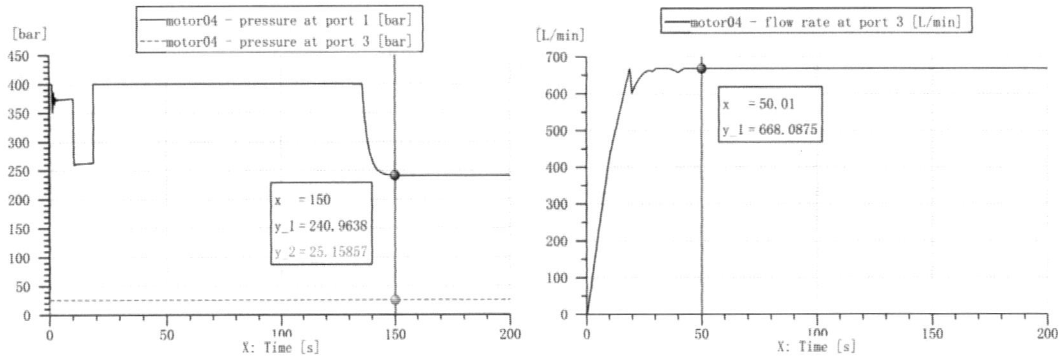

(a) (b)

Figure 5. Profiles of (a) the pressure and (b) the output flow of the hydraulic motor during acceleration of the tamping machine.

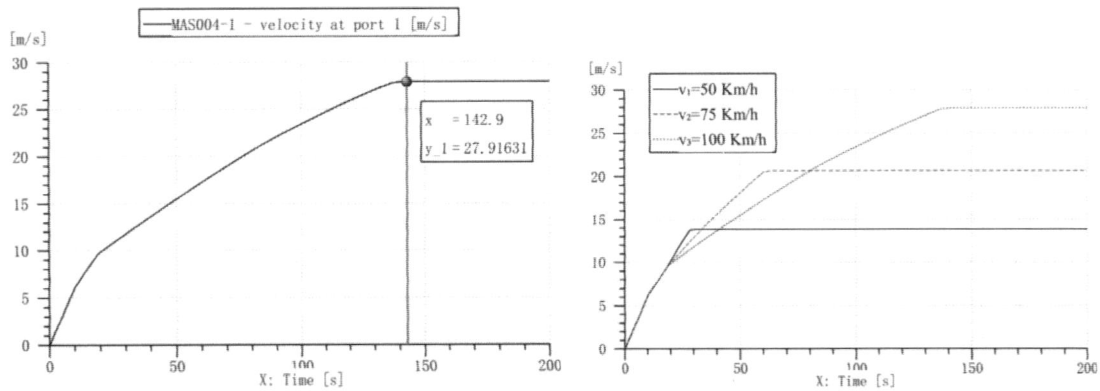

Figure 6. (a) The maximum speed of the tamping machine; (b) Acceleration performance of the tamping machine at various operation speeds.

4. Conclusions

- During volumetric speed control of the closed-type hydrostatic propulsion drive system, the tamping machine will get a maximum speed of 100 km/h for mobility, the acceleration periods are 32 seconds, 65 seconds and 143 seconds or so when the vehicle is accelerated to 50 km/h, 75 km/h and 100 km/h, respectively.
- The tamping machine with a closed-type hydrostatic propulsion drive system has got a 35-100 km/h scope for high-speed mobility, and the acceleration performance is also satisfactory.
- The mathematical model established and results obtained in this work will be instructive for further parameter sensitivity analysis and tamping machine design optimizations.

Acknowledgements
The authors thank financial support from the National Natural Science Foundation of China (NSFC) under Grant No. 11572123, the Hunan Provincial Natural Science Foundation under Grant No. 2017JJ4015 and the Research Fund for High-level Talent of Dongguan University of Technology under Project No. GC200906-30.

References
[1] Kache M 2014 Investigation an all-hydraulic hybrid system for diesel-hydraulic rail cars *European Transport Research Review* (6):181–189
[2] Zhou Y Z 2011 *Research on the fluid power drive system of a continuous tamping machine* (Master's Degree Thesis, Changsha: Central South University, China)
[3] Wang X 2008 *Research on the hydrostatic transmission running system of 08-32 tamping machine* (Master's Degree Thesis, Changsha: Central South University, China)
[4] Famurewa S M, Xin T, Rantatalo M and Kumar U 2015 Optimisation of maintenance track possession time: A tamping case study. *Proceedings of the Institution of Mechanical Engineers Part F Journal of Rail & Rapid Transit* **229**(1): 12–22
[5] Aursudkij B 2007 *A laboratory study of railway ballast behaviour under traffic loading and tamping maintenance* (Ph.D Thesis, Nottingham: University of Nottingham, UK)
[6] Dasgupta K 2000 Analysis of a hydrostatic transmission system using low speed high torque motor *Mechanism and Machine Theory* **35**(10) 1481–1499
[7] Wang X, Hu J K and Long P 2008 Research on the acceleration model of hydraulic driving vehicle *Mechanical research and application* **21**(4) 28–29, 31
[8] Long P 2008 *A study on the acceleration performance of a military utility vehicle with hydrostatic propulsion drive* (Master's Degree Thesis, Changsha: Central South University, China)
[9] Yan C G 2010 The analysis of 08-16 tamping machine's hydraulic drive system in the high-speed running *Equipment Manufacturing Technology* (10) 28–29
[10] Shi T L 2013 Simulation and improvement of the hydraulic driving system of tamper machine based on AMESim *J. Zhengzhou University* (*Engineering Science*) **34**(5) 89–92
[11] Hu Y, Hu J K and Fang J K 2014 Research on pressure stability of clamping hydraulic system in railway tamping device *J. Railway Science and Engineering* **11**(3) 146–150
[12] Ji Z H and Zhu F M 2011 Tamping machine fault diagnosis based on fuzzy inference. *Proceedings of the 2011 International Conference on Advances in Construction Machinery and Vehicle Engineering, Shanghai, China.*
[13] David A C 2009 *Automotive Engineering* Butterworth-Heinemann, Oxford, UK.

IOP Conf. Series: Materials Science and Engineering **616** (2019) 012015 doi:10.1088/1757-899X/616/1/012015

Joint Trajectory Time Optimization of Cobot Based on Particle Swarm Optimization

Cheng Zhenyi[1, *]

[1]School of Mechanical and Automotive Engineering, Shanghai University Of Engineering Science.Shanghai, China

[*]E-mail:447337499@qq.com

Abstract: In order to improve the working efficiency of the robot and make the welding quality of the part of skin group improved greatly. We Researched on the end trajectory of the Cobot. In the application scenario of Friction Stir Welding (FSW) test bench, an algorithm of joint space trajectory is proposed according to the kinematics characteristics of the robot and the practical application needs.Based on particle swarm optimization (PSO), the motion trajectory of robot joint is fitted by high order polynomial interpolation. While ensuring the end trajectory, the stability of the joint motion is guaranteed. The simulation experiment is carried out by using MATLAB software. The joint space optimal motion planning is realized. The comparison algorithm is verified by experiments. According to the experimental results, the particle swarm optimization algorithm improves the efficiency of the cooperative robot significantly. It is helpful to the end stability of friction stir welding robot.

1. Introduction

In recent years, friction stir welding (FSW) has developed rapidly in the world. Under the background of China's industry 4.0, major research institutes have begun to vigorously develop the technology of FSW[1]. Our school's Intelligent Robot Research and Development Center focuses on the development of friction stir welding robots. When studying the related control problems, it encounters the problems of jitter and uneven axial force [2, 3]. After referring to the feed-forward control and other solutions [4], it decides to optimize the joint arms of the cooperative robot first. Robot trajectory planning is the key of robot control. At present, there are two optimization directions: one is to optimize the time, the other is to optimize the energy of the system. Compared with the working efficiency of the robot, the friction stir welding robot is not sensitive to the change of system energy, so the motion time is chosen as the optimization object. In order to improve the performance of the robot, reduce the jitter of the robot in work and the change of the axial force in welding. In this paper, particle swarm optimization (PSO) is used to optimize the trajectory of the robot joint.

There are several ways to optimize the time of trajectory planning, such as genetic algorithm. At present, most of the research on robot joint trajectory at home and abroad adopt genetic algorithm [5]. Particle swarm optimization is still in the exploratory stage, but it also has a considerable research foundation [6]. At the same time, it has many shortcomings and limitations.

This paper chooses UR5 cooperative robot as the research object, and studies the trajectory planning time of point-to-point (PTP). In the simulation experiment, the end trajectory of the robot for friction stir welding arc workpiece was optimized by using high order polynomial interpolation and particle swarm optimization. The practicability and expected results of the research results are verified

IOP Conf. Series: Materials Science and Engineering **616** (2019) 012015 doi:10.1088/1757-899X/616/1/012015

by the combination of actual test and simulation experiments. Experiments show that this method can not only optimize the trajectory planning time, shorten the time of robot joint movement, but also can be applied in practice. For friction stir welding test, it can greatly shorten the welding cycle, improve work efficiency and increase the life of the stirring head.

2. Kinematics Analysis of Collaborative Robot

UR5 cooperative robot is used in this experiment, which is a 6-DOF robot. According to the specifications, the size diagram (Figure 1) is obtained. According to the D-H parameters (shown in Table 1), the robot model can be constructed, and the forward and inverse solutions and singular values [7] can be obtained.Forward and Inverse kinematics.According to the manual of the Cobot UR5(Figure 2), the DH parameters of the robot in table 1 are obtained.

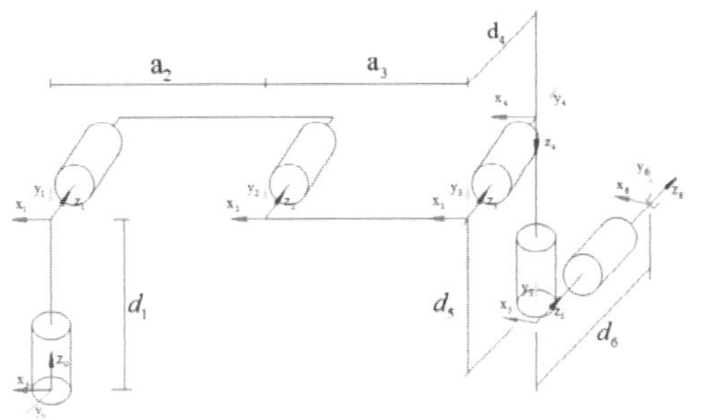

Figure 1. UR5 Robot size drawing

Figure 2. Cobot

Table 1 D-H parameters of the UR5

Connecting rod	Torsion Angle	Around corner	Distance	Length
n	α_i / rad	θ_i / rad	d_i / mm	a_i / mm
1	π/2	θ_1	89.459	0
2	0	θ_2	0	425
3	0	θ_3	0	392.5
4	π/2	θ_4	109.15	0
5	-π/2	θ_5	94.65	0
6	0	θ_6	82.3	0

The kinematics principle and Simulation of cooperative robot are analyzed [8]. Combining with MTLAB robot simulation library, the forward and inverse kinematics solutions of cooperative robot can be obtained. In order to plan the joint space motion of the robot, it is necessary to find the position points of the joint motion. According to the inverse kinematics characteristics of the robot, the position points of the joint motion can be obtained.

2.1Planning of End Position

According to the actual welding requirements of the workpiece, an arc trajectory of FSW is planned based on high order interpolation [9]. In case of fixed end trajectory, there are innumerable inverse solutions. According to the simulation experience of the inverse solution of the joint robot [10], the solutions of the four path points of the joint are obtained. According to Cartesian transformation of solve function , the definition of coordinate system and worktable coordinate system can be applied to basic inverse kinematics. The base is the first joint point. The solution is shown in Table 2 below.

IOP Conf. Series: Materials Science and Engineering **616** (2019) 012015 doi:10.1088/1757-899X/616/1/012015

Table 2 The Inverse kinematics of each joint at the path point

Joint/Path Point	1	2	3	4
1	87.55°	90.51°	120.62°	171.23°
2	-67.52°	-67.02°	-67.04°	-46.14°
3	82.26°	77.45°	79.09°	54.46°
4	-193.77°	-190.6°	-190.5°	-190.5°
5	-54.21°	-57.72°	-86.10°	-86.33°
6	-81.72°	-80.21°	-81.72°	-81.72°

3. Experimental design based on particle swarm optimization

3.1. Solution of Position, Velocity and Acceleration

There are many interpolation methods for trajectory planning of joint points, such as high-order polynomial interpolation, low-order polynomial interpolation, Hermite interpolation, spline interpolation and so on. However, the high order polynomial interpolation may be unstable; the low order polynomial interpolation has poor smoothness, and the first derivative does not exist near the interpolation node; the piecewise three order Hermite interpolation has better smoothness than the low order polynomial, but it does not have the second order smoothness in the interpolation interval. In this paper, the high order spline interpolation function is used to interpolate the control points and optimize the joint trajectory of the cooperative robot based on friction stir welding.

High-order spline interpolation is used to interpolate the spatial motion of robot joints. In the experiment, the path of joint space motion is divided into three intervals. Let the position of joint motion be, and the time of different intervals be. The relationship between joint spatial motion position and time is as follows:

$$\begin{cases} q_1 = a_{13}t_1^3 + a_{12}t_2^2 + a_{11}t_1 + a_1 \\ q_2 = a_{25}t_2^5 + a_{24}t_2^4 + a_{23}t_2^3 + a_{22}t_2^2 + a_{21}t_2 + a_{20} \\ q_3 = a_{33}t_3^3 + a_{32}t_3^2 + a_{31}t_3 + a_{30} \end{cases} \quad (1)$$

According to the above formula, the formulas for calculating the velocity and acceleration of each point in the motion of each joint of the robot can be obtained. The velocity is as follows:

$$\begin{cases} \dot{q}_1 = 3a_{13}t_1^2 + 2a_{12}t_1 + a_{11} \\ \dot{q}_2 = 5a_{25}t_2^4 + 4a_{24}t_2^3 + 3a_{23}t_2^2 + 2a_{22}t_2 + a_{21} \\ \dot{q}_3 = 3a_{33}t_3^2 + 2a_{32}t_3 + a_{31} \end{cases} \quad (2)$$

The acceleration is shown as follows:

$$\begin{cases} \ddot{q}_1 = 6a_{13}t_1 + 2a_{12} \\ \ddot{q}_2 = 20a_{25}t_2^3 + 12a_{24}t_2^2 + 6a_{23}t_2 + 2a_{22} \\ \ddot{q}_3 = 6a_{33}t_2 + 2a_{32} \end{cases} \quad (3)$$

According to the known conditions, the following three groups of formulas can be obtained from the path points passed at different times and the motion state of the robot joints at different times.Let the value of four path points and the motion time of three intervals be.The formulas for solving parameters in the first interval are as follows:

$$\begin{cases} Q(1) = a_{10} \\ Q(2) = a_{13}t(1)^3 + a_{12}t(1)^2 + a_{11}t(1) + a_{10} \\ 0 = a_{11} \\ 0 = 2a_{12} \end{cases} \tag{4}$$

The formulas for solving the parameters of the second interval are as follows:

$$\begin{cases} Q(2) = a_{10} \\ Q(3) = a_{25}t(2)^5 + a_{24}t(2)^4 + a_{23}t(2)^3 + a_{22}t(2)^2 + a_{21}t(2) + a_{20} \\ a_{21} = 3a_{13}t(1)^2 + 2a_{12}t(1) + a_{11} \\ 2a_{22} = 6a_{13}t(1) + 2a_{12} \\ a_{31} = 5a_{25}t(2)^4 + 4a_{24}t(2)^3 + 3a_{23}t(2)^2 + 2a_{22}t(2) + a_{21} \\ 2a_{32} = 20a_{25}t(2)^3 + 12a_{24}t(2)^2 + 6a_{23}t(2) + 2a_{22} \end{cases} \tag{5}$$

The formulas for solving the parameters of the third interval are as follows:

$$\begin{cases} Q(3) = a_{30} \\ Q(4) = a_{33}t(3)^3 + a_{32}t(3)^2 + a_{31}t(3) + a_{30} \\ 0 = 3a_{33}t(3)^2 + 2a_{32}t(3) + a_{31} \\ 0 = 6a_{33}t(3) + 2a_{32} \end{cases} \tag{6}$$

According to the above formulas, the parameters of higher order polynomials can be obtained. Thus, the trajectory formula, velocity formula and Acceleration Formula of joint space motion can be obtained. The position, velocity and acceleration at any time can be calculated according to the formula.

3.2 Optimization of particle swarm optimization

In order to optimize joint space trajectory planning based on PSO, several examples need to be selected as optimization objects. In the experiment, a three-dimensional example is formed by the time taken to complete the motion of three intervals, and 20 groups of particles are generated randomly. The mean of the particles is between. The fitness function of the example is shown in equation (7):

$$fitness = t_1 + t_2 + t_3 \tag{7}$$

The fitness value of particles is calculated according to fitness function. The smallest fitness of a single particle is selected as the local optimal particle and the smallest fitness of all particles is selected as the global optimal particle.

In addition to the fitness function, the parameters (8) in the updated particle velocity formula need to be determined. The inertia weight, maximum and minimum, decreases as the number of iterations increases, as shown in formula (10). Learning factors. The random factors are random numbers between them, which are generated by the function of MATLAB.

$$v[m] = w \bullet v[m] + c_1 r_1 (pBest[m] - present[m]) + c_2 r_2 (pBest[m] - present[m]) \tag{8}$$

$$present[m] = present[m] + v[m] \tag{9}$$

$$W = W_{\max} - (W_{\max} - W_{\min})(n / N) \tag{10}$$

In this experiment, besides the general constraints, the constraints of joint space motion must be satisfied, that is, the joint motion speed and acceleration can not exceed the prescribed maximum speed and acceleration. From the formula (11), the maximum velocity and large acceleration of different particles can be obtained.

$$\{|V| \le V_{max}; |a| \le a_{max} \tag{11}$$

Particle swarm optimization (PSO) with constraints is used to optimize the time of joint space trajectory, and the shortest time of joint motion is obtained.In order to realize the optimization process, we first verify whether the particle meets the constraints in Formula (11). Satisfaction is calculated by formula (7), otherwise it is infinite. The fitness values of each particle are preserved. By comparing the fitness values of 20 particles, the minimum fitness value is selected, that is, the optimization is completed, as the global optimal particle. Through N iterations, the particle is updated continuously, even the fitness value, and the optimal particle is selected to obtain the optimal solution under constraints, that is, the shortest time of trajectory planning.

4. Simulation

In this paper, PSO based high order polynomial interpolation method is used to optimize the joint trajectory of cooperative robots. The joint trajectory is smooth and stable.

According to the beam-saving condition of the robot, the position-time relationship of the joint motion is obtained as shown in Figure. 3. From the comparison of Figure 4 and Figure 5, it is obvious that polynomial of degree five interpolation can make the trajectory curve more flexible, especially the acceleration improvement at the end of the robot.

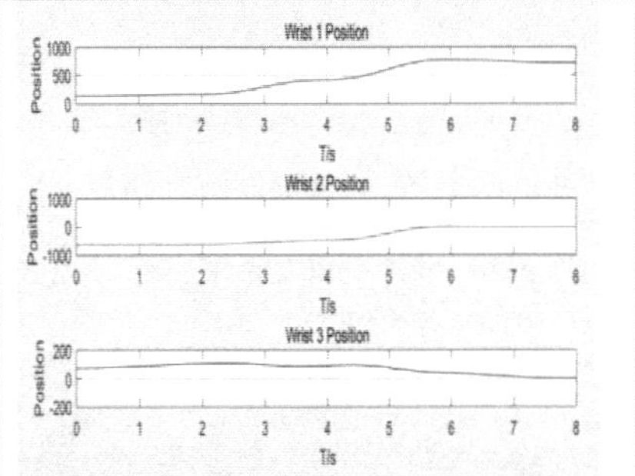

Figure 3. Relation between position and time of wrist 1, 2 and 3

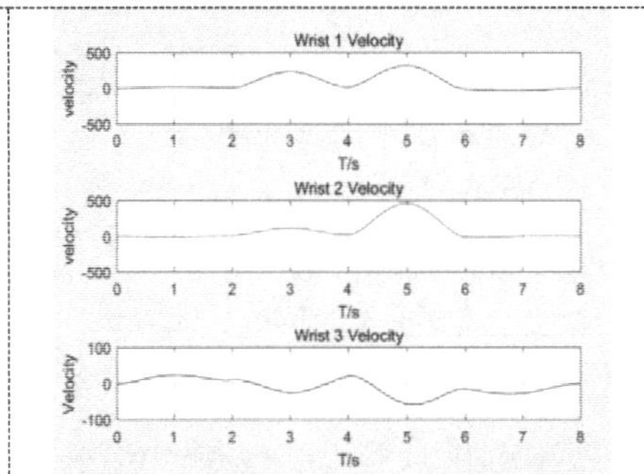

Figure 4. Relation between velocity and time of wrist 1, 2 and 3

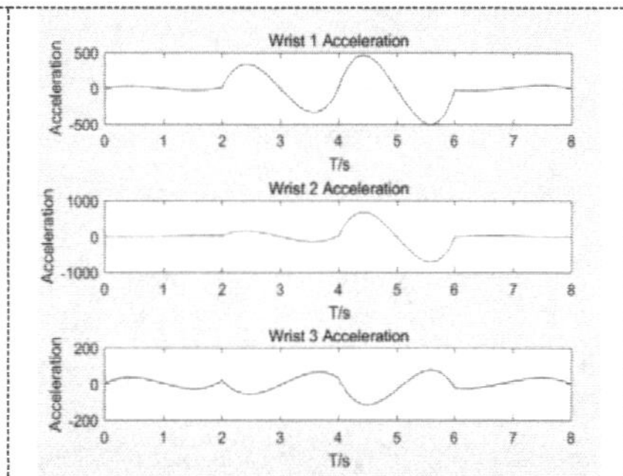

Figure 5. Relation between acceleration and time of wrist 1, 2, 3

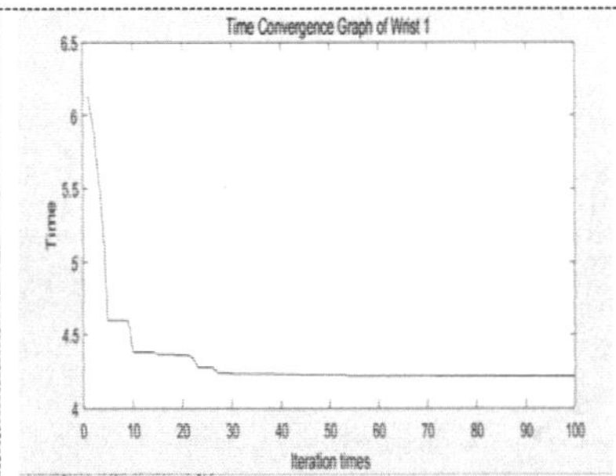

Figure 6. Convergence figure of Wrist 1

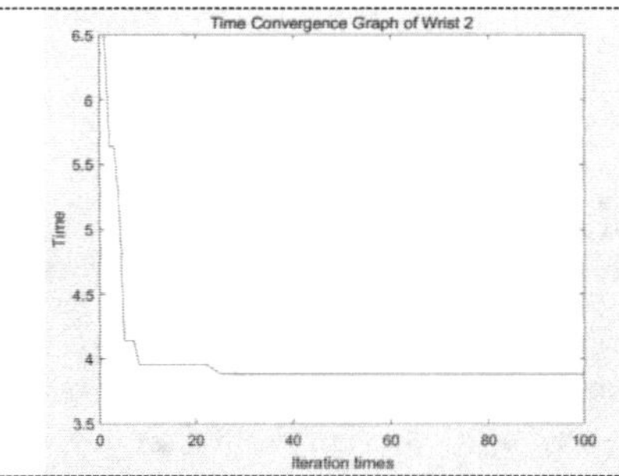

Figure 7. Convergence figure of Wrist 2

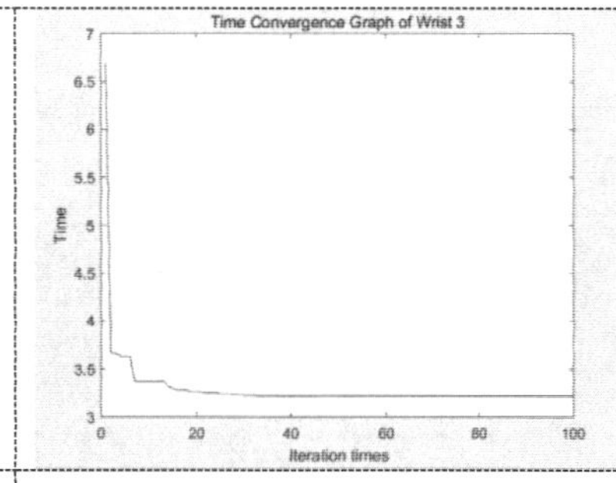

Figure 8. Convergence figure of Wrist 3

It is easy to find that Figures 6, 7, and 8 represent the iteration process. After about 30 iterations, the time tends to be stable. About a third of the time can be saved.

As can be seen from Table 3, the initial time to complete the specified trajectory is 6 seconds. After optimization, the time consumed by the robot can be shortened to about four seconds. This method not only ensures the smooth motion of the robot joint, but also improves the efficiency of the robot and enhances the axial force stability of friction stir welding.

Table 3. Joint optimization results.

Time Gap	T1	T2	T3	Total time
Initial time	2	2	2	6
Optimization 1	1.44	1.36	1.46	4.26
Optimization 2	1.46	1.36	1.45	4.27
Optimization 3	1.45	1.38	1.39	4.32
Optimization 4	1.47	1.36	1.81	4.64
Optimization 5	1.45	1.36	1.49	4.30

5. Conclusion

The trajectory of the robot optimized by high-order interpolation and particle swarm optimization has good smoothness, greatly shortens the movement time, improves the efficiency of friction stir welding robot, and enhances the stability of welding.

In the experiment, the robot will have a short standstill after reaching a posture. I think the next work direction is the feed forward control of the robot's terminal trajectory to further improve the efficiency of the cooperative robot.

References

[1] Zhang hao, huang yongde. Research status of friction stir welding technology and process suitable for robot welding [J]. Materials review, 2018, 32(1):128-134.

[2] Wan minhong, zhou weijia. Design and motion control of high-precision and heavy-duty friction stir welding robot [J]. Robot, 2018, 40(6):817-824.

[3] Sun hongyu, zhou qi. Analysis of axial pressure change in friction stir welding [J]. Thermal processing technology, 2018(1):206-209.

[4] Tao zhengxin. Design and research of motion controller for friction stir welding robot [D]. Yanshan university, 2016.

[5] Xiong zhengwei. Trajectory planning of three-dimension feeding manipulator based on b-spline and multi-objective genetic algorithm[J].Mechanical transmission,2018,v.42; No.257(5):125-128.

[6] Feng Bin.Robot joint space based on particle swarm algorithm the optimal trajectory planning[J]. Journal of combination machine tools and automatic processing technology,2018, No.531(5):6-9.

[7] Guo xiaobao. Kinematics and singularity analysis of UR robot [J]. Equipment machinery, 2017(1).

[8] Ge jianbing, zhai xueqin. Kinematics simulation of robot based on MATLAB [J]. Machine design and manufacturing, 2008(9):168-169.

[9] Han jiang. Joint trajectory planning algorithm for industrial robots based on hybrid interpolation [J]. China mechanical engineering, 2018, 29(12):1460-1466.

[10] Lu jia-hao.Research on joint robot motion simulation based on MATLAB Robotic Toolbox [J]. Machine tool & hydraulics, 2017(17).

IOP Conf. Series: Materials Science and Engineering **616** (2019) 012016 doi:10.1088/1757-899X/616/1/012016

Study on Cure Deformation of Metal/CFRP Laminated Composites

Xueshu Liu[1*], Jinlei Li[1], Jianhua Lei[2], Qi Wang[2] and Qian Gao[2]
[1] School of Automotive Engineering, Dalian University of Technology, Dalian 116024, China;
[2] Aerospace Research Institute of Materials & Processing Technology, Beijing 100076, China

[*]Corresponding author's e-mail: liuxs@dlut.edu.cn;

Abstract. During the curing process of composite materials, the differences of the heat transfer ability and mechanical properties of material in various directions result in residual stresses and curing deformation in the composite after curing. Curing deformation has adverse effect on the mechanical performance of composites. The development of internal stress during curing may cause matrix cracking, delamination, warpage and rebound deformation of composites. In this paper, numerical simulation is used to predict the curing deformation of components, thereby reducing production costs and improving production efficiency as well. According to the curing kinetic equation, the HETVAL subroutine was firstly developed to calculate the temperature field and the curing degree field of the curing process. UMAT and UEXPAN subroutines were then applied to solve the mechanical constants and thermal strain and curing shrinkage strain during the curing process. After the proposed model was verified, the deformation of the metal composite co-cured components was studied. It was found that when the thickness ratio of metal to CFRP is one to one the composite gains the greatest geometrical deviation after curing.

1. Introduction

Composite materials have been widely used in many fields due to their high specific strength, large specific modulus, fatigue resistance and corrosion resistance. However, due to the thermal expansion and contraction effect of the material itself during the curing process of the composite material, the chemical shrinkage effect of the matrix and the thermal expansion coefficient of the composite material and the mould do not match, which inevitably causes deformation after moulding, especially large size and complicated structure member. Curing deformation can cause assembly stress, reduce structural strength, shorten fatigue life, and even lead to component scrapping. In order to ensure the quality of component manufacturing while minimizing cost, it is important to use numerical simulation to predict curing deformation. In the early research, Cowley [1] approximated the curing deformation of the orthogonally laid strip laminates to the cylindrical shape, and applied the classical laminate theory to estimate the curvature deformation of the cylinder. Yi [2] et al. used nonlinear finite element method to calculate the residual stress during the cooling process of composites. Kappel [3] reported a preliminary experimental study on the mechanical interaction and spring deformation of unidirectional, crossed and quasi-isotropic laminates with C-type and L-type specimens. Zhang [4] proposed a modified viscoelastic constitutive model for predicting curing residual stress of polymer matrix composites. Ersoy [5] predicted the spring of AS4/8552 composite C-section parts. Hörberg [6] introduced the experimental results and analysis of the influence of laminate thickness on the elastic deformation and shape deformation of thermosetting composite L-shaped steel. Brauner [7] presented a simulation method for analysis of induced deformation of composite multi-segment flaps.

This paper aims to study the curing process and the deformation of composite materials by simulation. Using the material customization function in the finite element software ABAQUS, a subroutine for calculating the curing exotherm and curing degree as well as the mechanical constant and strain was developed. Then, the correctness of the model is verified by comparison with the examples given in the literature. Finally, the model was used to study the geometrical deformation of metal/CFRP composites.

2. Theoretical basis of numerical simulation

2.1. Thermal-chemical theory

The curing process of the thermosetting resin matrix composites is a simultaneous physical and chemical process of heat conduction and curing reaction, the essence is a transient heat conduction problem with a nonlinear internal heat source. Wherein, the internal heat source is derived from the chemical reaction exotherm of the resin matrix. In this paper, a three-dimensional heat conduction model is used to calculate the temperature distribution, and the curing reaction kinetic equation is introduced to calculate the internal heat source.

$$\rho_p c_p \frac{\partial T}{\partial t} = \frac{\partial}{\partial x}(k_{xx}\frac{\partial T}{\partial x}) + \frac{\partial}{\partial y}(k_{yy}\frac{\partial T}{\partial y}) + \frac{\partial}{\partial z}(k_{zz}\frac{\partial T}{\partial z}) + q \tag{1}$$

where ρ_p is the density of composite material; c_p is the specific heat capacity of the composite; k_{xx}, k_{yy}, k_{zz} are the thermal conductivity of the composite in the x, y and z directions respectively; t is the curing time; T is the transient temperature at the current moment; q represents the heat generation rate of the curing reaction, which can be expressed as:

$$q = \rho_r (1 - V_f) H_r \frac{d\partial}{dt} \tag{2}$$

where ∂ is the degree of cure; ρ_r is the density of the resin; V_f is the volume fraction of the fiber; H_r is the total heat released by the resin per unit mass in the curing process; $\frac{d\partial}{dt}$ represents the rate of curing reaction and the specific form depends on the type of resin. In this paper, using the improvement equation of the curing mechanics of polyester resin proposed by Johnston [8] is as follows:

$$\frac{d\partial}{dt} = \frac{K\partial^m (1-\partial)^n}{1 + e^{C\{\partial - (\partial_{CO} + \partial_{CT} T)\}}} \tag{3}$$

$$K = Ae^{-\Delta E/RT} \tag{4}$$

where A is the frequency factor; ΔE is the activation energy; C , ∂_{CO} , ∂_{CT} are the diffusion constant; R is the gas constant; m and n are the reaction series.

2.2. Determination of mechanical properties of resin

During the curing process, the fiber hardly changed in volume during the whole process, and the resin undergoes a liquid-gel state-glass state, and the mechanical properties vary greatly. This paper uses Johnstonl's improved model to describe the change in mechanical properties of the resin during curing.

$$\begin{cases} E_r = E_r^0, (T^* < T_{C1}^*) \\ E_r = E_r^0 + \dfrac{T^* - T_{C1}^*}{T_{C2}^* - T_{C1}^*}(E_r^\infty \text{-} E_r^0), (T_{C1}^* < T^* < T_{C2}^*) \\ E_r = E_r^\infty, (T^* > T_{C2}^*) \end{cases} \tag{5}$$

$$v_r = \frac{E_r^\infty - E_r(1 - 2v_r^\infty)}{2E_r^\infty} \tag{6}$$

$$T^* = (T_g^0 + a_{Tg}\partial) - T \tag{7}$$

where E_r , v_r are the instantaneous modulus of elasticity and poisson's ratio of resin respectively; E_r^0 is the elastic modulus of resin before curing; E_r^∞ is the elastic modulus of fully cured resin; v_r^∞ is the poisson's ratio of fully cured resin; T_{c1}^*, T_{c2}^* are constants, indicating the beginning and end of glass transition; T^* is the difference between resin temperature and instantaneous glass transition temperature; T_g^0 is the glass transition temperature before curing; a_{Tg} is a constant, determined by experiment; ∂ is the degree of cure.

2.3. Curing shrinkage strain

During the curing process, the volume shrinkage of the resin is also one of the reason for the composite deformation to cause residual stress, which is related to the polymerization reaction of the resin under the temperature load. Assuming that the dimensions of the cube micro-elements in the three directions are l_1, l_2, l_3 and the variation are Δl_1, Δl_2, Δl_3 ,the volume change of the micro-element is:

$$\Delta V = l_1 l_2 \Delta l_3 + l_2 l_3 \Delta l_1 + l_1 l_3 \Delta l_2 + l_1 \Delta l_2 \Delta l_3 + l_2 \Delta l_1 \Delta l_3 + l_3 \Delta l_1 \Delta l_2 + \Delta l_1 \Delta l_2 \Delta l_3 \tag{8}$$

Volume change rate is:

$$\Delta v = \frac{\Delta V}{V} = \varepsilon_1 + \varepsilon_2 + \varepsilon_3 + \varepsilon_1\varepsilon_2 + \varepsilon_1\varepsilon_3 + \varepsilon_2\varepsilon_3 + \varepsilon_1\varepsilon_2\varepsilon_3 \tag{9}$$

Since the resin is isotropic, the shrinkage strain of the resin is:

$$\varepsilon_m^{sh} = \sqrt[3]{1 + \Delta v} - 1 \tag{10}$$

The chemical shrinkage strain of the three main directions of the composite material can be calculated by the following formula:

$$\begin{cases} \varepsilon_1^c = 0 \\ \varepsilon_2^c = \varepsilon_3^c = \left(\sqrt[3]{1+\Delta v} - 1\right)\left(1 - V_f\right)\left(1 + v_r\right) \end{cases} \tag{11}$$

where V_f is the fiber volume content; v_r is the resin Poisson's ratio. Since the 1 direction is constrained by the fiber and the strain is small, it is considered that the contraction strain in the 1 direction is 0.

2.4. Thermal strain

During the cooling phase of the curing process, thermal deformation occurs inside the laminate due to the inconsistent thermal expansion coefficients of the composite material in all directions. The formula for calculating the thermal strain of a composite material is:

$$\begin{cases} \varepsilon_1^T = \beta_1 \Delta T \\ \varepsilon_2^T = \varepsilon_3^T = \beta_2 \Delta T = \beta_3 \Delta T \end{cases} \tag{12}$$

where β_1 , β_2 , β_3 are coefficients of thermal expansion in the direction of the composite material 1,2 and 3; ΔT is the temperature difference.

3. Numerical simulation

In order to verify the algorithm, a laminated plate of AS4/8552 material is used to simulate the curing process. The laminate dimensions are shown in figure 1, where L=300mm, W=30mm, B=1mm. Single layer thickness is 0.125mm and layup sequence is $[0_4/90_4]$.The thermochemical parameters of AS4/8552 materials and the mechanical properties are shown in reference [9].

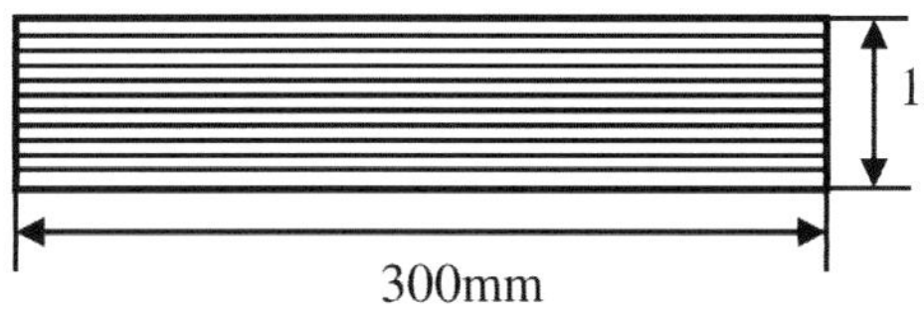

Figure 1. Dimension diagram of laminate

3.1. Temperature and degree of cure

In this paper, the ABAQUS three-dimensional finite element simulation analysis software is used to simulate the temperature field and curing degree field of the laminate. At the same time, the HETVAL subroutine is used to calculate the curing exotherm of the resin matrix, and the temperature data is transmitted to ABAQUS for update. Finally, the curve of temperature and curing degree changing with time in the curing process of the center point of the laminates is obtained, and compared with the results in the literature, as shown in figure 2.

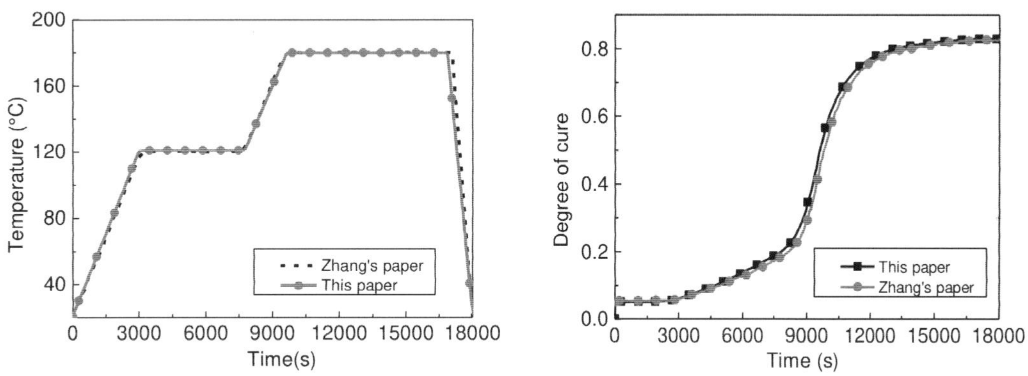

(a)Comparison curve of temperature with time (b)Comparison curve of curing degree with time

Figure 2. Comparison curve of temperature and curing degree at the center point of the laminate

From the figure 2, the settlement result of this paper is basically consistent with the temperature and curing degree change curve calculated in literature [11] , in which the maximum relative error of temperature is 0.9% and the maximum relative error of curing degree is 1.2%, within an acceptable range. Therefore, the program and calculation result in this paper are correct.

3.2. Curing deformation

Based on the calculation results of temperature and curing degree, curing deformation of laminates is simulated. According to the experimental results in literature [10] and the calculation results in literature [11] show that according to the $[0_4/90_4]$ laid the curing deformation of unsymmetrical laminated is near to the cylindrical shape, with the curvature of the cylindrical to represent the deformation degree of laminated plates. The curvature calculation method as shown in figure 3, where L is half of the chord length of the deformed cylinder, and U is the maximum deviation displacement in the Z direction.

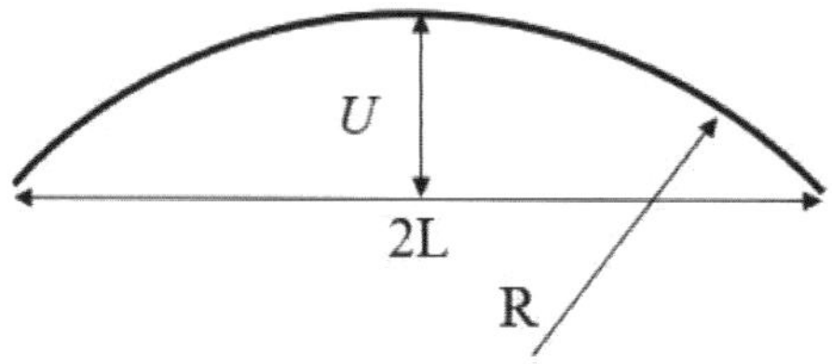

Figure 3. Diagram of curvature calculation

The formula for calculating the cylinder curvature is:

$$t = \frac{1}{R} = \frac{2U}{L^2 + U^2} \tag{13}$$

Figure 4 is a deformation diagram of the asymmetrically laminate obtained after curing, and the curvature of the cylinder is calculated and compared with the curvatures in [10] and [11], as shown in table 1.

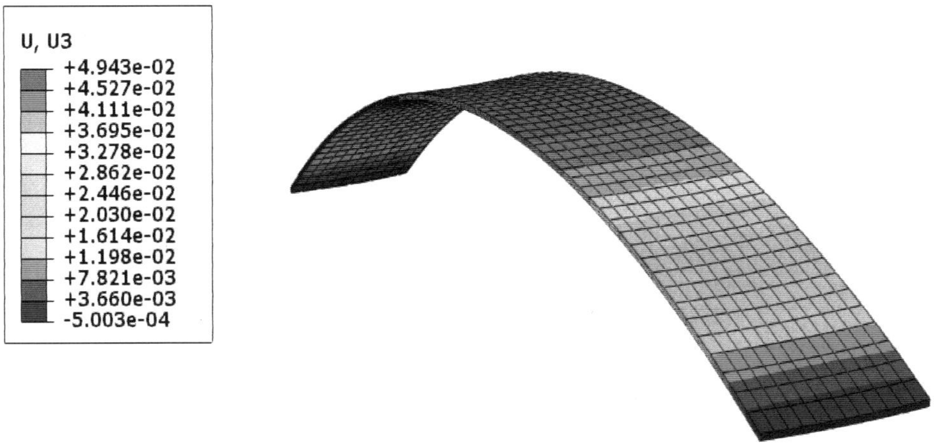

Figure 4. Deformation cloud diagram of asymmetric laminate

Table 1. Comparison of the calculation results in this paper with the literature

	t / m^{-1}	error/%
Experimental results in reference [10]	4.07	0
Calculation results in reference [11]	4.37	+7.37
The calculation results in this paper	3.98	-2.2

The error between the calculation result and the experiment is 2.2%, which verifies the correctness of the simulation method and the program. Figure 5 shows the displacement change in the Z direction during the curing of the laminate. It can be seen from figure 5 that the laminate has two obvious deformations during the curing process. The first deformation is caused by the curing shrinkage of the resin, and the second deformation is caused by the thermal expansion and contraction caused by the temperature drop of the laminate. At about 9000s, the displacement has a negative value, that is, the laminate has a downward bend. This is because the gel point of the selected material is 0.651. When the degree of cure reaches 0.651, the temperature is still rising. At this time, the phenomenon of curing shrinkage and thermal expansion occurs at the same time. The deformation caused by thermal expansion is greater than the deformation of the curing shrinkage, resulting in deformation.

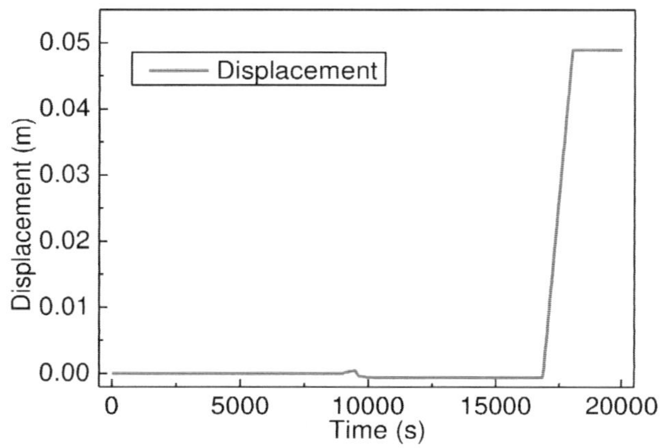

Figure 5. The displacement of the center point of the laminate in the Z direction

4. Deformation analysis of co-curing of metal and composite laminates.

The fiber metal laminate is co-cured based on the epoxy resin and the metal. During the curing process, different materials are bonded to each other, and no subsequent bonding process is needed. This simplifies the process and ensures that pure composite materials will not lose very important strength and stiffness when their properties are damaged. AS4/8552 composite materials and aluminum are co-cured to study the deformation results of dimension and layup sequence.

4.1 Finite element model

The dimensions of the metal aluminum and composite materials are all based on 300 mm × 30 mm × 1 mm, and the finite element model is shown in figure 6. The ABAQUS is used to establish a three-dimensional finite element model. The thermal conduction unit DC3D8 is used to simulate the temperature field and the curing degree field. The calculation results are imported into the stress analysis model, and the three-dimensional solid element C3D8R is used to simulate the residual stress and curing deformation. On this basis, the deformation laws of metal and composite materials with thickness ratios of 3 mm, 2 mm and 1 mm, that is (metal thickness: composite thickness) of 3:1,2:1,1:1,1:2,1:3 and composite lamination angles of 0°,45°and 90°are studied respectively.

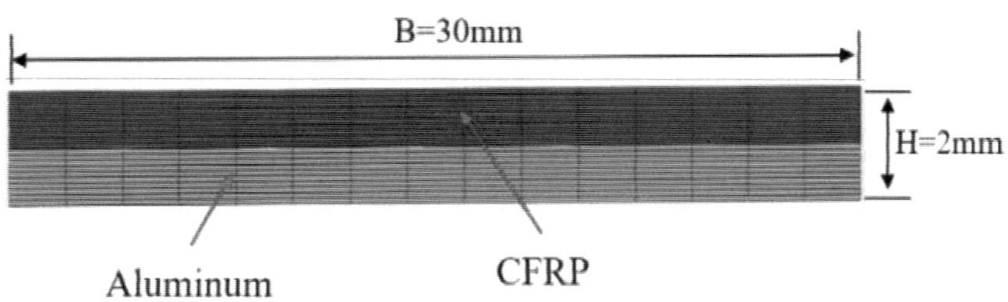

Figure 6. Co-curing Model of Metals and Composites

4.2 Result analysis

The simulation of the curing process of the metal-composite with different thickness ratios is carried out, and the displacement of the Z direction in the three cases of the ply angles of 0°, 45° and 90° is obtained, as shown in figure 7.

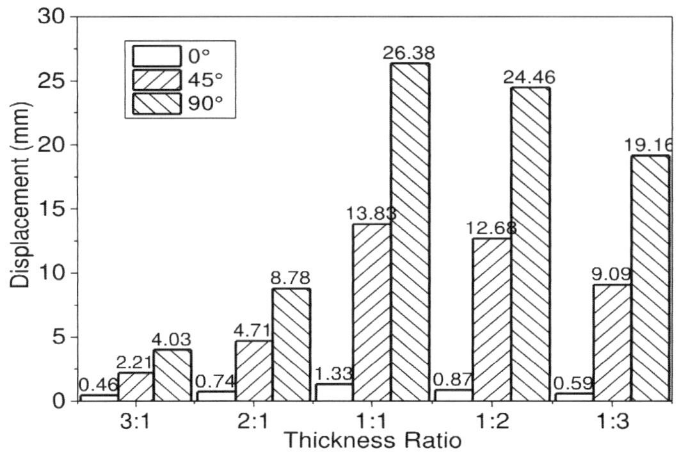

Figure 7. Comparison of displacement in the Z direction with different thickness ratios

It can be seen from the figure that the deformation amounts corresponding to different thickness ratios are different. Within the three ply-angles, the amount of curing deformation shows a certain regularity. When the thickness ratio of the metal to the composite material is 1:1, that is, the thickness of the metal and the composite material is the same, the deformation amount of the laminate in the Z direction is the largest. Keeping the thickness of the composite material constant and increasing the thickness of the metal, the corresponding deformation amount is gradually reduced. Under the same conditions, the thickness of the metal is kept constant, and when the thickness of the composite is increased, the corresponding deformation amount also tends to decrease. At the same time, it can be seen that increasing the thickness of the metal is less than increasing the thickness of the composite during the curing process. This is because the metal has only thermal expansion and contraction during the curing process, and the coefficient of thermal expansion is larger than that of the composite. The chemical reaction does not occur in the whole process, and the recovery ability after deformation is strong. Therefore, in the manufacture of the laminate, appropriately increasing the content of the metal can effectively reduce the deformation of the member after the completion of the curing.

5. Conclusion

(1) Based on the time-varying characteristics of the mechanical parameters of composite materials, a three-dimensional finite element calculation model of composite materials is established by using mathematical models including thermochemical model, resin curing shrinkage model and residual stress model, which can be used to simulate the curing process of composite materials.

(2) In comparison with the results of the references, it is proved that the algorithm developed for calculating the curing process of the composite material is reasonable and effective.

(3) Based on the algorithm proposed in this paper, the metal composite co-cured components are simulated and proved that when the thickness of metal and composite materials is the same, the deformation after curing is the largest. Appropriate increase of metal thickness can reduce deformation to a certain extent.

References

[1] Kevin-D Cowley and Beaumont Peter-W-R 1997 J. The measurement and prediction of residual stresses in carbon-fibre/polymer composites: Composites Science and Technology,57(11): 1445-1455.

[2] Sung Yi, Chian Kerm-Sin and Hilton Harry-H 2002 J. Nonlinear Viscoelastic Finite Element Analyses of Thermosetting Polymeric Composites during Cool-Down after Curing:Journal of Composite Materials, **36**(1): 3-17.

[3] Erik Kappel 2016 J. Forced-interaction and spring-in – Relevant initiators of process-induced

distortions in composite manufacturing: Composite Structures, Volume **140**, Pages 217-229..
[4] J-T Zhang, Zhang M and Li S-X 2016 J. Residual stresses created during curing of a polymer matrix composite using a viscoelastic model: *Composites Science and Technology*, Volume **130,** Pages 20-27.
[5] Nuri Ersoy, Garstka Tomasz and Potter Kevin 2010 J. Modelling of the spring-in phenomenon in curved parts made of a thermosetting composite: *Composites Part A: Applied Science and Manufacturing*, **41**(3): 410-418.
[6] Erik Hörberg, Nyman Tonny and Åkermo Malin 2019 J. Thickness effect on spring-in of prepreg composite L-profiles – An experimental study: *Composite Structures*, Volume **209,** Pages 499-507.
[7] Christian Brauner, Bauer Stefan and Herrmann Axel-S 2014 J. Analysing process-induced deformation and stresses using a simulated manufacturing process for composite multispar flaps: *Journal of Composite Materials*, **49**(4): 387-402.
[8] Andrew Johnston 1998 M. An Integrated Model of the Development of Process-Induced Deformation in Autoclave Processing of Composite Structures.
[9] Yizhi Fu and Shenyan Chen 2012 A. Analysis of process-induced deformation of complicated composite curved surface: *63rd International Astronautical Congress 2012 (IAC 2012)* Volume **8** of **14**. Pages 6223.
[10] M Gigliotti, Wisnom M-R and Potter K-D 2003 J. Development of curvature during the cure of AS4/8552 [0/90] unsymmetric composite plates: *Composites Science and Technology*, **63**(2): 187-197.
[11] Jikui Zhang, Li Zhengneng and Guan Zhidong 2007 J. Influence of degree of cure and cure shrinkage on the finial deformation of the unsymmetric composite laminates: *Fuhe Cailiao Xuebao/Acta Materiae Compositae Sinica*, **24**(2): 120-124.

IOP Conf. Series: Materials Science and Engineering **616** (2019) 012017 doi:10.1088/1757-899X/616/1/012017

Performance Analysis of Drilling Test of Aramid Fiber Composite

Jin lei[1, *] and Qi jing[2]

[1] Liaoning Provincial College of Communications
[2] Tianjin hydraulic (machinery) group co. LTD

[*]Jason_sy_j@aliyun.com

Abstract. Aramid fiber-reinforced polymer (AFRP) is a typical difficult material to machine. It is easy to occur defects such as ablation and burr in the drilling process due to the serious non-uniformity and heterotrophy of these material properties, as well as low heat transfer coefficient and high fibre toughness of the aramid fibre, which remarkable affects the machining quality and assembly accuracy. In this paper, three types of drills with different cutting edge shapes were compared to analyse the influence of drilling on cutting force, cutting temperature and machining quality. The results showed that the tool geometry has a great influence on the drilling performance of AFRP. Due to poor drilling quality, 8-face drill and twist drill are not suitable for drilling AFRP drilling tools. The machining quality of Brad & Spur drill is relatively good, which can meet the drilling requirements of AFRP.

1. Introduction

Aramid fiber-reinforced polymer (AFRP) is an advanced materials with typical high difficult, specific strength material, specific stiffness, low density and designability [1,2]. For its exceptional shock resistance and low density, the materials are widely applicated in aerospace, armored weapons, electronic vehicles and human body protection and other fields [3-5]. However, due to the characteristics of low inter laminar strength and low thermal conductivity of aramid fiber composite materials, under the coupling action of cutting force and cutting heat, defects such as burr, tear and layering are easy to occur in machining process, which will affect the geometric error of workpiece and assembly accuracy. Especially, drilling is a semi-closed machining mode, and the cutting area temperature is extremely high during machining, which makes it difficult to dissipate the processing heat in time, easily causing ablation on the hole wall [6,7].

In particular, drilling is a semi-closed machining mode, and the cutting area temperature is extremely high during machining, which makes it difficult to dissipate the processing heat in time, easily causing ablation on the hole wall. studied The experimental study on the perforation of the kevlar fiber was operated by Hemant C using laser processing. The result showed that the AFRP can significantly inhibit burrs and tears on the inlet and outlet surfaces, but ablative defects occur in the inlet and outlet holes due to the high-energy heat generated by the highly focused laser beam [8]. A.Díaz-Álvareza analyzed effect of standard twist bit and three-point bit on drilling of Arlington composite material. The empirical formulas of the damage area and axial force at the inlet and outlet of the material are established through variable parameter experiments [9]. Sinan L et al, compared the effects of feed velocity on borehole axial force with and without support ring, as the upper and lower support ring is added when the twist drill machining AFRP plate. It was found that the lamination, tear and burr defects on the inlet and outlet surfaces were the least when the support ring size was larger bit diameter [10].

In this paper, the three-point drill bit, eight-face drill bit and twist drill bit are used as drilling tools to study the influence on the processing performance of AFRP, providing some references for the high quality and efficient processing technology of arrayon composite materials.

2. Experiments

2.1. Test materials and cutting tools

The experimental composite material was f12/602 arrayon fiber reinforced epoxy resin base composite plate material, F12 the fiber model, the weaving direction was 0/90°, and the single-layer fiber thickness was 0.13mm. The curing temperature of epoxy resin was 130°C, and the glass conversion Tg was 150°C. The laminates were 5.8±0.1mm thick and the fiber mass fraction was 62±2wt%. After molding, it is cut into 150mm×100mm specification for convenient testing. The specific physical properties are shown in table 1.

In this experiment, three kinds of drills with the same blade diameter of 6mm were used for drilling research, the maford-207 diamond-coated three-point drill, the kyocera-120 diamond-coated octahedral drill and the kyocera-101 titanium nitride coated standard twist drill, as Figure 1 shown. In the test, cutting parameters were adopted as spindle rotation speed of 4000r/min and feeding speed of 40mm/min.

Table 1. Mechanical properties of AFRP laminates

Materials	Volume density (g.cm$^{3)}$	Compression strength (MPa)	Extension strength (MPa)	Bending strength (MPa)	Shear strength (MPa)	Heat conductivity coefficient (W/(m·K))	Weaving direction
F12/602	1.44	171	726	398	38.4	0.35	0/90°

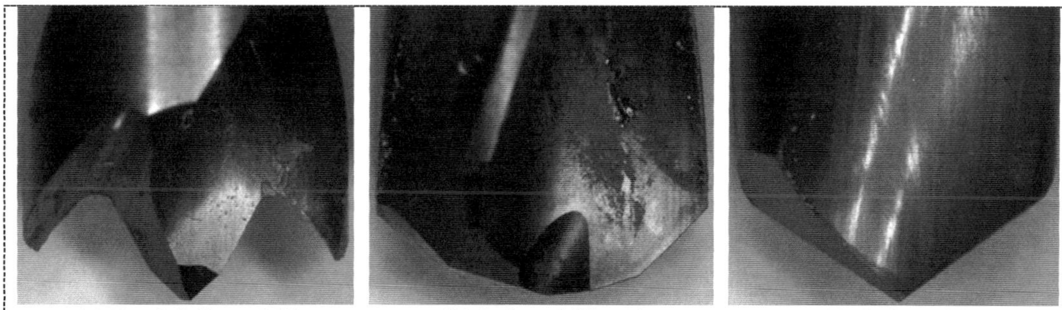

(a) Brad & Spur drill (b) 8-face drill and (c) Twist drill

Figure 1. Drills in experiment.

2.2. Experiment devices

The machine tool used in the drilling test of AFRP was KVC850M of the long march three-axis vertical machining center. The cutting force was collected by piezoelectric dynamometer, LN5861 charge amplifier and ADLINK usb-1901 acquisition card. Drilling temperature is completed by manual thermocouple and data acquisition card. The thermocouple adopts OMEGA-K thermocouple, measuring range -200~500°C, accuracy 0.1°C. Before the test, a 1mm micro drill bit was used to conduct embedded hole processing of AFRP. The location of embedded hole was 0.2mm away from the wall of the processed hole. The thermocouple was encapsulated in the material with cryogen.

3. Result and analysis

3.1. Cutting force

The axial force of the three types of bits routinely process AFRP laminated platesc were shown as Figure 2. During drilling, because the drill is composed of two symmetrical cutting edges, the radial force generated by the cutting edges can cancel each other under normal circumstances, resulting in a small torque generated by drills. In the direction of feed, the axial force is larger due to the extrusion of the transverse edge and the cutting and friction of the main cutting edge. a raft of studies suggest that axial force during mechanical drilling of laminated composite materials is the main factor affecting lamination, tear and burr defect [11-12]. Therefore, only three kinds of axial forces generated by bit cutting are discussed.

The axial force values of conventional machining of brad & spur drill, 8-face drill and twist drill are respectively 41.8N, 104.7N and 51.8N, as shown in Figure 2. The axial force of each cutter will rise sharply, when it is just cut into the material axial. The maximum value of the force appears after cutting a certain distance inside the material. And then, the axial force decreases gradually during the cutting process. According to the characteristics of drilling, the axial force can be divided into three stages.

(1)A~B Drill transverse edge, main cutting edge and secondary cutting edge cut into the material stage

Each cutting edge of the drill bit gradually extrudes the cut material, and the axial force increases sharply. As the cutting edge of the twist drill and the eight-sided drill continuously ACTS on the material surface, the axial force in this stage presents an increased state.

(2)B~C Drill transverse edge, main cutting edge and secondary cutting edge all work together stage

Each cutting edge of the drill is fully involved in cutting, and the axial force continues to increase and then reaches the peak. Due to the U-shaped groove structure between the middle sharp Angle and the outer edge sharp Angle of the three-pointed drill, the axial force at this stage increases for a period of time and then reaches the peak. After the peak value, the axial force presents a steady downward trend due to the screw Angle of the drill.

(3)C~D Drill cutting edge, main cutting edge and secondary cutting edge cutting material stage

Each cutting edge of the drill gradually moves away from the material and the force acting on the workpiece decreases sharply until it no longer exerts force on the workpiece, and the value of axial force returns to 0. Due to the lack of support at the outlet, the axial force fluctuates slightly at that time.

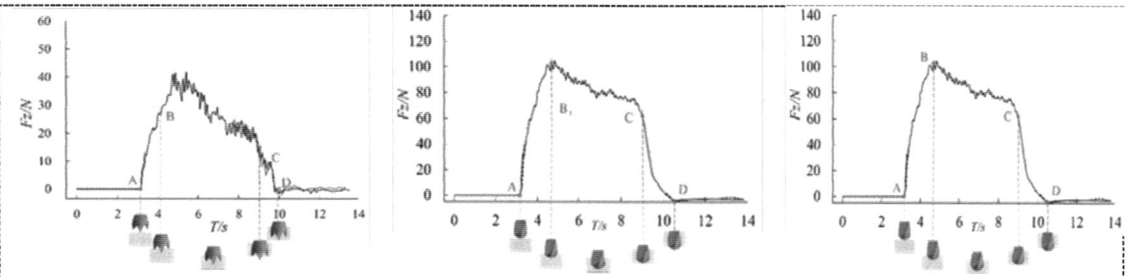

(a) Brad & Spur drill (b) 8-face drill and (c) Twist drill

Figure 2. Evolution of the Thrust force with the drill movement.

3.2. Cutting temperature

The maximum temperature value of hole wall during AFRP machining is shown in Figure 3. The temperature of brad & spur drill, 8-face drill and twist drill are respectively 234.7°C, 285.4°Cand 195.7°C. Due to the low thermal conductivity of aramid fiber composites (0.35W/m•K), a large amount of cutting heat generated during drilling could not be diffused in time. When the cutting zone temperature exceeds the resin base glass conversion temperature Tg, the resin base softened, degraded and even completely carbonized, causing serious ablative defects in the inlet and outlet and pore wall. Cutting heat energy reduces the bond strength of resin, which will lead to the fiber being easily pulled out from the resin matrix under the action of cutting force, leading to the emergence of burr defects. On the one hand, the reduction of resin bond strength will reduce the export stiffness of composite workpiece, and make the uncut material heave to the exit direction during the cutting out of the material, so as to inhibit the emergence of export layering. On the other hand, the reduction of resin bond strength will reduce the bonding strength between layers, easy to produce stratification.

The twist drill bore the lowest temperature as shown in Figure 3. The main source of cutting heat is the work done by material resisting cutting force. The 8-face drill produces the most cutting force and thus the most heat and temperiment. The brad & spur drill has the smallest cutting force, but because of the U-groove shape of the cutting edge, the area of contact between the tool and the workpiece material is larger, the material produces more friction, generates more heat, resulting in a high cutting temperature.

3.3. Surface quality of machined holes

AFRP is a highly heterogeneous material with obvious anisotropy. As the characteristics of high toughness aramid fiber and low affinity with resin base, it is easy to form such phenomena as fiber pulling out and matrix and fiber defusing under the action of thermal/force coupling. During mechanical drilling, the fibers on the surface of material inlet and outlet are difficult to be cut due to lack of support, resulting in burr defects, as shown in Figure 4. In the dry cutting state of the 8-facee drill, there are a lot of fibers on the inlet and outlet surface that have not been cut off. At the same time, the excessive cutting temperature caused obvious burns on the fibers on the material surface, and almost all the resins at the hole wall have been heavily carbonized.

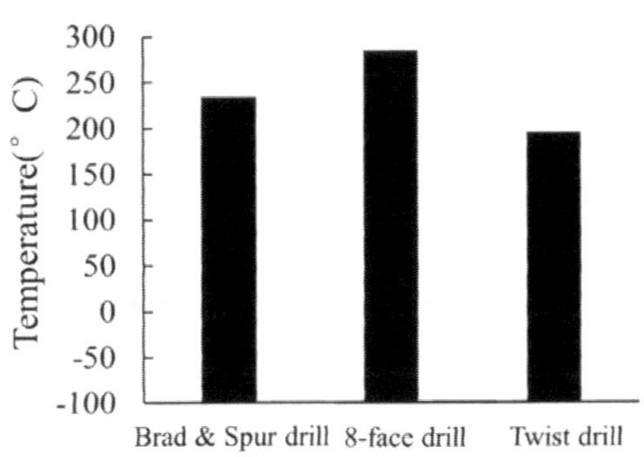

Figure 3. Temperature of drilling AFRP.

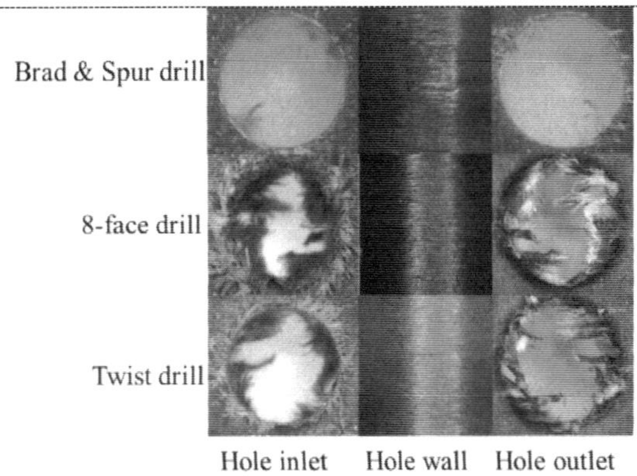

Figure 4. The surface morphology of drilling AFRP.

According to the observation of hole wall, resin ablative phenomenon increases obviously with the increase of bit axial feed. Significant carbonization has been observed at the outlet. The reason is the cutting edge of the drill bit continuously cuts the material, under the semi-closed environment, the cutting heat is continuously low and accumulates and increases, resulting in the cutting temperature increasing with the feed of the drill hole, resulting in more serious material burns at the outlet than at the inlet.

Compare the other two drills, under the dry cutting condition of the brad & spur drill, the burrs at the inlet and outlet are greatly reduced, and there are almost no long burrs at the inlet and only a few remaining burrs at the outlet. The reason is the special external geometric edge shape of the three-point drill, as shown in Figure 5. Different from the extruded cutting from the center of the hole to both sides of the Fh 'produced by the twist drill and the 8-face drill, the brad & spur drill, due to the existence of external edge, first cuts the outer fiber of the hole, and applies the horizontal component force Fh' on the material to the center of the circle, thus forming a cutting state from the edge of the hole to the center of the hole.

No force is produced to extrude the fiber on the edge of the hole, reducing the burr caused by the fiber on the edge of the hole due to lack of support. And in the hole wall observation can be, three-pointed drill will also produce ablative resin base, resin base is heavily carbonized, at the same time in the hole wall produced serious fiber pulling defects.

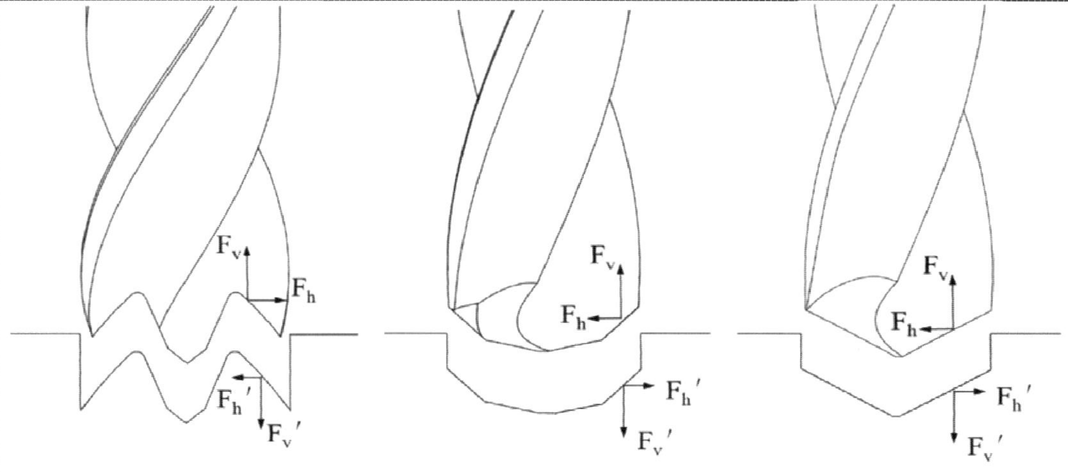

(a) Brad & Spur drill (b) 8-face drill and (c) Twist drill

Figure 5. Model of the drilling force.

4. Conclusion

The influence of cutting tool geometry on the drilling performance of array on fiber composites is obvious. The 8-face drills, which have multiple cutting edges, will generate greater cutting force and cutting heat when drilling, and the quality after drilling is poor, which is not suitable for AFRP. The twist drills produce less cutting heat, less burrs at the cutting inlet than the 8-face drills,, and less surface burns. The cutting force produced by the brad & spur drill is small, and the hole wall may have defects such as fiber pulling out, but the import and export quality is relatively good.

References

[1] Kaiyan Sun, Jun Li. Research on drilling method of aramid fiber composites [J]. Engineering Plastics Application,2009,37(3):37-39
[2] Martin D, Rafael W, et al. Modeling of drilling assisted by cryogenic cooling for higher efficiency[j].CIRP Annals - Manufacturing Technology 2014 (63) 73–76
[3] Li Guo. Research on high performance lightweight armor material processing technology [D]. Nanjing: Nanjing University of Science and Technology,2006.
[4] Abrao A M, Faria P E, Rubio J C C, et al. Drilling of fiber reinforced plastics: a review[J]. Journal of Materials Processing Technology. 2007, 186(1-3): 1-7
[5] Xia T,Kaynak Y, Arvin C et al. Cryogenic cooling-induced process performance and surface integrity in drilling CFRP composite material. The International Journal of Advanced Manufacturing Technology, 2016,82: 605-616
[6] Yini Zhang. Study on thermal behavior of aramid fiber composites during processing [D]. Dalian University of Technology, 2016
[7] Guiyu Li.Mechanism and process parameter optimization of laminate composite drilling defects [D]. Shandong University, 2011
[8] Hemant C , Dilpreet S et al machining of Kevlar fiber reinforced laminates e Effect of

polyetherimide versus polypropylene matrix[J]. Composites Science and Technology2016 (134) 267-274

[9] A Díaz-Álvareza, M Rodríguez-Millána et al Experimental analysis of drilling induced damage in aramid composites[J]. Composite Structures,2018

[10] Sinan L Tao Y Investigation of hole quality during drilling of KFRP based on the interaction between collars and cutter[J]. Int J AdvManufTechnol 2018 (95):4101–4116

[11] Abrao A M,Faria P E, Rubio J C C, et al. Drilling of fiber reinforced plastics: a review[J]. Journal of Materials Processing Technology. 2007, 186(1-3): 1-7

[12] Martin D, Rafael W, et al. Modeling of drilling assisted by cryogenic cooling for higher efficiency[j].CIRP Annals - Manufacturing Technology 2014 (63) 73–76

IOP Conf. Series: Materials Science and Engineering **616** (2019) 012018 doi:10.1088/1757-899X/616/1/012018

Technological Rule of Ultrasonic Assisted Magnetic Abrasive Finishing of Titanium Alloy

T Y Jiang, F J Ma*, X Liu, Y Liu, S F Zhang and Z H Sha

School of Mechanical Engineering, Dalian Jiaotong University, No. 794, Huanghe Road, Shahekou District, Dalian 116028, China

*The corresponding author's e-mail address: mafj@djtu.edu.cn

Abstract. The influence of machining time, ultrasonic amplitude, spindle speed and machining gap on surface roughness and material removal rate were studied and analysed through the experiment of UAMAF of titanium alloy. The results show that the surface roughness decreases rapidly in the beginning and then gradually stabilizes as the machining time increases. The surface roughness reduces from the original Ra0.94μm to Ra0.10μm, when the machining time reaches 30min that is optimum machining time. The material removal rate drops approximately linearly with the increase of machining time. The surface roughness decreases significantly and the material removal rate increases rapidly as the ultrasonic amplitude rises. The surface roughness reduces to Ra0.12μm and the material removal rate increases to 30.5mg/h, when the ultrasonic amplitude is up to 14μm. The surface roughness decreases in the beginning and then increases, and the material removal rate rises in the beginning and then falls, as the spindle speed increases. The surface roughness achieves a minimum of Ra0.11μm and the material removal rate reaches a maximum of 22.74mg/h, when the spindle speed is 1400r/min. The surface roughness drops in the beginning and then rises, and the material removal rate increases in the beginning and then decreases, as the machining gap enlarges. The optimum machining gap is about 1.25mm. In this case, the surface roughness achieves a minimum of Ra0.16μm, and the material removal rate reaches a maximum of 17.7mg/h.

1. Introduction

Titanium alloy with high strength, high temperature resistance, corrosion resistance, good fatigue properties and fracture toughness has been widely used in key parts of aviation, aerospace, shipping and energy fields [1]. Extremely high requirements for finishing quality and efficiency are required in precise finishing of titanium alloy parts. However, titanium alloy is typical difficult-to-machine material because of its small elastic modulus, high viscosity and poor process performance. Therefore, the quality and efficiency in finishing of titanium alloy have yet to be improved.

For achieving high-quality and high-efficiency finishing, a variety of finishing technologies that compounds with multiple energies have been applied to the finishing of difficult-to-machine materials such as titanium alloy, aluminium alloy and superalloy. ultrasonic assisted magnetic abrasive finishing (UAMAF) can significantly improve the finishing quality and efficiency of difficult-to-machine materials among these technologies through the organic compounding of ordinary magnetic abrasive finishing and ultrasonic vibration [2-3]. The finishing principle of UAMAF is shown in Figure 1. Ultrasonic vibration is applied on magnetic pole to make magnetic abrasive impact workpiece surface at a velocity of *Vu* with a high-frequency, on the basis of rotating at a speed of *Vs*. The action of magnetic abrasive on workpiece is high-frequency and high-speed impacting besides cutting. The amount of magnetic abrasives in contact with the surface of workpiece per unit time is increased. Finally, high-quality and high-efficiency finishing of difficult-to-machine materials is achieved [4]. Therefore, many researches on UAMAF have been carried out by many scholars.

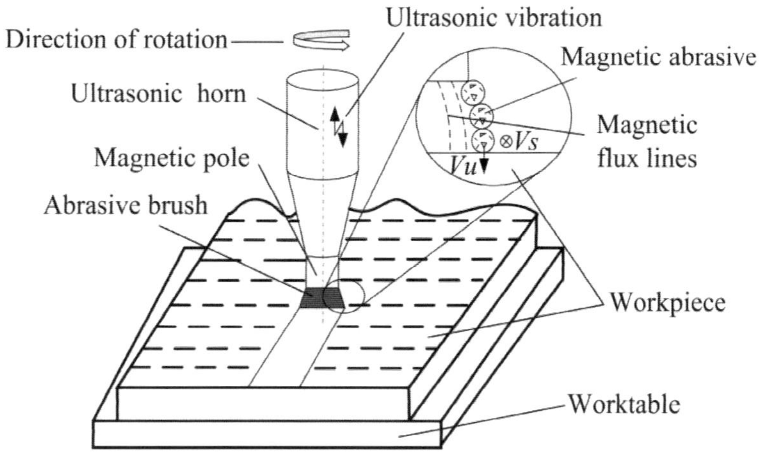

Figure 1. Finishing principle of UAMAF.

Y P Lu et al. [5] proposed to attach ultrasonic vibration to magnetic abrasive finishing. It is revealed that enhancing the cutting ability of the abrasive brush is critical to improve the efficiency of magnetic abrasive finishing by analysing mathematical removal model of traditional magnetic abrasive finishing. J Ma et al. [6] used the self-developed UAMAF setup to carry out single factor test on the finishing of 45# steel. The influence of magnetic induction on the quality of finishing was analysed. Surface roughness reduces significantly as magnetic induction increases. The surface roughness decreases from the original Ra1.35μm to Ra0.11μm, when the magnetic induction is 0.9T. The experiment of UAMAF of the inner wall of TA18 titanium alloy tube was carried out by Y Tan et al. The results show that the surface roughness of TA18 tube reduces from the original Ra1.2μm to Ra0.07μm after finishing for 50min with the vibration frequency of 19kHz and the amplitude of 10μm [7]. H Yun et al. conducted UAMAF test on Al_2O_3 ceramic tubes. It is found that the increase of vibration frequency and spindle speed can improve the material removal rate of UAMAF and reduce the surface roughness. The optimum surface roughness is Ra0.03μm, and the maximum material removal rate is 48.5mg/h after finishing for 50min with the vibration frequency of 19kHz. The surface roughness decreases to Ra0.03μm, and the material removal rate increases by 17.5mg/h, when the spindle speed increases from 600r/min to 2000r/min [8]. The comparative test of ordinary magnetic abrasive finishing and UAMAF of AISI52100 hardened steel with hardness HRC61 was conducted by R Mulik et al. The results show that the finishing quality of UAMAF is significantly improved, and the surface roughness can be as low as Ra0.022μm [9-10]. UAMAF has advantages in achieving high-quality and high-efficiency finishing as shown in above studies. But relatively few studies on the influence of machining parameters on surface roughness and material removal rate in UAMAF of difficult-to-machine alloys such as titanium alloy and superalloy were researched. Therefore, more and more comprehensive and in-depth studies on UAMAF of titanium alloy are necessary to be carried out.

A systematic technological test of UAMAF of titanium alloy is conducted in this study. Surface roughness and material removal rate of UAMAF of titanium alloy materials under different finishing parameters are measured by surface roughness meter, ultra-depth 3D microscope and precision analysis scale. The effects of machining time, ultrasonic vibration and other finishing parameters on UAMAF are studied and analysed.

IOP Conf. Series: Materials Science and Engineering **616** (2019) 012018 doi:10.1088/1757-899X/616/1/012018

2. Experiment setup and conditions

The test of UAMAF is carried out on the experiment setup as shown in Figure 2. It is built by the ultrasonic vibration system integrated on a three-axis CNC machine tool. The resonant frequency of ultrasonic vibration system is 21.91kHz and the amplitude is continuously adjustable between 5μm and 15μm. The magnetic pole is NdFeB magnet of type N38 with the diameter of 6mm and the length of 75mm. The coolant is Castrol 9930 water soluble cutting fluid with a concentration of 8%. The abrasive is iron-based diamond magnetic abrasive with the average particle diameter of 300μm prepared by chemical composite plating. Its surface topography is shown in Figure 3. The other finishing parameters used in the test are shown in Table 1.

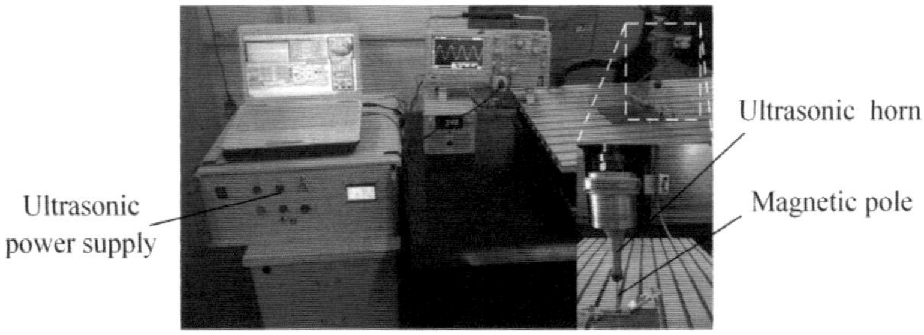

Figure 2. Experiment setup of UAMAF.

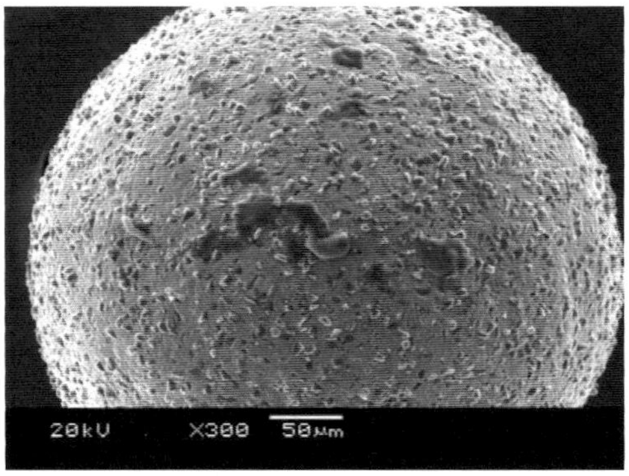

Figure 3. Surface topography of magnetic abrasive.

Table 1. Finishing parameters.

Finishing parameters	value
Spindle speed (r•min^{-1})	1000
Feed rate (mm•min^{-1})	20
Machining gap (mm)	1.25
Finishing time (min)	20
Finishing length (mm)	10
Amplitude (μm)	10
Ultrasonic frequency(kHz)	21.91

The workpiece material is TC4 titanium alloy in the test of UAMAF. The initial surface of the workpiece is obtained by milling. The milling parameters are that the spindle speed is 2500r/min, tool diameter 20mm, feed per tooth 0.05mm and cutting depth 0.1mm. The coolant is water-based lubricating cutting fluid. The surface roughness after milling is about Ra0.94μm.

The average value of three measurements that are detected by Talysurf PGI840 Surface Roughness Profiler before and after finishing is taken as the surface roughness of the workpiece.

The material removal amount of finishing is measured by Sartorius CP225D precision analytical balance. The workpiece before and after finishing needs to be cleaned with an ultrasonic cleaner for 20 minutes and dried with electric thermostatic drying oven for 15min at 150°C. The material removal rate is calculated by

$$Q = \frac{M_1 - M_2}{t} \tag{1}$$

where M_1 is the weight of the workpiece before finishing(mg); M_2 is the weight of the workpiece after finishing(mg); t is the machining time(h).

3. Results and discussion

3.1. Effect of machining time on surface roughness and material removal rate

The machining time in the test is 10min, 20min, 30min, 40min and 50min respectively and the other finishing parameters are shown in Table 1. Single factor test for UAMAF is conducted. The effect of machining time on surface roughness and material removal rate is shown in Figure 4.

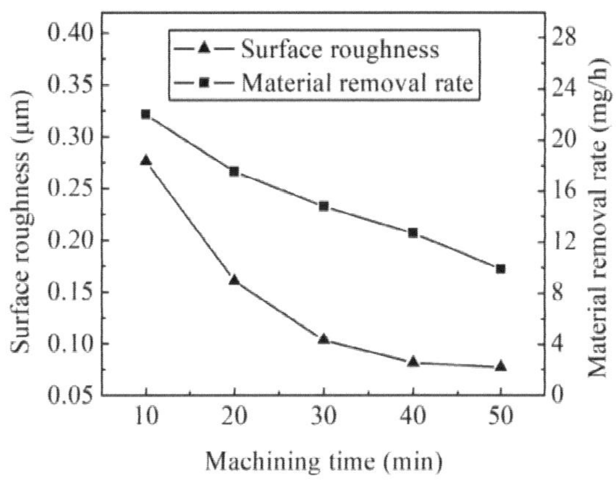

Figure 4. Effect of machining time on surface roughness and material removal rate.

It can be seen from Figure 4 that the surface roughness decreases as the machining time increases. The surface roughness declines from the original Ra0.94μm to Ra0.10μm, after finishing for 30min. But the surface roughness of workpiece is decreasing more and more slowly as the machining time goes further. The surface roughness reduces from Ra0.10μm to Ra0.08μm only, after finishing for another 20min.

The reason for this phenomenon is that UAMAF has vertex effect. Original surface of workpiece is obtained by milling. The difference of height between peak and valley of surface is comparatively large, and the magnetic abrasive has a higher pressure on the peak. Therefore, workpiece will be removed more quickly at the peak, resulting in a faster decrease in surface roughness. In addition, high-frequency and high-acceleration impact of ultrasonic vibration causes the magnetic abrasive to perform ultrasonic shot peening while performing cutting removal. Plastic flow is produced from the wave peak of workpiece surface to flow into the valley and fill it. However, the difference of height between peak and valley becomes smaller, and the vertex effect is weakened as the finishing progresses. The entire plane begins to be finished by the magnetic abrasive, when the groove between adjacent peaks can be machined, after finishing for a certain period of time. Therefore, the surface roughness decreases more and more slowly.

It also can be seen from Figure 4 that the material removal rate drops approximately linearly with the increase of machining time. The material removal rate is 10.5 mg/h at 50 min, which is 55% lower than that of 10 min. The reason for this result is that some diamond grains on magnetic abrasives are worn or fall off as the machining time increases, and the machining performance of magnetic abrasives is reduced. As a result, the material removal rate decreases approximately linearly.

3.2. Effect of ultrasonic amplitude on surface roughness and material removal rate

The ultrasonic amplitude in the test is 6μm, 8μm, 10μm, 12μm and 14μm respectively and the other finishing parameters are shown in Table 1. Single factor test for UAMAF is conducted. The effect of ultrasonic amplitude on surface roughness and material removal rate is shown in Figure 5.

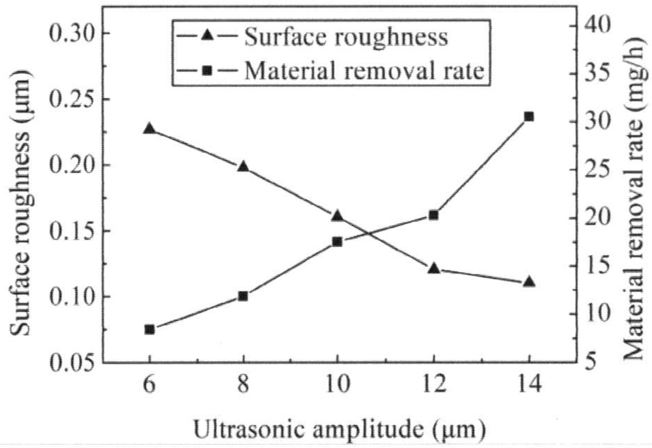

Figure 5. Effect of ultrasonic amplitude on surface roughness and material removal rate.

As can be seen from Figure 5, the surface roughness decreases rapidly and the material removal rate increases remarkably as the ultrasonic amplitude increases. The surface roughness reduces from Ra0.26μm to Ra0.12μm, and the material removal rate increases from 7.6mg/h to 30.5mg/h, when the amplitude increases from 6μm to 14μm.

The reason for this phenomenon is that an increase of ultrasonic impulse is caused by the increase of ultrasonic amplitude, which results ultrasonic force to rise linearly. The cutting force of a single magnetic abrasive on workpiece consists of magnetic field force and ultrasonic force. The magnetic field force remains unchanged during finishing. Therefore, the cutting force of UAMAF rises linearly with the increase of ultrasonic amplitude, which improves the ability of magnetic abrasive to cut into workpiece. Workpiece surface material is removed quickly and the material removal rate rises. The impact energy between magnetic abrasives and workpiece surface are increased due to the increase of the ultrasonic amplitude at the same time. Plastic flow is enhanced from the wave peak to the valley of workpiece surface. The randomness of the magnetic abrasive trajectory is also increased, resulting in complex interleavings of the cutting trajectories between adjacent abrasives. Therefore, the surface roughness reduces significantly.

3.3. Effect of spindle speed on surface roughness and material removal rate

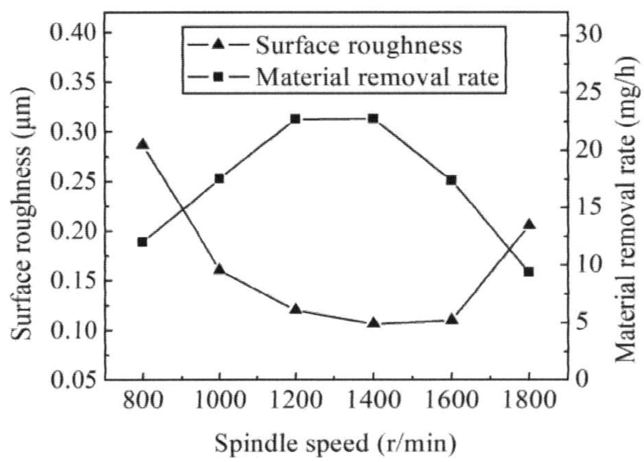

Figure 6. Effect of spindle speed on surface roughness and material removal rate.

The spindle speed in the test is 800r/min, 1000r/min, 1200r/min, 1400r/min, 1600r/min and 1800r/min respectively and the other finishing parameters are shown in Table 1. Single factor test for UAMAF is conducted. The effect of spindle speed on surface roughness and material removal rate is shown in Figure 6.

It can be seen from Figure 6 that the surface roughness decreases and the material removal rate increases in the beginning of finishing as the spindle speed increases. The surface roughness reaches the lowest Ra0.11μm, and the material removal rate achieves the highest 22.74mg/h, when the spindle speed is 1400r/min. The surface roughness rises rapidly and the material removal rate drops significantly, when the spindle speed increases further. The surface roughness increases to Ra0.21μm, and the material removal rate reduces to 9.85mg/h, which is 57% lower than 22.74mg/h, when the spindle speed is 1800r/min.

The reason for this phenomenon is that the cutting length per unit time is increased due to the increase of spindle speed. So the surface roughness of the workpiece reduces and the material removal rate rises. The centrifugal force of the magnetic abrasive in the "abrasive brush" is enhanced, as the spindle speed increases. However, the magnetic abrasive will break away from the magnetic field force and fly out of the finishing area along the tangential direction of spindle rotation, when the centrifugal force is greater than the magnetic field force, resulting in a decrease of the amount of the abrasives involved in finishing in the "abrasive brush". Therefore, the surface roughness increases and the material removal rate decreases, when the spindle speed increases continuously.

3.4. Effect of machining gap on surface roughness and material removal rate

The machining gap in the test is 0.75mm, 1mm, 1.25mm, 1.5mm and 1.75mmrespectively and the other finishing parameters are shown in Table 1. Single factor test for UAMAF is conducted. The effect of machining gap on surface roughness and material removal rate is shown in Figure 7.

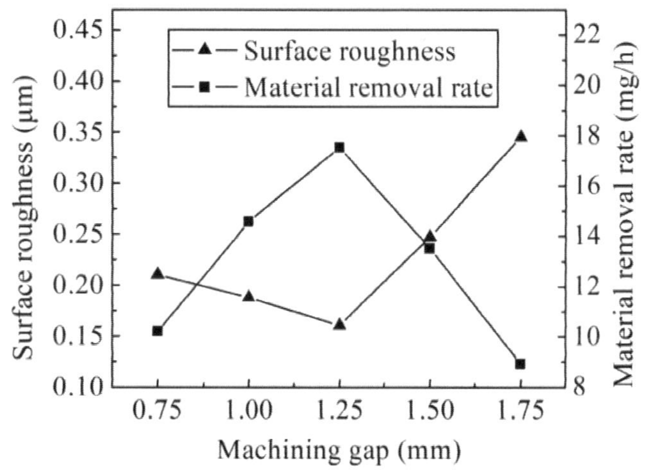

Figure 7. Effect of machining gap on surface roughness and material removal rate.

As can be seen from Figure 7, the surface roughness decreases and the material removal rate increases as the machining gap enlarges in the range from 0.75mm to 1.25mm. The surface roughness reaches the lowest Ra0.16μm, and the material removal rate achieves the highest 17.7mg/h, when the machining gap is 1.25mm. However, the surface roughness rises and the material removal rate drops rapidly, when the machining gap is greater than 1.25mm. The surface roughness increases from the lowest Ra0.16μm to Ra0.35μm, and the material removal rate reduces from 17.7mg/h to 8.8mg/h, when the machining gap increases from 1.25mm to 1.75mm.

The reason for this phenomenon is that magnetic abrasives are layered evenly along magnetic flux lines in machining gap. The magnetic pole collides with the magnetic abrasive of the uppermost layer, then ultrasonic vibration energy is transmitted layer-by-layer, and to the workpiece through the lowermost layer of the magnetic abrasive finally in UAMAF. Therefore, the loss energy in ultrasonic vibration transmission increases linearly as the machining gap enlarges. However, the amount of magnetic abrasives in the machining gap increases at the same time. Magnetic abrasives are caused to collide with each other by ultrasonic vibration, resulting renewability and fluidity of the magnetic abrasive are enhanced, so the finishing ability is improved. Therefore, the surface roughness decreases, and the material removal rate increases as the machining gap enlarges in the range from 0.75mm to 1.25mm. Magnetic induction intensity in the machining gap decreases and the loss energy of ultrasonic vibration transmission rises further as the machining gap continues to enlarge, so the total cutting force composed of magnetic force and ultrasonic force decreases quickly. Therefore, the surface roughness increases and the material removal rate reduces as the machining gap increases from 1.25mm to 1.75mm.

4. Conclusions

A systematic experiment of UAMAF of titanium alloy is carried out in this research. The influence of machining time, ultrasonic amplitude, spindle speed and machining gap on surface roughness and material removal rate are studied and analysed. The conclusions are as follows:

1) Due to the vertex effect of UAMAF, the surface roughness decreases rapidly as the machining time increases. The surface roughness declines from the original Ra0.94μm to Ra0.10μm, after finishing for 30min. However, the surface roughness of workpiece is decreasing more and more slowly, because the vertex effect is weakened as the machining time goes further. The surface roughness reduces from Ra0.10μm to Ra0.08μm only, after finishing for another 20min. Therefore, to ensure the finishing quality, the optimal machining time is 30min.

2) Some diamond grains on magnetic abrasives are worn or fall off, so the finishing performance of magnetic abrasive is reduced, and the material removal rate is decreased approximately linearly as machining time increases. The material removal rate is 10.5mg/h at 50min, which is 55% lower than that of 10 min.

3) The cutting force of UAMAF is increased and the ability of magnetic abrasives to cut into workpiece is improved with the increase of ultrasonic amplitude. Workpiece surface material is removed quickly and the material removal rate rises. Plastic flow from the wave peak to valley of workpiece surface is enhanced and the randomness of the magnetic abrasive trajectory is increased due to the increase of the ultrasonic amplitude at the same time. So the surface roughness reduces quickly with the increase of ultrasonic amplitude. The surface roughness reduces to Ra0.12μm, and the material removal rate is up to 30.5mg/h, when the amplitude increases to 14μm. Therefore, to ensure the finishing efficiency and surface quality, high amplitude is used as much as possible.

4) The surface roughness of workpiece drops, and the material removal rate rises, because the cutting length per unit time is increased with the increase of spindle speed in the range from 800r/min to 1400r/min. However, the surface roughness increases rapidly, and the material removal rate decreases significantly, because more and more magnetic abrasives will break away from the magnetic field force and fly out of the finishing area and the amount of magnetic abrasives involved in finishing is reduced gradually as the spindle speed continues to increase. Therefore, a reasonable spindle speed is 1400r/min. In this case, the surface roughness achieves the lowest Ra0.11μm, and the material removal rate reaches the highest 22.74mg/h.

5) The amount of magnetic abrasives in the machining gap increases as the machining gap enlarges in the range from 0.75mm to 1.25mm. Fluidity of abrasives is enhanced by ultrasonic vibration, which reduces the surface roughness, and increases the material removal rate gradually. However, magnetic induction intensity in the machining gap decreases and the loss energy of ultrasonic vibration transmission rises further as the machining gap continues to enlarge, which increases the surface roughness and reduces the material removal rate. Therefore, the optimum machining gap is about 1.25mm. In this case, the surface roughness reaches a minimum of Ra0.16μm, and the material removal rate achieves a maximum of 17.7mg/h.

Acknowledgments
The authors would like to acknowledge the financial support for this research from National Natural Science Foundation of China (No. 51505057), Scientific Research Project of Education Department of Liaoning Province (No. JDL2017033), Natural Science Foundation of Liaoning Province (No. 20170540113) and High-level Talents Innovation Support Project of Dalian (No. 2017RQ049).

References
[1] Lu Y and Li XY 2012 Current status and development trend of machining of titanium alloy for aerospace. *J. Aeron Manuf Technol.* **14** 55-57
[2] Chen Y, Liu ZQ and Wang XK 2012 Ultrasonic vibration assisted magnetic grinding process research. *J. T Chin Soc Agric Mach.* **19** 294-298
[3] Jiao AY, Quan HJ and Chen Y 2013 Experimental study on ultrasonic magnetic composite grinding of titanium alloy cone holes. *J. Chin J Mech Eng-en.* **19** 114-119
[4] Ma FJ, Tao DS, Gong C, Kang RK, Zhang SF and Sha ZH 2016 Technology of magnetic abrasive finishing in machining of difficult-to-machine alloy complex surface. *J. Journal of Hebei University of Science and Technology.* **37** 449-456
[5] Lu YP, Ma J and Zhang JQ 2007 Ultrasonic magnetic abrasive composite grinding technology. *J. Journal of China Coal Society.* **5** 552-556
[6] Ma J 2007 Study on ultrasonic magnetic composite grinding of mold surface. *J. electromachining & Mould.* **3** 59-62
[7] Lu YP, Ma J and Zhang JQ 2007 Ultrasonic magnetic abrasive composite grinding technology. *J. Journal of China Coal Society.* **5** 552-556
[8] YUN H, HAN B and CHEN Y 2016 Internal finishing process of alumina ceramic tubes by ultrasonic-assisted magnetic abrasive finishing. *J.International Journal of Advanced Manufacturing Technology.* **85** 727-734
[9] Mulik R and Pandey P 2010 Mechanism of Surface Finishing in Ultrasonic-Assisted Magnetic Abrasive Finishing Process. *J. Advanced Manufacturing Processes.* **25** 1418-1427
[10] Mulik R and Pandey P 2011 UAMAF of hardened AISI 52100 steel using unbonded SiCabrasives. *J. International Journal of Refractory Metals and Hard Materials.* **29** 68-77

IOP Conf. Series: Materials Science and Engineering **616** (2019) 012019 doi:10.1088/1757-899X/616/1/012019

Study on the Interaction between Vibration Mode and Radiated Sound of Conical Shell

Wu Jiang-Hai*, **An Fang, Duan Yong and Sun Yu-Dong**

China Ship Scientific Research Centre, National Key Laboratory on Ship Vibration & Noise, Shanshui east road 222, Wuxi,China,214082

*CSSRC_WJH@163.com

Abstract. Conical shell is one of the most common structures in engineering, especially in airplane and ship structures. In order to reduce its vibration and radiation sound, which were widely concerned, this paper used BEM method to analysis the vibration characteristic ,radiated sound and the interaction between them. Three different models were established to study the difference between conical shell in the air and under water. The numerical results demonstrate that the existence of water has great influences on the frequency and amplitudes of the radiated sound power. Due to the low density, the natural frequency of conical shell dominants the peak frequency of radiation sound in the air. And the contribution of each vibration mode to the radiation sound different from each other. The results of this study can provide some valuable suggestions for the position selection of active control actuators on conical shell.

1. Introduction

Conical shell structure is widely used in aerospace vehicles and submarines, and its vibration and noise characteristics are the focus of research.. Irie[1-2] et al. used transfer matrix method to study the frequency characteristics of isotropic conical shells under nine boundary conditions and different thickness forms. Tong[3] studied the free vibration of orthotropic conical shells under various boundary conditions. The displacement solution was obtained by power series method, and good accuracy was obtained. Mecitonglu[4] studied the free vibration of ribbed isotropic conical shells, and obtained a simple orthotropic conical shell model by means of the average energy method. Lam[5] et al. studied the free vibration of isotropic truncated conical shells under simply supported boundary conditions. Wang Anwen[6] proposed a theoretical equation with matrix for axial symmetric deformation of cones. Cui Weicheng[7-8] et al. deduced the exact solution of stress calculation for conical shells from the basic differential equation of rotating shells, and converted the basic Flugge differential equation for solving the moment problem of cylindrical shells into the second-order ordinary differential equation with complex constant coefficients, which can be used for the boundary effect of conical shells. Luo Dongping[9] used shell theory and transfer matrix method to study the free vibration of ring-stiffened cylindrical shells. Based on Flugge shell theory, Chen Meixia[10] et al. used power series method to study the vibration characteristics of circular ribbed conical shells in water under different boundary conditions, and studied the influence of fluid load, ring rib, half-cone angle and boundary conditions on their vibration characteristics.

In order to offer the actuator position and control frequency of vibration active control on conical shell. This paper presents the coupled vibro-acoustic characteristics of conical shell by using the BEM

IOP Conf. Series: Materials Science and Engineering **616** (2019) 012019 doi:10.1088/1757-899X/616/1/012019

method.

2. Vibration theory of conical shell

Conical shell as shown in figure1, the radius of small and large end part are r_1 and r_2, half cone angle is α. The stress-displacement relationship is shown in the following equation:

$$\begin{cases} \varepsilon_x = \dfrac{\partial u}{\partial x} \\ \varepsilon_\theta = \dfrac{1}{r}\dfrac{\partial v}{\partial \theta} + \dfrac{\sin\alpha}{r}u + \dfrac{\cos\alpha}{r}w \\ \varepsilon_{\theta x} = \dfrac{\partial v}{\partial x} - \dfrac{\sin\alpha}{r}v + \dfrac{1}{r}\dfrac{\partial u}{\partial \theta} \\ k_x = -\dfrac{\partial^2 w}{\partial x^2} \\ k_\theta = \dfrac{\cos\alpha}{r}\dfrac{\partial v}{\partial \theta} - \dfrac{1}{r^2}\dfrac{\partial^2 w}{\partial \theta^2} - \dfrac{\sin\alpha}{r}\dfrac{\partial w}{\partial x} \\ k_{x\theta} = -\dfrac{2\sin\alpha\cos\alpha}{r^2}v + \dfrac{\cos\alpha}{r}\dfrac{\partial v}{\partial x} + \dfrac{2\sin\alpha}{r^2}\dfrac{\partial w}{\partial \theta} - \dfrac{2}{r}\dfrac{\partial^2 w}{\partial x \partial \theta} \end{cases} \tag{1}$$

Where ε_x, ε_θ are tensile strain, $\varepsilon_{x\theta}$ is shear strain. k_x, k_θ are curvature change induced by tension, $k_{x\theta}$ is curvature change induced by torsion.

The stress-strain relationship is shown in the following equation:

$$\begin{cases} N_x = B(\varepsilon_x + \nu\varepsilon_\theta) \\ N_\theta = B(\varepsilon_\theta + \nu\varepsilon_x) \\ N_{x\theta} = N_{\theta x} = \dfrac{B(1-\nu)}{2}\varepsilon_{x\theta} \\ M_x = D(k_x + \nu k_\theta) \\ M_\theta = D(k_x + \nu k_x) \\ M_{x\theta} = M_{\theta x} = \dfrac{B(1-\nu)}{2}k_{x\theta} \end{cases} \tag{2}$$

Where $N_x, N_\theta, N_{x\theta}$ and $N_{\theta x}$ are shell surface force, $M_x, M_\theta, M_{x\theta}$ and $M_{\theta x}$ are shell bending moment. $B = \dfrac{Eh}{1\text{-}\nu^2}$ is the tensile stiffness, $D = \dfrac{Eh^3}{12(1-\nu^2)}$ is the torsional stiffness, E is the Modulus of elasticity and h is the thickness of conical shell.

Accordingly, the vibration equation of a conical shell is:

$$\begin{cases} \dfrac{\partial N_x}{\partial x} + \dfrac{1}{r}\dfrac{\partial N_{\theta S}}{\partial \theta} + \dfrac{\sin\alpha}{r}(N_x - N_\theta) + \rho h\omega^2 u = 0 \\ \dfrac{\partial N_{x\theta}}{\partial x} + \dfrac{1}{r}\dfrac{\partial N_\theta}{\partial \theta} + \dfrac{2\sin\alpha}{r}\dfrac{N_{\theta x}}{x} - \dfrac{\cos\alpha Q_\theta}{r} + \rho h\omega^2 u = 0 \\ \dfrac{\cos\alpha}{r}N_\theta + \dfrac{\sin\alpha}{r}\dfrac{\partial Q_\theta}{\partial \theta} + \dfrac{\partial Q_x}{\partial x} + \dfrac{\sin\alpha}{r}\dfrac{Q_x}{x} + \rho h\omega^2 u = 0 \end{cases} \tag{3}$$

IOP Conf. Series: Materials Science and Engineering **616** (2019) 012019 doi:10.1088/1757-899X/616/1/012019

Where ρ is the density of material, Q_x, Q_θ are the shear forces, they are defined as follow:

$$\begin{cases} Q_x = \dfrac{\partial M_x}{\partial x} + \dfrac{\sin\alpha}{r} M_x + \dfrac{1}{r}\dfrac{\partial M_{\theta x}}{\partial \theta} - \dfrac{\sin\alpha}{r} M_\theta \\ Q_\theta = \dfrac{\partial M_{x\theta}}{\partial x} + \dfrac{2\sin\alpha}{r} M_{\theta x} + \dfrac{1}{r}\dfrac{\partial M_\theta}{\partial \theta} \end{cases} \tag{4}$$

The vibration equation of a conical shell can be written in the form of a matrix.

$$\begin{bmatrix} L_{11} & L_{12} & L_{13} \\ L_{21} & L_{22} & L_{23} \\ L_{31} & L_{32} & L_{33} \end{bmatrix} \begin{Bmatrix} u \\ v \\ w \end{Bmatrix} + \rho h \omega^2 \begin{Bmatrix} u \\ v \\ w \end{Bmatrix} = 0 \tag{5}$$

The free vibration equation of truncated conical shells has the following basic properties:

1) Because of the existence of the half cone angle α, the tension-bending coupling term appears in the constitutive equation of the shell, which makes the order of the equation 1, 3 and 5 different. Moreover, the coefficient term of equation 1 is variable, so it is difficult to obtain its exact solution. General equation 5 is solved numerically;
2) When $\alpha \to 0$, equation3 can converted to the vibration equation of an isotropic cylindrical shell with radius R, while $\alpha \to \dfrac{\pi}{2}$ equation 3 can converted to the vibration equation of circular plate with radius R;
3) boundary conditions must be used to solve equation 5:
   Rigid fixed boundary: $u = v = w = \partial w / \partial r = 0$
   Simply supported edges boundary: $N_x = w = M_x = v = 0$.

3. Analysis of vibration mode of conical shell

The conical model in this paper adopts the typical truncated cone model. The dimensions are shown in the figure below. The upper half radius is $R_1 = 0.09m$, the lower half radius $R_2 = 0.5m$, the half cone angle α=21.04° and the bus length is L=1.13m. The whole model is made of isotropic uniform material with thickness of 20 mm, E=2.1e^{11}, γ=0.3. The mesh size of the model is 0.05m.

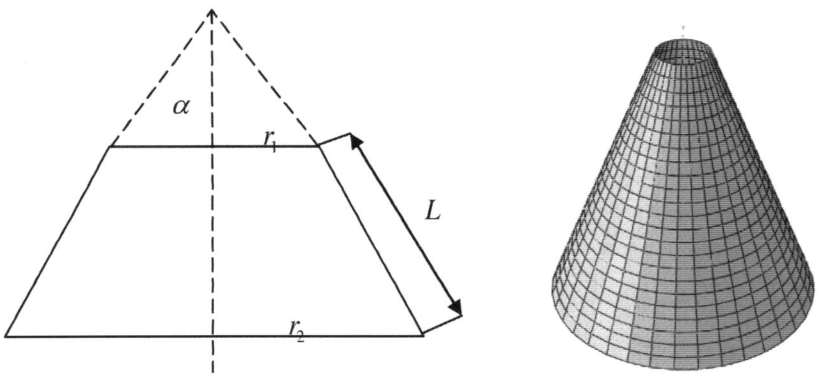

Figure 1. Conical shell model

The first 10 vibration modes of conical shell are studied in Figure 2.

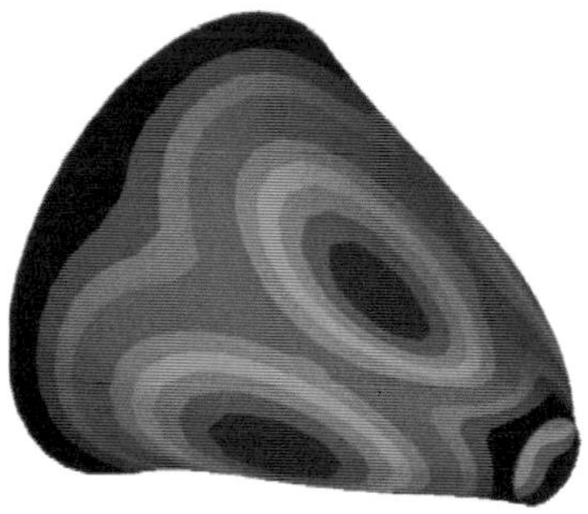

Mode 1 Frequency=589 Hz

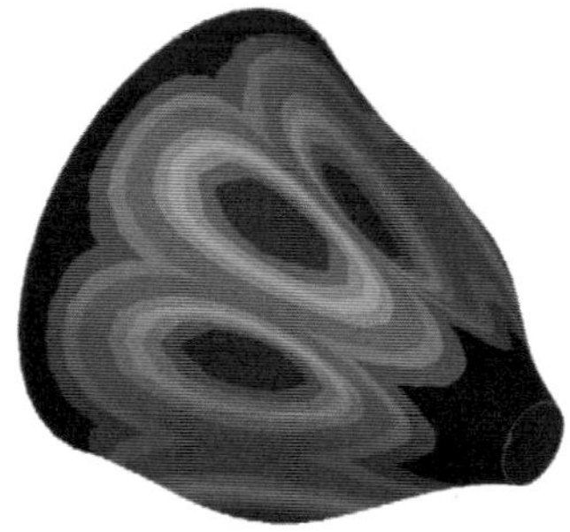

Mode 2 Frequency=613Hz

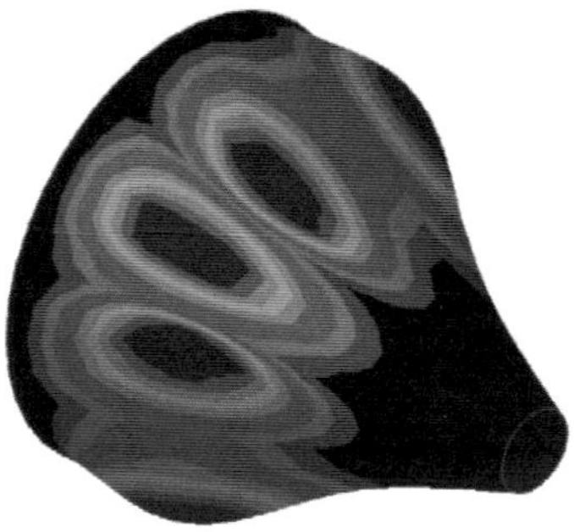

Mode 3 Frequency=773Hz

Mode 4 Frequency=967Hz

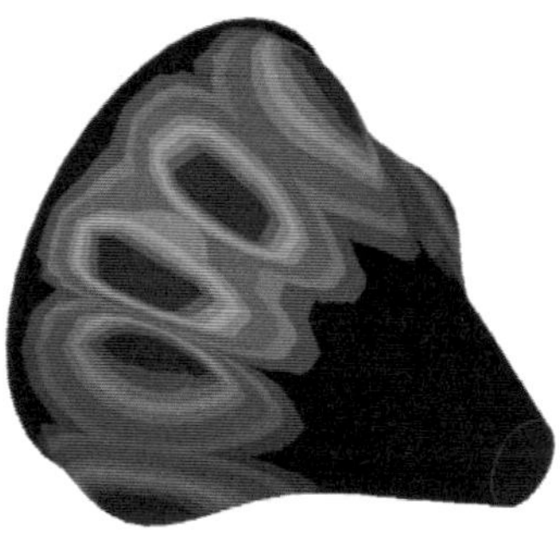

Mode 5 Frequency=1004 Hz

Figure 2. Vibration modes of conical shell

Because of the symmetry of the structure, actually there are five vibration modes. The first, second, third and fifth modes are consistent with the breathing modes of cylindrical shells. With the increase of frequency, the number of "bulges" on the conical surface increases, and the fourth mode is the bending mode of the whole model. The graph of vibration modes are as follow.

4. Radiated sound and mode participation factor

The radiated sound of conical shell is calculated by using the BEM method and the two radiated sound power results(in air and in water) are compared in Figure 3 and Table 1, it is easy to found that the frequency of radiated sound power consist with the natural frequency of conical shell. Because the density of water is much higher than that of air, the vibration sound radiation of structure in water is different from that in air. The effect of water on conical shell should not be neglected. Different from the result in air, the frequency in water is lower, and the frequency is misplacement with the natural frequency. The amplitude in air are close to 85dB, while the amplitude in water different from 89.85dB to 66.75dB.

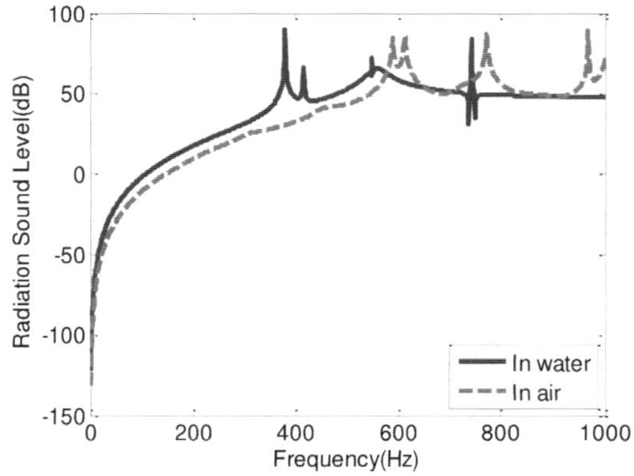

Figure 3. Radiated sound power

Table 1. Frequency of vibration mode and radiated sound power

Mode frequency	Radiated sound power in air	Radiated sound power in water
589Hz	589Hz,84.87dB	379Hz,89.85dB
613Hz	613Hz,86.35dB	415Hz,66.75dB
773 Hz	773Hz,87.46dB	548Hz,72.43dB
967 Hz	967Hz,88.96dB	744Hz,83.92dB
1004.43 Hz		

In order to effectively suppress the peak value of each order radiated sound power, four peak values of radiated sound power are analyzed by modal participation. Figure 4 is a diagram of modal participation factors of each order, and Figure 5 is a cloud of modal participation factors of each order. From Figure 4, it is clear that there are two dominant modes for each order peak. 589Hz is dominated by the first and second order vibration modes, 613Hz is dominated by the third, fourth order vibration modes, 773Hz is dominated by the fifth and sixth order vibration modes, 967Hz is dominated by the seventh and eighth order vibration modes. Although the structure has a high degree of symmetry, the first and second order frequencies are identical with the mode shapes, but due to the location of the excitation load, the contribution rate of the first two order vibration modes to the first peak of the radiated acoustic power is deviated. The first radiation peak is affected by the first and second order vibration modes, but it can be clearly seen from the nephogram of modal participation factors that the first order vibration mode has greater effect. The seventh order vibration mode participation factor is the smallest, and the first, fourth, fifth and eighth modes participation factor are the largest.

Consistent with the analysis in air, the parameters of each mode at the peak of underwater radiation are analyzed. Because of the symmetry of the structure, the first and second order dominant 379Hz peak, the third and fourth order dominant 415Hz peak, the fifth and sixth order dominant 548Hz peak, and the ninth and tenth order dominant 744Hz peak. As can be seen from the above figure, the influence of the 7 and 8 order modes on the underwater radiation of the cone is relatively small. The first, fourth, fifth and tenth order vibration modes have great influence on underwater radiated noise, so active control should be carried out for these fourth order vibration modes.

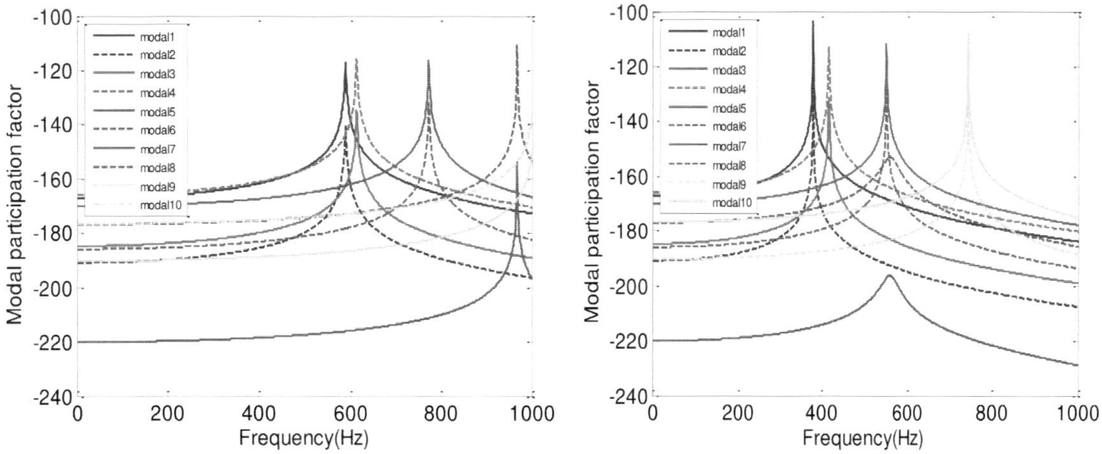

In air In water

Figure 4. Modal participation factor

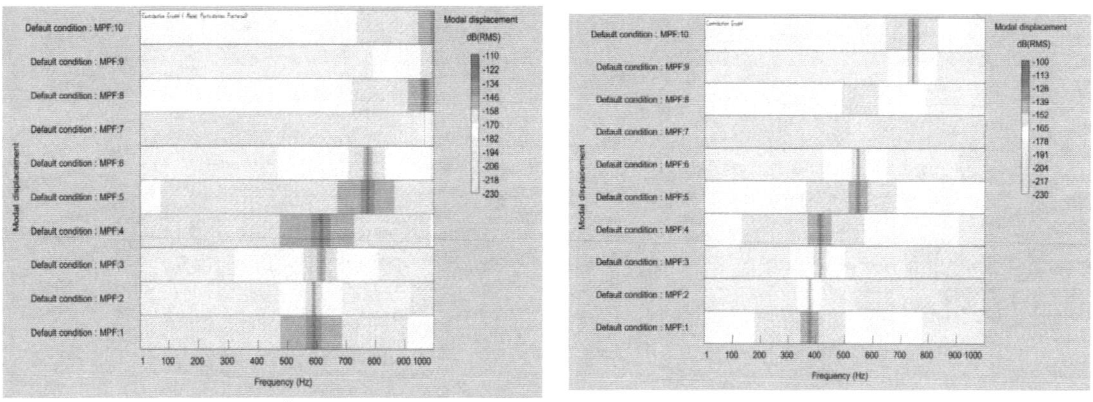

In air In water

Figure 5. Modal Participation Factor Nephogram

The position of active actuator is related to the effect of active control. Thus, the optimization of actuator position has always been a research hotspot in active control.In this paper, a 360 degree monitoring point is set at the center of the conical geometric model to calculate the acoustic directivity of the acoustic field radiated by the model under the excitation load. The load acts on the position of 270 degrees around the circle. The acoustic directivity at the frequencies corresponding to the four main radiated sound powers is shown in Figure 6.

As can be seen from the above figure, if the first peak value wants to be suppressed, the actuator can be distributed along the circular direction of the cone with a distance of 90 degrees, starting at 0 degrees; if the second peak value wants to be suppressed, the actuator can be distributed along the circular direction of the cone with a distance of 60 degrees, starting at 30 degrees; if the third peak wants to be suppressed, the actuator can be distributed along the circular direction of the cone with a distance of 45 degrees and a starting angle of 0 degrees. If the fourth peak value wants to be suppressed, the actuator can be distributed along the circular direction of the cone with a distance of 180 degrees. It is no difficult to find that the directivity is highly symmetrical at the peak of the radiated sound power and is consistent with the vibration mode of the structure.

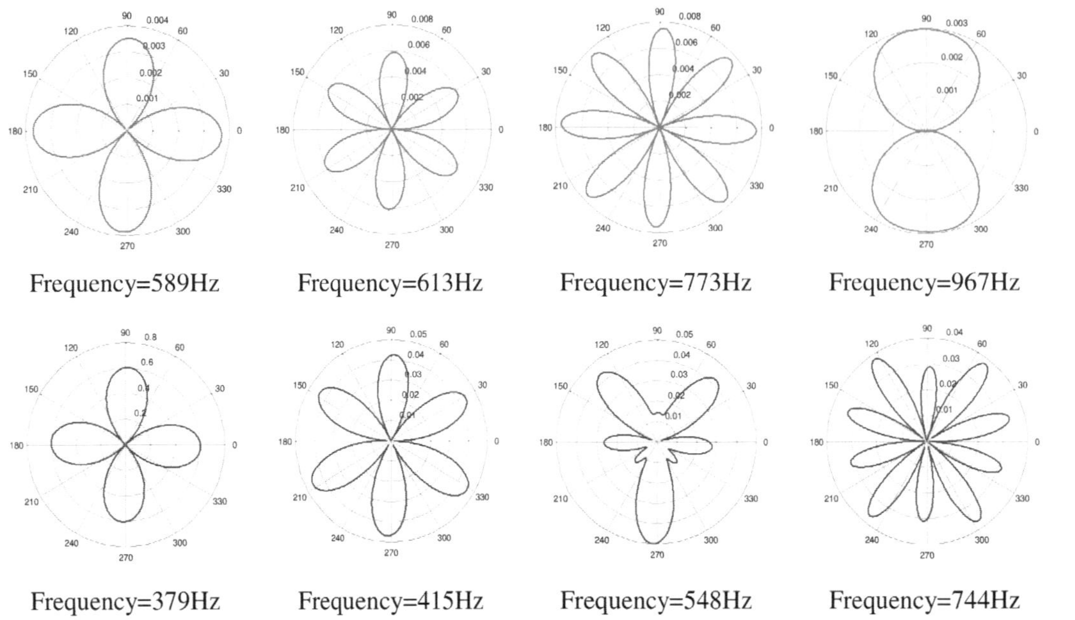

Figure 6. Acoustic Directivity of Radiated Sound Field

5. Analysis of conical shell with stiffness

The more reality conical shell is ring stiffened conical which is common used in engineering. In this section, the stiffened conical shell model is analyzed. Keep the amplitude and position of the excitation force unchanged. Size of the T shape beam is $80mm \times 80mm \times 10mm \times 10mm$.

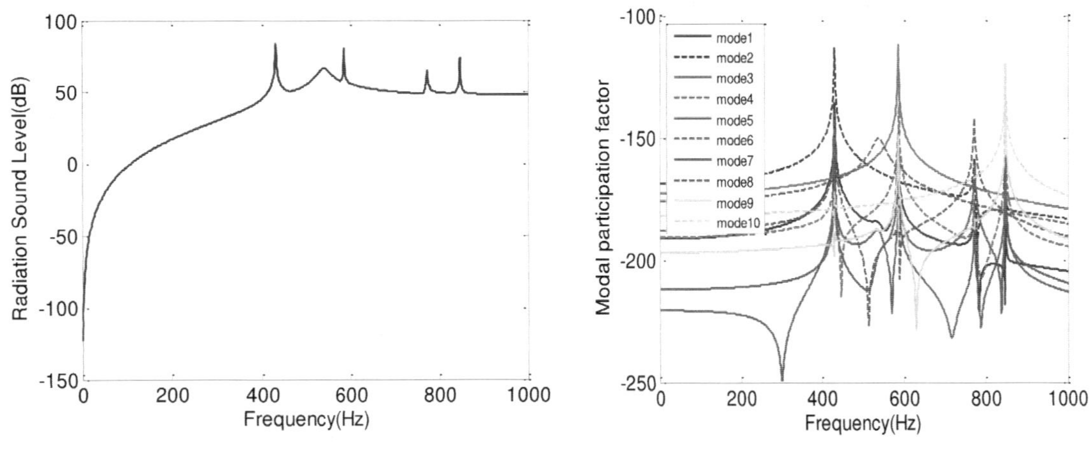

Figure 7. Radiated sound power and mode participation of ring stiffened conical shell

Different from the typical conical shell mentioned above, the each mode participation factor of ring stiffened conical shell no longer has a single peak frequency. Each mode contains several frequencies and becomes more complicated. concrete modal participation factors are shown in the table 2 and figure 7.

It is clear from the table below that the 430Hz of radiated sound power dominated by mode2, the 586Hz dominated by mode3, the 771Hz dominated by mode8, the 848Hz dominated by mode9. Among all modes, the first, fourth, fifth and seventh order vibration modes participate in all four peaks of radiated sound power, and the tenth order mode only affects the fourth peak.

Table 2. Mode participation of ring stiffened conical shell

Participation factor	430Hz	586Hz	771Hz	848Hz
Mode-1	-135.6	-146	-177.2	-169.1
Mode-2	-113.5	/	-158.2	/
Mode-3	/	-111.9	/	-151.6
Mode-4	-152.7	-136.2	-175.6	-148.6
Mode-5	-157.6	-166.6	-188.2	-163..8
Mode-6	/	-151.6	-164.2	/
Mode-7	-162.4	-152.6	-160.9	-150.6
Mode-8	-140.3	-177.2	-141.9	/
Mode-9	/	-148.7	/	-120
Mode-10	/	/	/	-125.5

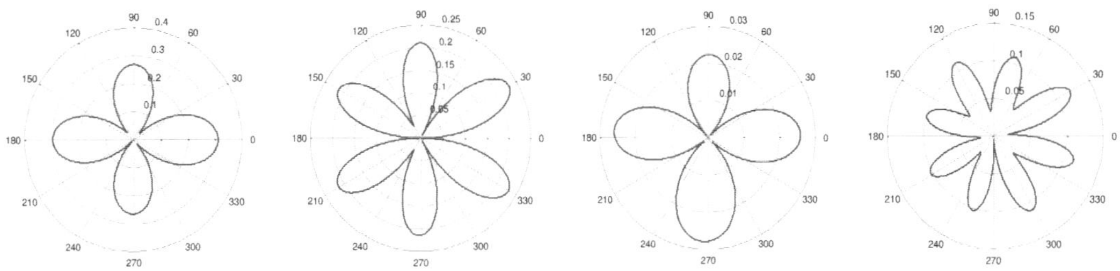

Frequency=430Hz Frequency=586Hz Frequency=771Hz Frequency=848Hz

Figure 8. Acoustic Directivity of conical shell

It is obviously found that the acoustic directivity diagram of ring stiffened conical shell(in Figure 8) are highly symmetrical. So the active actuators should be equality distributed around the circumference.

6. Conclusion

The following results have been obtained through the calculation:1) The peak frequency of radiated sound power in the air is consistent with the natural frequency of the conical shell itself;2) The frequency of underwater radiated sound power is lower than the vibration natural frequency of the conical shell;3) In both media, the first-order vibration model has an effect on the radiated sound power of the conical shell, so this mode can be selected to active control;4) The acoustic directivity of the main peak position of the radiated sound power of the conical shell is generally symmetrical, so the active control actuator can refer to the vibration mode of the structure and distribute periodically along the circumferential direction;5) The ribbed can change the sound and radiation efficiency of vibration radiation.

Reference

[1] Irie T , Yamada G , Kaneko Y . Free vibration of a conical shell with variable thickness[J]. Journal of Sound & Vibration, 1982, 82(1):83-94.

[2] Irie T , Yamada G , Tanaka K . Natural frequencies of truncated conical shells[J]. Journal of Sound & Vibration, 1984, 92(3):447-453.

[3] Tong L . Free vibration of orthotropic conical shells[J]. International Journal of Engineering Science, 1993, 31(5):719-733.

[4] Mecitoglu Z . Vibration characteristics of a stiffened conical shell[J]. Journal of Sound and Vibration, 1996, 197(2):191-206.

[5] Lam K Y , Hua L . Vibration analysis of a rotating truncated circular conical shell[J]. International Journal of Solids & Structures, 1997, 34(17):2183-2197.

[6] Wang Anweng, Guo Rixiu. Asymptotic Solution of the Second Degree for the Equation of

IOP Conf. Series: Materials Science and Engineering **616** (2019) 012019 doi:10.1088/1757-899X/616/1/012019

Conical Shells[J].Journal of Naval Engineering University, 1993(2):4-12.

[7] Cui Weicheng, Pei Junhou, Zhang Wei. A Comparison of Equivalent Cylinder Solution with Accurate Solution for Conical Shells [J]. Journal of Ship Mechanic, 2000(4):33-42.

[8] Cui Weicheng, Pei Junhou, Zhang Wei. Simplified Solution Having the Same Accuracy as Thin Walled Shell Theory for Conical Shells [J]. Journal of Shanghai Jiaotong University, , 2002, 36(1):125-129.

[9] Luo Dongping, Zhao Yuxi. Analysis of Free Vibration Characteristics of Ring-Ribbed Conical Shells [J]. Journal of Vibration and Shock, , 1990(4):64-69.

[10] Chen Meixia, Deng Naiqi, Zhang Chong. Vibration of submerged and ring stiffened conical shells[J]. Journal of Vibration and Shock, 2014, 33(14):25-32.

IOP Conf. Series: Materials Science and Engineering **616** (2019) 012020 doi:10.1088/1757-899X/616/1/012020

Development of Explosion-containment Vessel with 3 kg TNT Equivalent

Kan Zhou[1,a], Ge Huang[1], Kai Fang[1], Huang Huang[1] and Haijun Bai[1]

[1]Institute of Chemical Defense, Beijing, China

[a]517525860@qq.com

Abstract. In this paper, the design theory of explosion-containment vessel is studied, the charge ratio distance, incident overpressure and reflection overpressure acting on the inner wall of the explosion-containment vessel are calculated when 3kg TNT is exploded in it, the transient load generated in the explosion is converted into equivalent static load by the method of dynamic coefficient, and the main wall thickness of 3kg TNT equivalent explosion-containment vessel is determined. A 3kg TNT equivalent capsule explosion-containment vessel is designed and tested the explosion performance, the explosion performance test results show that the design of the 3kg TNT equivalent explosion-containment vessel is reasonable, safe and reliable, and it is of great practical significance to apply it to the transportation and destruction of weapons and ammunition and emergency treatment.

1. Introduction

Explosion-containment vessel is a kind of special sealed pressure container, which can effectively restrain the explosion inside, so that the shock wave and debris generated by the explosion are limited inside the container [1]. It can realize effective close protection for test equipment and personnel, and test and observe the process of explosion.

Since 1945 when Los Alamos laboratory developed the world's first explosive container, many countries have developed explosive containers with TNT equivalent from several kilograms to dozens of kilograms or even hundreds of kilograms [2]. Due to different application purposes, the structure of explosive containers is also diverse, common spherical, cylindrical and composite containers. Such containers are mainly used for explosive tests in scientific research departments, storage and transfer of nuclear and chemical weapons, prevention of environmental pollution caused by radioactive or toxic wastes and other harmful materials, and transportation and destruction of expired weapons and ammunition.

In recent years, explosion emergencies are on the rise all over the world. In order to prevent suspected explosives carried by terrorists from endangering social security, explosion-containment vessel need to be used to deal with suspected explosives. Therefore, it is very important to study explosion-containment vessel to deal with emergencies [3].

2. Design of explosion-containment vessel

Baker et al. [4-10] have done a lot of work on the basic theory of explosion-containment vessel. Explosion-containment vessel requires small volume and light weight, and has a large anti-explosion ability, its size and wall thickness is of great significance. The most fundamental difference between an explosion-containment vessel and a general pressure container is that its internal function is transient. After the explosion occurred in the center of the explosion-containment vessel, a shock wave propagates in all directions is formed. The shock wave reaches the inner wall of the explosion-containment vessel and forms a reflection shock wave, which is transient load acting on the inner wall of the vessel. Therefore, the strength of the transient load is determined before the design of the explosive container, and then it is converted into the equivalent static load. When the equivalent static load is determined, the main wall thickness of the vessel will be designed based on the equivalent static load.

2.1. selection of explosive container materials
The structural dimensions of the 3kgTNT equivalent explosion-containment vessel are preliminarily determined as shown in figure 1. The explosion-containment vessel designed in this paper is capsule shaped and consists of tank body, hemispherical head, flange and nozzles.

It is mainly used for the transportation of expired weapons and suspicious explosive items. The steel required to be selected has certain strength. In addition, welding and other conditions should be considered.

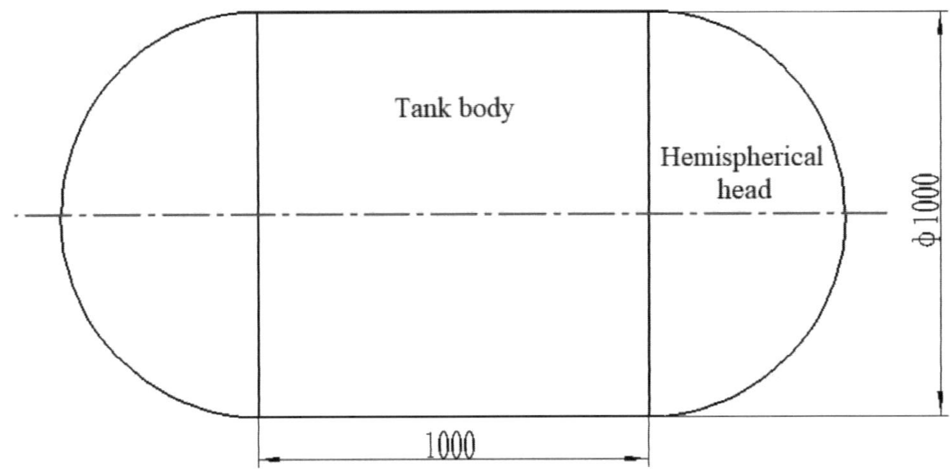

Figure 1. Dimensions of explosion-containment vessel.

2.2. Calculation of transient loads
There are many methods to calculate the explosive transient load inside the explosion-containment vessel, but each method has certain preconditions.

The explosion load is the basis of the strength calculation of the explosion-containment vessel. The dynamic response parameters of the explosion-containment vessel under the action of blast wave mainly include: incident overpressure, reflection overpressure and action time of overpressure. These parameters are generally determined by empirical formulas based on a large number of experiments.
The capsule shaped explosion-containment vessel designed in this paper has a radius of 500mm, a length of 2000mm, the charge is 3kg TNT.

2.2.1. Determination of charging ratio distance. The charging ratio distance can be determined by equation (1).

$$\overline{R} = R / \sqrt[3]{W} \text{ (unit: m)} \tag{1}$$

Where, R is the radius of the explosion-containment vessel (unit: m), W is the TNT equivalent (unit: kg). Put $R = 0.5m$, $W = 3kg$ into the equation (1), to get $\overline{R} = 0.35$.

2.2.2. Determination of incident overpressure. The incident overpressure Δp_1 can be calculated by K-G (Kinney and Graham) formula.

$$\Delta p_1 = 808 \times p_0 \frac{1+(\overline{R}/4.5)^2}{\left[1+(\overline{R}/0.048)^2\right]^{1/2}\left[1+(\overline{R}/0.32)^2\right]^{1/2}\left[1+(\overline{R}/1.35)^2\right]^{1/2}} \tag{2}$$

Where, p_0 is the atmospheric pressure. Put $\overline{R} = 0.35$, $p_0 = 1.01325\times10^5\, pa$ into the equation (2), to get $\Delta p_1 = 7.22 Mpa$.

2.2.3. Determination of reflection overpressure. When the shock wave acts on the container wall, it can be considered as the shock wave acting on the rigid solid wall. Therefore, reflection overpressure Δp_2 is calculated by the equation (3).

$$\Delta p_2 = 2\Delta p_1 + 6\Delta p_1^{\ 2} / (7 p_0 + \Delta p_1) \tag{3}$$

Where, p_0 is the atmospheric pressure. Put $\Delta p_1 = 7.22 Mpa$, $p_0 = 1.01325\times10^5\, pa$ into the equation (3), to get $\Delta p_2 = 53.93 Mpa$.

2.3. Determination of equivalent static load

The calculation methods of equivalent static load mainly include the empirical formula method, hongdu method and dynamic coefficient method. This paper adopts the dynamic coefficient method.

2.3.1. Determination of dynamic coefficient. According to literature [11], the calculation formula of action time of reflection overpressure is shown below.

$$t_1 = \eta R / \sqrt{Q_0} \tag{4}$$

Where, R is the radius of the container shell (unit: m), Q_0 is the detonation heat per unit mass (unit: J/kg), and η is the empirical coefficient. When the column is symmetric, $\eta = 0.5$;For TNT, $Q_0 = 4860874.8 J/kg$, put $R = 0.50m$ into the equation (4), to get $t_1 = 1.134\times10^{-4}$.

For a thin shell, if only the extended motion is considered, then the period of natural oscillation:

$$T = 2\pi R\sqrt{\rho / E} \tag{5}$$

Where, E is the elastic modulus, ρ is the density of shell material (unit: kg/m^3). $E = 210\times10^9\, pa$. $\rho = 7.8\times10^3 kg/m^3$. $R = 0.50m$, $T = 6.051\times10^{-4}$ s, $t_1 / T = 0.1874 < 3/8 = 0.375$ .According to literature [1], when $t_1 / T < 3/8$, dynamic coefficient C_d is selected according to the following formula:

$$C_d = \sqrt{\left(\frac{\omega t_1}{2}\right)^{-2} \sin^4 \frac{\omega t_1}{2} + \left(\frac{\sin \omega t_1}{\omega t_1} - 1\right)^2} \tag{6}$$

Where, ω is the natural vibration frequency of the explosion-containment vessel, put $\omega = 2\pi / T = 1.038\times10^4$, $\omega t_1 = 1.1771$ into the equation (6) to get $C_d = 0.567$.

2.3.2. Determination of equivalent static load. The calculation method of shell strength is basically the same as that used in the design of anti-explosion protection structure.The calculation formula of equivalent static load is

$$p_e = \Delta p_2 \times C_d \tag{7}$$

Put $\Delta p_2 = 53.93Mpa, C_d = 0.567$ into the equation (7) to get $p_e = 30.58Mpa$. In the structural design, the equivalent static load of 30.58Mpa is used for the main structure calculation.

2.4. determination of wall thickness of container body
The wall thickness of the vessel under static pressure is

$$\delta_1 \geq \frac{p_e D}{2[\sigma]\phi - p_e} \tag{8}$$

Where, p_e is the design pressure (unit: MPa), D is the inner diameter of the container (unit: mm), $[\sigma]$ is the allowable stress of the material under the design temperature, ϕ is the weld coefficient, and is generally taken as 1.

For material used in this paper, the safety coefficient is 2, $[\sigma] = 345Mpa$, put $p_e = 30.58Mpa$, $D = 1000mm$ into equation (8) to get the required wall thickness of the container $\delta_1 \geq 46.4mm$.

In actual engineering, from the perspective of safety, the steel plate with a thickness of 60mm is conservatively selected as the plate of the tank body, while the plate with a thickness of 66mm is selected as the upper and lower semicircle head.

3. Test verification of explosion-containment vessel

3.1. test plan
The 3kg TNT cartridge is bound into an approximate cube with insulating tape and placed in the center of the explosion-containment vessel as shown in figure 2. The cartridge block is fixed by hanging net and placed horizontally as a whole.

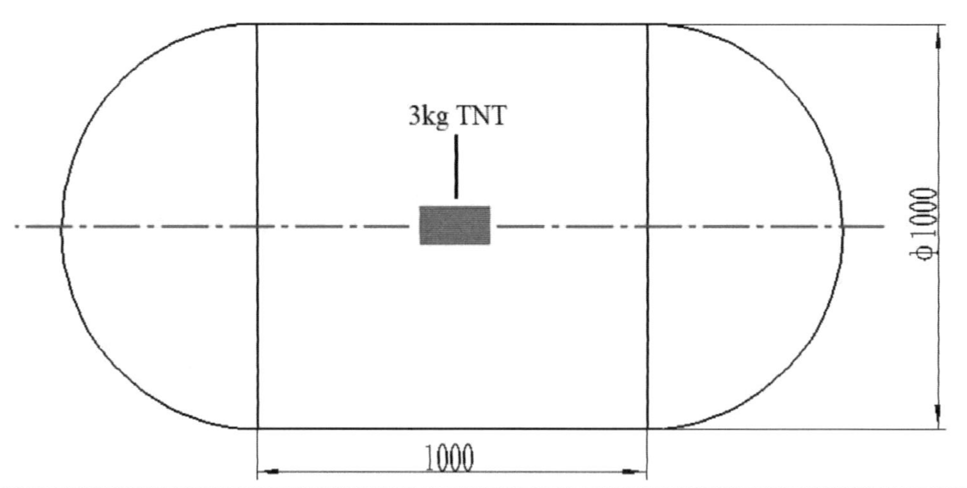

Figure 2. The position of 3kg TNT in the explosion-containment vessel.

After the explosion source in the explosion-containment vessel is detonated, the explosion test of the explosion-containment vessel shall be judged as qualified according to the following standards:

- no spatter should be generated outside the explosion-containment vessel
- no penetrable holes or cracks shall occur in the explosion-containment vessel
- the explosion-containment vessel can be opened and closed normally

3.2. test results
After the explosion test, the appearance of the explosion-containment vessel is detected, the tank body has no penetrable holes or cracks, the external parts are complete, and there is no shedding, damage, deformation or other phenomena, and there is no spatter flying out from the TNT cartridge explosion.

After the explosion-containment vessel is opened, there is no TNT residue and powder in the vessel and the switch mechanism can work normally. In conclusion, the design method of the explosion-containment vessel is safe and reliable.

In addition, this explosion-containment vessel has been carried out more than 10 times of scientific research explosion test so far, and the operation is in good condition. The switch sealing mechanism of this explosion-containment vessel can be fully automatic operation. The successful development of this vessel can be used for reference in the development of similar explosion-containment vessel in the future.

4. Conclusions
The excellent performance of explosion-containment vessel makes it widely used in the national defense industry, scientific research, and public safety. In this paper, the engineering design method of the capsule type explosion-containment vessel is introduced, and it has been verified by experiment. This design method can be briefly summarized as follows:

(1) Determine the shape, inner diameter and material of the explosion-containment vessel according to the working condition;

(2) Calculate the transient load acting on the inner wall of the explosion-containment vessel, including incident overpressure and reflection overpressure;

(3) The dynamic coefficient is determined by the reflection overpressure action time and the natural vibration circular frequency of the explosion-containment vessel, and then the static load is obtained;

(4) The main wall thickness of the explosion-containment vessel is designed based on the static load.

The technical measures that should be given priority in the development of explosion-containment vessel are introduced above. In addition, safety measures should be considered during the long-term operation of explosion-containment vessel. For example, ultrasonic inspection should be carried out regularly for explosion-containment vessel, and detailed archival records of cracks or defects should be established to ensure that no cracking accidents occur during the operation of containers.

References
[1] Ba-yi H 2000 The Engineering Design Method of Explosion-Containment Vessel and Its Application. *Pressure Vessel.* **17**(2) 39–41
[2] Wen-hui Z Hong-lu X Jun-wan H 1996 The research advances in the dynamics of explosive chambers. *Advances in Mechanics.* **2**(1) 68–78
[3] Zhuo-Ping D and Yu-Bin L 2003 Design of Blast Chamber for Break out Affair and Its Life Span Evaluation. *Chinese Journal of High Pressure Physics.* **17**(4) 295–300
[4] Baker W E 1960 The Elastic Plastic Response of Thin Spherical Shells to Internal Blast Loading. *J. Appl. Mech.* **27** 139–144
[5] Duffey T 1972 An Elastic Viscoplastic Solution for Impulsively Loaded Rings. *Int. J. Solids. Struct.* **8** 913–921
[6] Belov A I, Belyaev V M and Kornilo V A 1985 Calculation of Wall Loading Dynamics in a Spherical Combustion Chamber. *Combustion Explosion and Shock Waves.* **21(6)** 132–135

[7] Baker W E, Hu W G and Jackson T R 1966 Elastic Response of Thin Spherical Shell to Axisymmetric Blast Loading. *J. Appl. Mech.* **33** 800–805
[8] Karpp R 1983 Responses of Containment Vessels to Explosive Blast Loading. *J. of Pressure Technology.* **105**(2) 23–28
[9] Baker W E et al. 1976 Design Study of An Expermental Blast Chamber. *The Shock and Vibration.* **46**(3) 227–250
[10] Silvestrov V V 1994 The response of real containment vessel to internal detonation loading. *Combustion explosion and shock waves.* **30**(4) 148–156
[11] Shi-da Z 1989 Blast Chamber. *Explosion and Shock Waves.* **9** 85–96

Parametric Analysis of Wave-shaped Metal Sealing Ring Based on Hierarchical Solution Method

Zhang miaotian[1,a], Suo shuangfu[2], Feng yu[1], Shi jianwen[2] and Meng guoying[1]

[1] School of Mechanical Electronic & Information Engineering, China University of Mining and Technology Beijing, China
[2] School of Mechanical Engineering, Tsinghua University, Beijing, China

[a] Email: zhangmiaotian1231@163.com;

Abstract. The wave-shaped metal sealing ring is widely used in the aerospace field, whose failure will directly affect the flight safety and even cause major personal accidents and economic losses. The performance optimization problem of the wave-shaped seal is a multi-objective optimization problem, and the multiple targets are affected by the geometric design factors, and are not completely independent of each other. In this paper, for the sealing problem under the severe conditions of aero-engine, the corrugated metal sealing ring is taken as the research object. The W-shaped metal sealing ring is selected as an example, and the finite element model is established by using ANSYS software. Sealing, stability and resilience are chosen as indicators to evaluate the sealing performance. The multi-objective optimization problem is transformed into a single-objective optimization problem by hierarchical solution method, and the sealing structure is parameterized. The optimization results of the geometric parameters of the sealing ring are highly consistent with the product manual of the sealing ring. The correctness and feasibility of the optimization design method based on the analysis of three aspects of performance and the hierarchical solution for dimension reduction are verified.

1. Introduction

With the vigorous development of aerospace technology at home and abroad, the development of aerospace technology and the manufacture of large aircraft are the needs of China's entry into an innovative country[1]. Seals are key basic components in many high-tech fields (such as the aerospace industry, ergonomics industry)[2-3]. It has been shown in the literature that whenever the seal leakage is reduced by 1%, the engine thrust can be increased by 1% and the fuel consumption rate can be reduced by 0.1%. Therefore, the United States has set a 60% reduction in flow system leakage in its national engine development plan to improve the overall performance of the engine[4]. In the 1980s, the aerospace powerhouse headed by the United States invested a large amount of resources in aerospace sealing technology research, and elastic metal sealing is one of the research results[5]. Domestic research started relatively late. Since the 21st century, the state has studied key basic components such as seals as the focus of research on aero-engine technology[6].

There is a lack of systematic methods for the active optimization design of seal rings in China. The finite element software is used to try to optimize the design of the sealing ring, but the optimization design goal is not closely related to the performance of the sealing ring. The result is quite different from the foreign product manual. The research in this paper is mainly aimed at the sealing problem under the severe conditions of aero-engines.

IOP Conf. Series: Materials Science and Engineering **616** (2019) 012021 doi:10.1088/1757-899X/616/1/012021

The performance optimization problem of the wave-shape seal is a multi-objective optimization problem, and the multiple targets are affected by the geometric design factors, and are not completely independent of each other. After a detailed analysis of the optimization problem of the wave-shaped seal, this paper finally chooses the method of layered solution and constructing the ranking optimization function to solve the multi-objective optimization problem.

2. Finite element model establishment

2.1 Structure and installation of corrugated metal seals

In the wave-shaped metal sealing ring, the C-shaped and W-shaped sealing rings are the most common. The C-shaped metal seal is easier to manufacture and has been widely used in the aerospace industry, the nuclear industry and the chemical industry. W-shaped seals have better compression resilience performance, and the permanent deformation after single compression rebound undergoing large deformation is small, even without permanent deformation. However, in China, the W-shaped seal is still in the research stage. Figure 1 Figure 2 is a schematic view showing the installation structure of the W-shaped seal ring. When installed, the seal is subject to a certain pre-deformation by the longitudinal pre-tightening compression force.

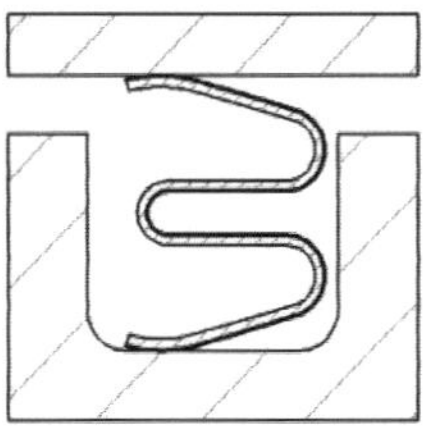

Figure 1. Before installation.

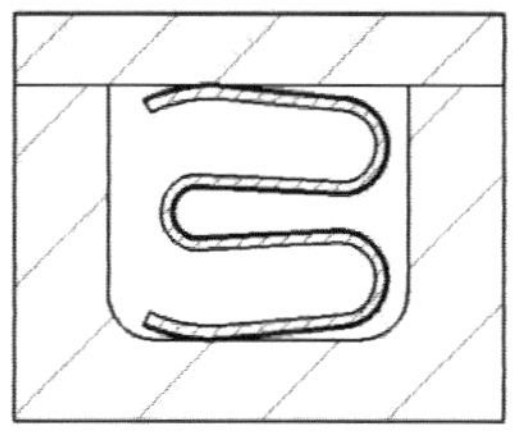

Figure 2. After installation.

2.2 Establishment of finite element model

The structure of the W-shaped metal seal is shown in Figure 1. Table 1 shows the initial structural parameters of the W-shaped seal ring.

Table 1. W-shaped metal sealing ring structure parameters.

Geometric parameter	Initial value (mm)
Outer diameter D	300
Wall thickness T	0.3
Wave height H	2.15
Free height FH	6.1
Contact surface radius of curvature R	3.05
Peak radius R1	0.75
Valley radius R2	0.9

The material of the W-shaped seal ring is made of nickel-based superalloy GH4169. The GH4169 has excellent comprehensive performance in the high temperature environment below 700°C, and the yield limit $\sigma_{0.2}$ at 650°C is in the forefront of common superalloys [7]. Table 2 shows the relevant parameters of GH4169.

Table 2. Mechanical properties of GH4169 material.

Temperature / °C	Modulus of elasticity / GPa	Yield strength / MPa
20	205	1030
650	153	865

The calculation model of the W-shaped seal is simplified[7-8], and the finite element calculation model is established in an axisymmetric manner and the upper and lower flanges are treated as pure steel bodies.The analysis unit type selects the PLANE182 unit, and the sealing ring is in contact with the upper and lower flanges to select the CONTACT 172 unit.

As shown in Figure 3, the Von-Mises stress distribution cloud diagram with an operating temperature of 650°C, a medium pressure of 3 MPa, and a compression of 20% was calculated. The maximum Von-Mises stress in the figure is 907.6 MPa, located at two valleys of the W-ring.

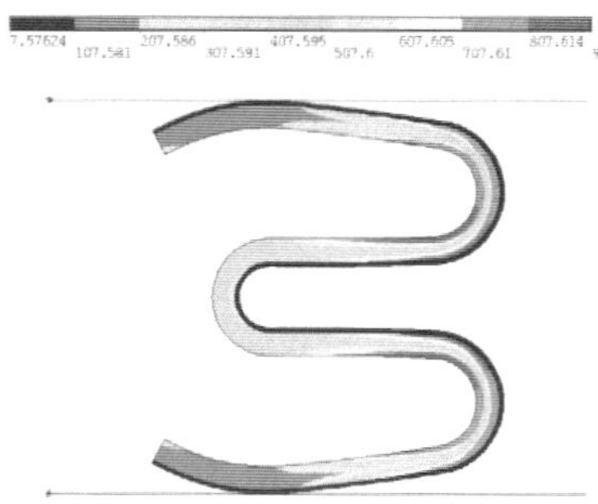

Figure 3. Von-Mises stress distribution of W-shaped seal.

2.3 Influence of structural parameters on sealing performance
According to the literature research, the comprehensive performance of the wave seal includes three aspects: resilience, sealing and stability, which are measured by the rebound rate, the maximum contact stress and the proportion of the shaped area.

2.3.1 Resilience. Figure 4 shows the change in rebound rate at different wall thicknesses. As can be seen from the figure, the smaller the wall thickness, the better the rebound resilience of the W-shaped seal. In the same wall thickness design, the higher the wave height, the better the rebound resilience. When the wall thickness is 0.2 mm, 0.25 mm, and 0.3 mm, there is a complete rebound, and in this region, the W-shaped seal has excellent resilience performance.

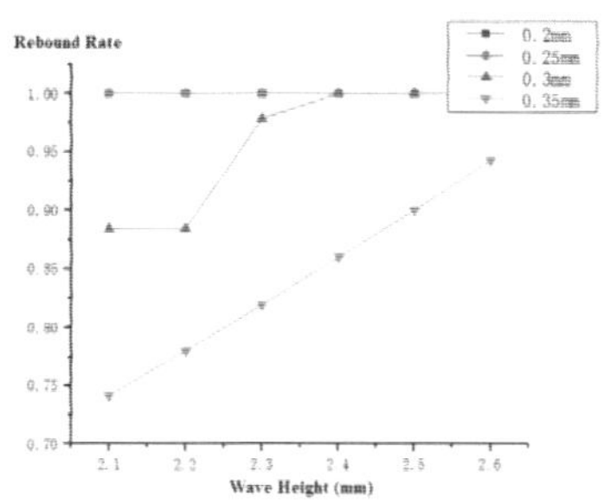

Figure 4. Rebound rate of different wall thickness waves.

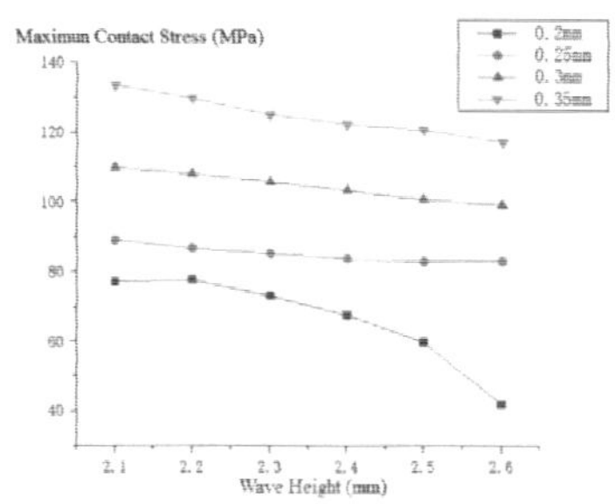

Figure 5. Maximum contact stress at different wall thickness waves.

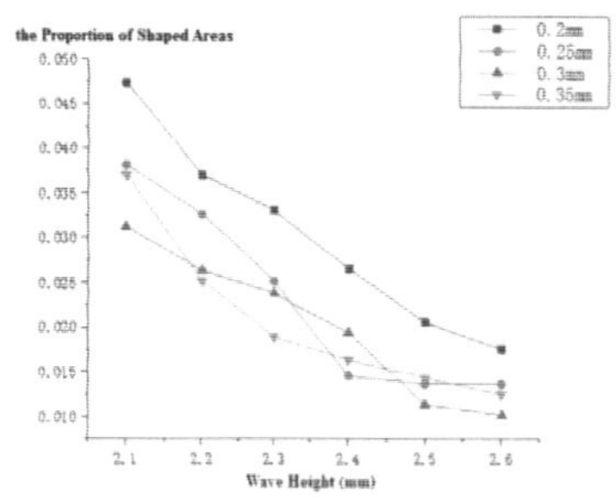

Figure 6. The proportion of shaped areas with different wall thicknesses.

2.3.2 Sealing. Figure 5 shows the maximum contact stress changes at different wall thicknesses. The maximum contact stress decreases as the wave height increases when the wall thickness is the same.

The maximum contact stress increases as the wall thickness increases when the wave height is the same.

2.3.3 Stability. Figure 6 shows the variation of the proportion of the shaped area under different wall thicknesses. It can be seen from the figure that when the wall thickness increases to a certain extent, it has a significant effect on the plastic deformation zone, and the proportion of the shaped area of the 0.35 mm wall thickness group is significantly higher than that of the other wall thickness groups. However, in the area with small wall thickness, the influence of wall thickness on the proportion of plastic deformation region could not be analyzed. As the wave height increases, the stability of the seal ring increases, and the proportion of the plastic deformation zone decreases.

By analyzing the influence of the structural parameters of the W-shaped seal on its comprehensive performance, it can be known that the influence of wall thickness and wave height on the comprehensive performance is complex, it is necessary to comprehensively consider the three factors of resilience, sealing and stability, and multi-objective optimization.

3. Parameter analysis of seal ring structure

3.1. Hierarchical Solution Method

According to the research results of the multi-objective optimization method of "Mathematical Planning", the hierarchical solution method is an effective method to convert multiple to fewer. The hierarchical solution method refers to: according to the optimization problem, the optimization sequence of the optimization target is discharged, and the first target $F_1(x)$ is first minimized to obtain the solution set x1. Then the solution of the next target $F_2(x)$ is performed in the solution set x1, so on and soforth, and finally the optimization result is obtained. In the optimization design problem of the wave-shaped seal ring, the rebound rate is determined to be the highest priority due to the leakage gap between the seal ring and the flange after the rebound failure. On the judgment of the subsequent sealing performance and stability performance, the objective function is solved by its single item ranking.

3.2. Boundary conditions and solutions

According to factors such as manufacturing process, manufacturing precision and dimensional design, a wall thickness of 0.05 mm and a wave height of 0.1 mm were selected as the minimum calculation interval. Xing Minjie[9] optimized the maximum contact stress under the condition of equivalent Von-mises stress, so that the sealing of the W-shaped seal is optimal. According to the feasible areas of the delineation factors, the feasible areas for setting the influencing factors are as follows:

$$\begin{gathered} 0.2 \le t \le 0.35 \\ 2.1 \le h \le 2.6 \end{gathered} \tag{1}$$

In the formula, t is the wall thickness of the sealing ring, the unit is mm; h is the wave height of the sealing ring, the unit is mm.

The structure objective function is as follows:

$$\begin{gathered} Z\,(t,h)\ \text{Optimal} \\ C\,(t,h)\ \text{Optimal} \\ S\,(t,h)\ \text{Optimal} \\ 0.2 \le t \le 0.35 \\ 2.1 \le h \le 2.6 \end{gathered} \tag{2}$$

In the formula, Z(t,h) is the proportion of the shaped area; C(t,h) is the rebound rate; S(t,h) is the maximum contact stress.

3.2.1 Rebound rate. If thc rcbound rate of the seal ring is not up to standard, there will be an interval between the seal ring and the flange, causing the internal high pressure medium to leak. In the JETSEAL product manual, the product's rebound rate is clearly described, and the product rebound rate of the high-pressure metal wave seal is 100%. Therefore, the objective function becomes:

$$\begin{gathered} C（t, h）=100\% \\ Z（t, h）\text{Optimal} \\ S（t, h）\text{Optimal} \\ 0.2 \le t \le 0.35 \\ 2.1 \le h \le 2.6 \end{gathered} \quad (3)$$

3.2.2 Maximum contact stress. Gong Xueting[10] conducted a leak test on the metal wave seal ring, and obtained a zero-leakage of the W-shaped seal ring with a wall thickness of 0.3 mm and a wave height of 3.5 mm in a medium pressure environment of normal temperature and 0-4 MPa. According to these conditions, the pressure is 1Mpa and 3Mpa, and the maximum contact stress is 79.5485Mpa and 105.377Mpa. The W-shaped seal under this condition is in a zero leakage state. That is, the objective function becomes:

$$\begin{gathered} \text{Min } Z（t, h） \\ C（t, h）=100\% \\ S（t, h）\ge 79.5485\text{Mpa}（p=1\text{Mpa}） \\ S（t, h）\ge 105.377\text{Mpa}（p=3\text{Mpa}） \\ 0.2 \le t \le 0.35 \\ 2.1 \le h \le 2.6 \end{gathered} \quad (4)$$

3.2.3 Proportion of shaping area. The optimal search of Z(t, h) is performed in the feasible region with a wall thickness of 0.05 mm and a wave height of 0.1 mm as the minimum calculation interval. The specific calculation results are shown in Tables 3 and 4.

Table 3. Results at 650°C, 1 Mpa, 20% compression.

Wall Thickness	Wave Height	Rebound Condition	Proportion of Shaping Area	Maximum Contact Load
0.2	2.2	Y	0	63.9846
0.2	2.4	Y	0	62.4157
0.2	2.6	Y	0	63.4863
0.25	2.2	Y	0	75.4104
0.25	2.4	Y	0.001155268	73.8492
0.25	2.6	Y	0.000693161	69.3513
0.3	2.2	N	0	99.4946
0.3	2.4	Y	0	91.2081
0.3	2.6	Y	0	84.9369
0.35	2.2	N	0.02310536	126.048
0.35	2.4	N	0.01155268	118.716
0.35	2.6	N	0	110.899

Table 4. Results at 650°C, 3 Mpa, 20% compression.

Wall Thickness	Wave Height	Rebound Condition	Proportion of Shaping Area	Maximum Contact Load
0.2	2.2	Y	0.036968577	77.7646
0.2	2.4	Y	0.02564695	67.6086
0.2	2.6	Y	0.017560074	41.9445

0.25	2.2	Y	0.032578558	86.7216
0.25	2.4	Y	0.014556377	83.6967
0.25	2.6	Y	0.012939002	83.057
0.3	2.2	Y	0.026340111	107.913
0.3	2.4	Y	0.01987061	103.24
0.3	2.6	Y	0.010166359	98.9253
0.35	2.2	N	0.025184843	129.747
0.35	2.4	Y	0.016404806	122.293
0.35	2.6	Y	0.012476895	117.041

According to the above calculation, the design with a wall thickness of 0.3 mm and a wave height of 2.2 mm was obtained as the optimal solution. This result is in good agreement with the wall thickness design results for high pressure seals (0.012 inch) in the JETSEAL product manual (wave height is not published in this product manual).

4. Conclusion

(1) Taking the W-shaped metal sealing ring as an example, the resilience, stability and sealing of the sealing ring are selected as the evaluation index of the sealing performance. Analyze the effect of different wall thickness and wave height on the rebound rate of the seal ring, the proportion of deformation of the shaped area and the maximum contact stress. It is concluded that the influence of wall thickness and wave height on the comprehensive performance is complex, and it is necessary to comprehensively consider the three factors of resilience, sealing and stability, and optimize it for multiple objectives.

(2) Selecting the rebound rate, the deformation ratio of the shaping area and the maximum contact stress as the optimization target of the W-shaped metal sealing ring, and applying the layered solving method to gradually transform the multi-objective optimization problem into a single-objective optimization problem. Through the dimension reduction solution, the optimal solution of the seal ring structure design under the required working conditions is obtained: the wall thickness is 0.3mm and the wave height is 2.2mm. The optimization results are in good agreement with the JETSEAL product manual. The correctness and feasibility of the optimization design method based on the analysis of three aspects of performance and the hierarchical solution for dimension reduction are verified.

Acknowledgments

The work decribed in this paper was supported by National Basic Research Program of China (973) (2014CB046404).

References

[1] Wen Jiabao 2008 *Let China's Big Plane Soar in the Blue Sky (Beijing: Defence Science & Technology Industry vol 05)*p 6-9

[2] Craggs A 1971 *The Transient Response of a Couple Plate-acoustic System using plate and acoustic finite elements (Nevada:Sound and Vibration vol 04)*p 509-528

[3] S Suzuki 1989 *Boundary Element Analysis of Cavity Noise Problems with Complicated Boundary (Nevada: Sound and Vibration vol 01)*p 147-153

[4] Hwang M, Pope A N, Shucktis B 1996 *Advanced Seals for Engine Secondary Flow Path (Reston: Journal of Propulsion and Power vol 04)*p 794-799

[5] Hu Guangyang 2012 *Application Research of Seal Technologies for Aeroengine (Shenyang:Aeroengine vol 03)*p 1-4

[6] Tan Songlin, Zhang Jinrong 2004 *The Application of Reliability Increasing in the Manned Spaceflight Rocket Engine (Xi'an:Journal of Rocket Propulsion,vol 4)*p 41-44

[7] Suo Shuangfu, Xing Minjie, Xue Qing, et al 2016 *Research on Effect Factors of Axial Stiffness of Metallic W-ring (Guangzhou: Lubrication Engineering vol 02)*p 14-17

[8] Wang Chenxi, Yang Yiyong, Suo Shuangfu, et al 2016 *Research on Compression-resilience and*

Sealing Performance of Metallic (Guangzhou:Lubrication Engineering vol 01)p 50-54
[9] Xing Minjie 2015 *Study on Sealing Performance and Leakage of Metal W-ring in the Aircraft Engine (Beijing: Beijing Institute of Technology)*
[10] Gong Xueting 2011 *Study on Characters of Metal W-ring Sealing used on Aircraft Engine (Beijing: Beijing University of Chemical Technology)*

IOP Conf. Series: Materials Science and Engineering **616** (2019) 012022 doi:10.1088/1757-899X/616/1/012022

Finite Element Analysis and Structure Optimization of the Middle Groove of Scraper Conveyor

Feng yu[1, a], Zhang Miaotian[1], Li Guoping[2] and Meng guoying[1]

[1] School of Mechanical Electronic & Information Engineering, China University of Mining and Technology Beijing, China
[2] China coal Zhangjiakou coal mining machinery co., Ltd Zhangjiakou, China

[a] Email: fengyu20062009@163.com;

Abstract. The middle groove is the most used and consumed component of the scraper conveyor. Its strength, rigid and wear resistance play an important role in the normal operation of the scraper conveyor. In this paper, the finite element analysis of the 1350 scraper conveyor is carried out of nonlinear finite element method, and the plastic deformation of the concave and convex ends of the groove is obtained during the tension and compression process. In order to prevent the phenomenon of the concave end of the groove, the paper proposes three methods to improve the performance of the middle groove. Through analysis and comparison, an optimization method is proposed to reduce the stress value during the tension and compression process by increasing the length of dumbbell rod concave end section and the diameter of the bell mouth, thus improving the reliability of the scraper conveyor.

1. Introduction

China is the world's largest coal-producing country and a major consumer of coal. Coal accounts for about 70% of China's energy industry and consumption structure, and is China's leading strategic energy source. The coal industry has become an important foundation for the rapid development of the national economy. The coal mine machinery and equipment that serves coal production will become the key to promoting China's coal technology advancement and industrial modernization [1]. The scraper conveyor is one of the most critical equipment in coal mining machinery. The middle groove is the main part of the scraper conveyor, and as a consumable part of the scraper conveyor, its reliability has an important influence on the life and reliability of the scraper conveyor [2].

The middle groove acts as a conveyor unit for the conveyor and its number determines the working length of the scraper conveyor. As the bearing part of the material, the middle groove is subjected to the intense friction of coal, scraper and chain during transportation, and is the component with the largest amount of consumption and consumption. The weight of the middle groove accounts for more than 70% of the total weight of the scraper conveyor. In the process of coal transportation, it not only has to bear the running load of the shearer, pushes and pulls the hydraulic support under the conditions of pushing and pulling, and also bears coal, but also the impact of the rock on the central trough, the pressing force, etc. Under the above conditions, the middle groove failure rate is extremely high, and the failure of the central trough directly leads to the failure of the scraper conveyor [3].

The finite element analysis of the middle groove with the dumbbell socket, the groove head and the shearer passing through the middle groove to study the reliability of the structure. On the basis of fully grasping the theory of structural optimization, we will further explore the application of structural optimization in practice to achieve the purpose of improving its reliability, and thus improve the reliability of the scraper conveyor [4].

2. Finite element model establishment

2.1 Establishment of finite element model

In this paper, the finite element analysis of the structure of the middle groove of the 1350 scraper conveyor is carried out. Modeling with Pro/E, using ANSYS software's powerful structural optimization function, optimizes the strength calculation size of the middle groove to reduce the stress level of the device. Homogenize the stress and improve the service life of the product to enhance the inheritance of the design, reduce the cost of the company's technical transformation and shorten the cycle of technological transformation [5].

The material properties of middle groove as shown in Table 1.

Table 1. Material properties of the middle groove.

Components	Material name	Young's modulus	Poisson's ratio	Yield strength	Tensile strength
Groove	30MnSiMo	2.02E5	0.3	766.07	862.03
Middle plate	NM360	2.02E5	0.3	950	1080
Bottom plate	NM360	2.02E5	0.3	950	1080
Concave end	20MnVB	2.02E5	0.3	885	1080
Convex end	20MnVB	2.02E5	0.3	885	1080
Others	Q345B	2.02E5	0.3	345	550

2.2 Finite element analysis of the groove in middle groove

The stress distribution diagram of the shovel groove is shown in Figure 1. The stress sipe of the groove is shown in Figure 2. The stress sipe of the groove is shown in Figure 3. The equivalent surface stress cloud is shown in Figure 4 The groove body is fully constrained on one side and the other side loads 720t at the dumbbell socket. In the stress cloud diagram, numerical specifications are made for the allowable stress of the material of 680 MPa and the allowable compressive stress of 1156 MPa. It can be seen from the figure that emerges to the action of 720t force, the dumbbell socket is only surface crushed, and the strength meets the requirements.

IOP Conf. Series: Materials Science and Engineering **616** (2019) 012022 doi:10.1088/1757-899X/616/1/012022

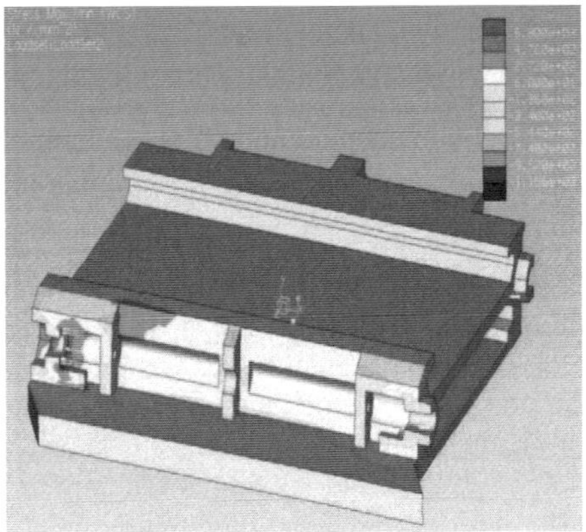

Figure 1. Schematic diagram of the shovel groove stress distribution.

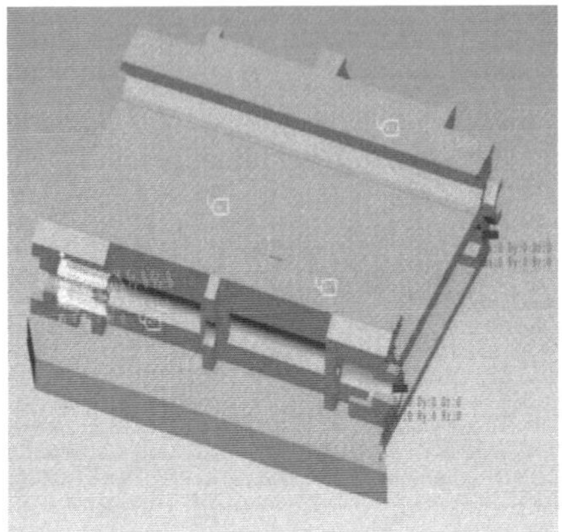

Figure 2. Stress loading diagram of the shovel groove.

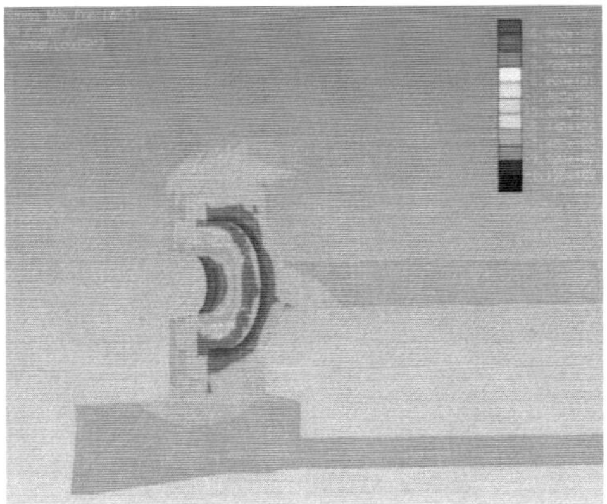

Figure 3. Stress slice diagram of the groove.

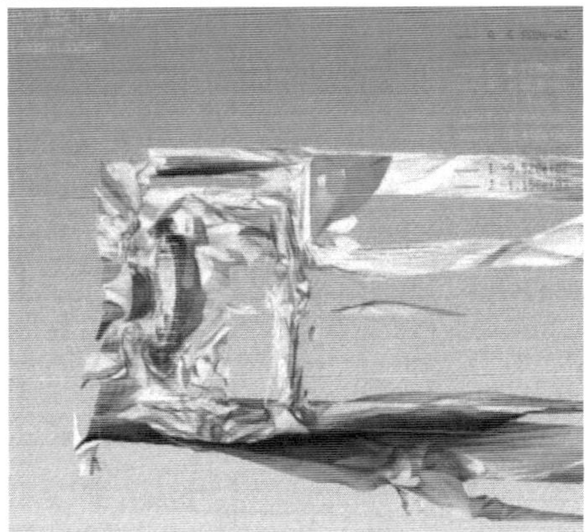

Figure 4. Equivalent surface stress cloud of the groove.

2.3Finite element analysis of the middle groove

The hydraulic support of the middle groove of the 1350 scraper conveyor is about 100t and the pulling force is about 150t. Theoretical analysis shows that when the frame is pulled, the concave head of the groove is the maximum force. Figure 5 shows the force of the middle groove when the frame is pulled: the horizontal pulling force acting on the ear is 150t, and the concave head of the groove and the concave head of the shovel groove are completely restrained. Figure 6 is a graph of stress distribution at The maximum stress in Figure 4 is 485.8 MPa, which is less than the allowable stress of material ZG30MnSi of 680 MPa. The under the action of 150t pull force, the strength of the concave head of the groove and the ear can Meet the demand.

IOP Conf. Series: Materials Science and Engineering **616** (2019) 012022 doi:10.1088/1757-899X/616/1/012022

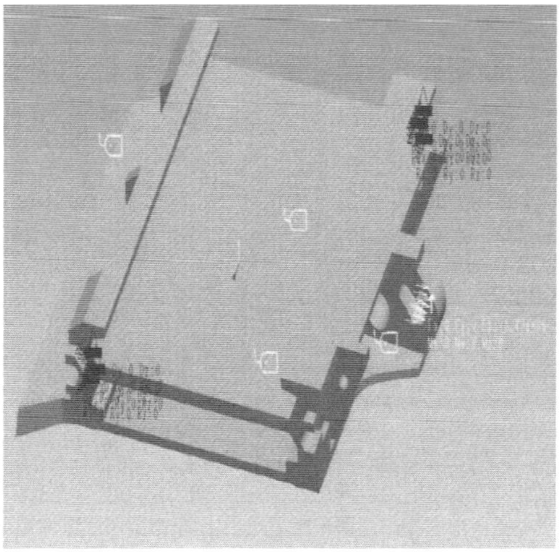

Figure 5. Force diagram of the middle groove when pulling the frame.

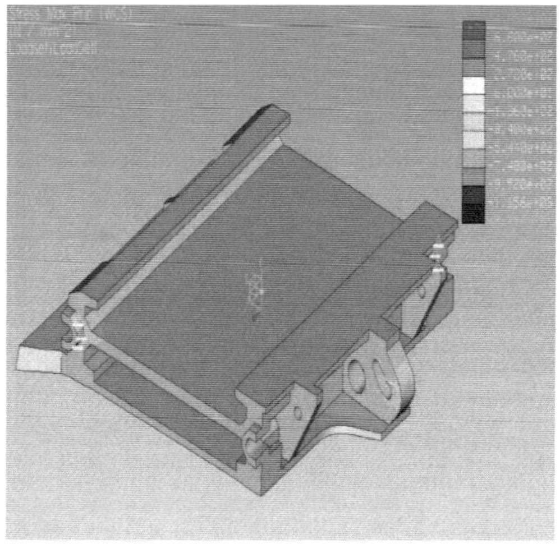

Figure 6. The distribution of the stress in the middle groove during the pulling of the frame.

2.4 Finite element analysis of the middle groove when the shearer passes

Take the worst working conditions for analysis: the middle groove is in a suspended state, the single sliding shoe of the shearer has a force of 100t on the shovel and the working area is 635mm × 135mm. Figure 7 is a force diagram of the shearer Figure 8 is a stress distribution diagram at this time. Figure 9 is a graph showing the stress distribution when the sliding shoe is moved to the middle of the blade.

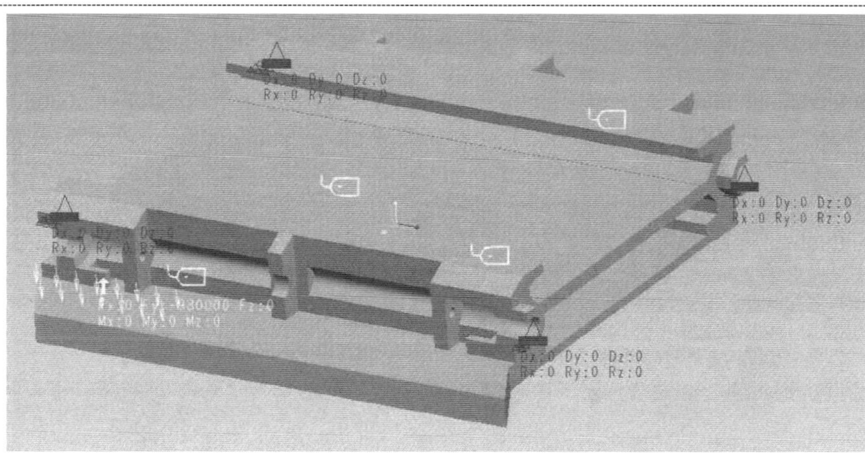

Figure 7. Force diagram of the sliding shoe just entering the middle groove.

IOP Conf. Series: Materials Science and Engineering **616** (2019) 012022 doi:10.1088/1757-899X/616/1/012022

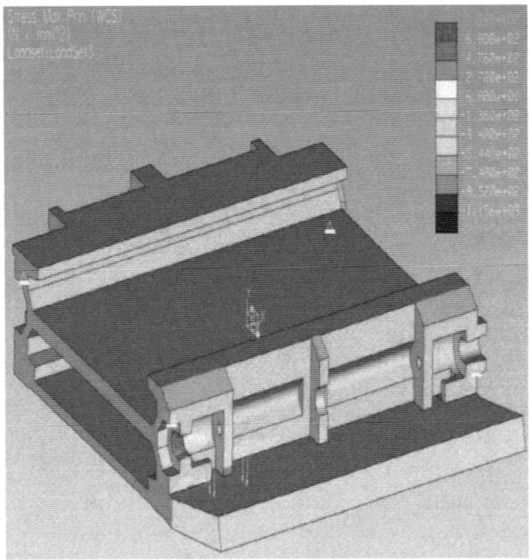

Figure 8. Stress distribution diagram of the coal mining machine sliding shoe just entering the middle groove.

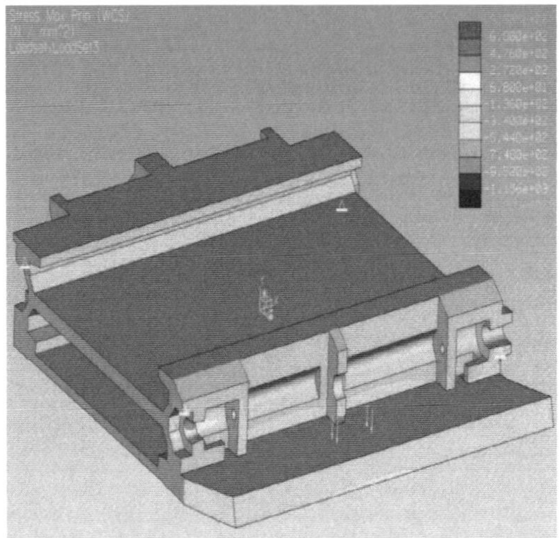

Figure 9. Stress distribution diagram of the sliding shoe going to the middle of the middle groove.

3. Structural optimization

3.1 Stress analysis of the concave and convex ends of the middle groove

The groove break is mainly the fracture of the shovel groove, and the fracture mode of the shovel groove is as shown in Figure 10: 1. occurs at the convex end; 2. occurs at the concave end. First, to reduce the stress value Of the concave end, only the stress value of the concave end can be reduced to avoid the fracture of the concave end. After repeated trials, it was determined that the two methods can reduce the stress value of the concave end: one is to increase The length of the dumbbell rod concave end; the other is to increase the diameter of the bell mouth.

Figure 10. Fracture form of the shovel groove.

After experimental comparison, the method of increasing the diameter of the bell mouth in the two methods is more effective than increasing the length of the dumbbell rod. If both methods are used simultaneous, the effect is better than using any single method. Figure 11 is a stress distribution diagram of the dumbbell rod section under the force of 720t in the middle groove of the 1350 scraper conveyor, and the maximum stress value is 950 MPa.

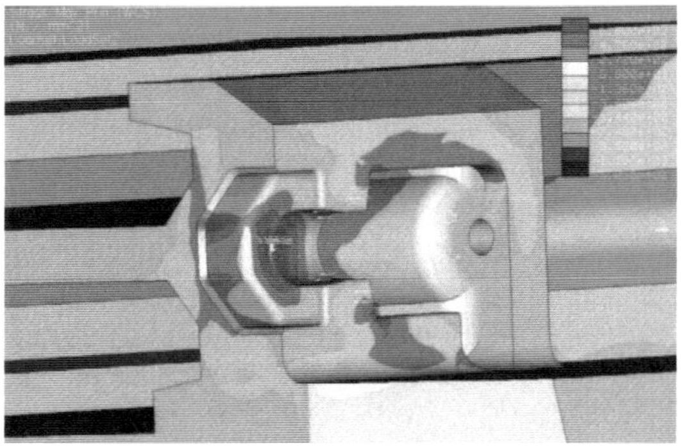

Figure 11. Stress distribution of the dumbbell rod under the force of 720t.

3.2 Structural optimization analysis

According to the method of increasing the length of the dumbbell rod concave end, the length is increased from 146 to 156. At this time, the stress condition of the dumbbell rod concave end is as shown in Figure 12, and the maximum stress value of The action rod section is 910 MPa. According to the method of increasing the diameter of the bell mouth, the diameter is increased from 84 to 114. At this time, the stress of the dumbbell rod concave end is as shown in Figure 13, and The maximum stress value of the dumbbell rod section is 710 MPa. Finally, the first and second methods are applied simultaneous, that is, the length of the concave end dumbbell rod is increased from 146 to 156, and the diameter of the bell mouth is Increased from 84 to 114. At this time, the stress condition of the concave end dumbbell rod section is as shown in Figure 14, and the maximum stress value of the dumbbell rod section at this time is 695 MPa.

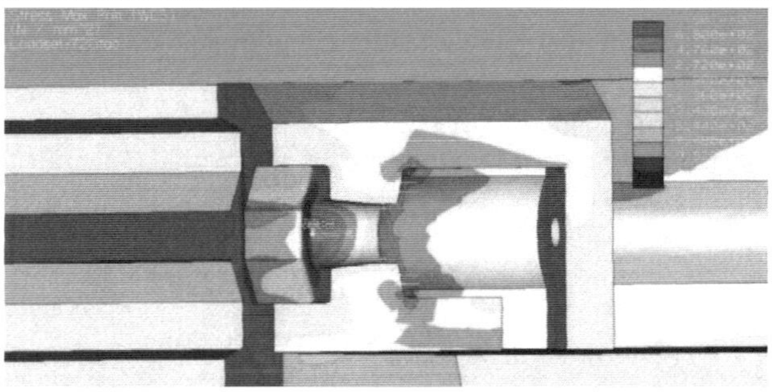

Figure 12. Increase the length of the dumbbell rod concave end.

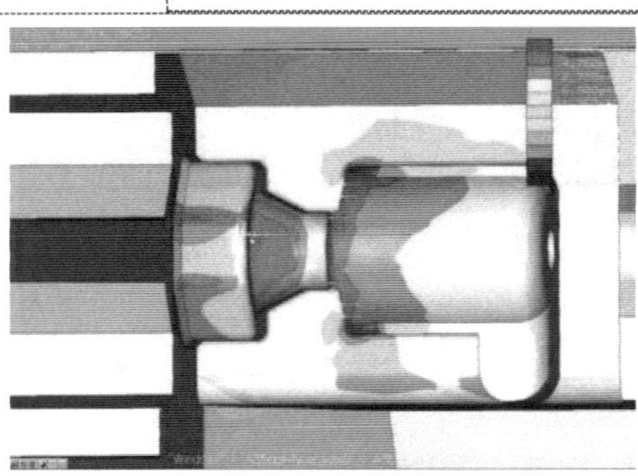

Figure 13. Increase the diameter of the bell mouth.

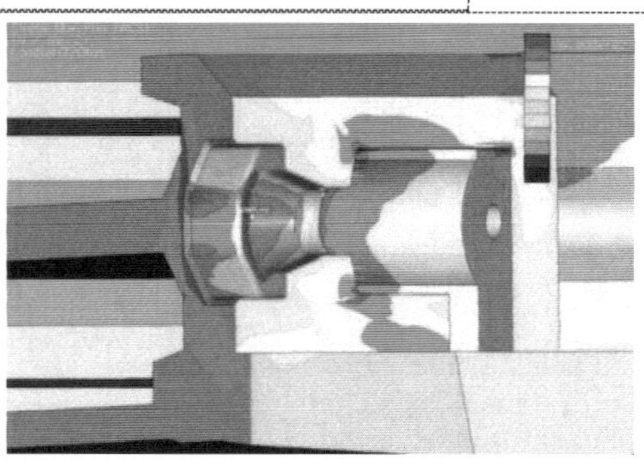

Figure 14. Two methods superimposed optimization.

IOP Conf. Series: Materials Science and Engineering **616** (2019) 012022 doi:10.1088/1757-899X/616/1/012022

It can be seen from the above analysis that the area exceeded 680 MPa is gradually reduced until it disappears. Table 2 shows the comparison between the original design and the three optimization schemes.

It can be seen from Table 2 that the percentage of stress reduction is the largest when the length of the dumbbell rod concave end and the diameter of the bell mouth are increased at the same time, and the optimization effect is the best.

Table 2. Comparison of analysis results of three optimization methods.

	Length of the dumbbell rod (mm)	Diameter of the bell mouth (mm)	Maximum stress value (MPa)	Percentage of stress reduction
The original design	146	84	950	0
Increase the length of the dumbbell rod concave end	156	84	910	4.21%
Increase the diameter of the bell mouth	146	114	710	25.26%
Two methods superimposed optimization	156	114	695	26.84%

4. Conclusion

In this paper, the nonlinear finite element method is used to model the middle groove. During the finite element analysis, it was found that the concave and convex ends of the groove had plastic deformation during the tension and compression process. In order to prevent the rupture of the concave end of the shovel groove, this paper proposes an optimization method to enhance the strength of the concave end of the shovel groove. According to the analysis, the length of the dumbbell rod and the diameter of the bell mouth are increased at the same time, and the optimization effect is best. The optimization improves the strength of the middle groove, effectively reduces the maximum stress value during the tension and compression process, maximizes the safety factor, and further improves the reliability of the scraper conveyor.

References

[1] Li Guixuan, Li Xinguo 2003 *Opportunities and challenges in revitalizing China's coal mining machinery (China Coal vol* 29: 02)p 8-10

[2] Jiang Lingyan 2007 *Technology status and development tendency of scrapper conveyor in mining face (Coal Science and Technology* vol 35: 08)p 102-106

[3] Yu Xueqian 2003 *Mine transportation equipment (China University of Mining and Technology Press).*

[4] Liu Pinqiang 2007 *The intensity analysis and optimization for the trough pf middle part (Hebei University of Technology)*

[5] Luo Qingji, Shi Guoxiang 2005 *Current status and development trend of fully mechanized mining face scraper conveyor (Colliery Mechanical & Electrical Technology* vol 05)p 54-57

IOP Conf. Series: Materials Science and Engineering **616** (2019) 012023 doi:10.1088/1757-899X/616/1/012023

The Dynamic Performance Study of Six-Dimensional Dynamic Force Generator

Cui Hongfei[1], Chen Quan[1], Li Wei[1], Tan Limeng[2], Jiang Mingyan[3,*], Song Deqiang[1]and Li Zheng[1]

[1]Beijing Oriental Metrology and Testing Institute, Beijing, China
[2]Beijing Space Vehicle General Design Department, Beijing, China
[3]School of Mechanical Engineering,Dalian University of Technology,Dalian,China

[*]1627568117@qq.com

Abstract: A six-dimensional dynamic force generating device is developed based on the dynamic performance calibration requirement of a six-dimensional force test platform based on the exciter force generating unit. The finite element software is used to perform modal analysis on the dynamic force generating device to obtain the natural frequency response of the device. Based on the finite element analysis results, the second-order transfer function at the sixth-order natural frequency point of the device is obtained, and the amplitude-phase frequency characteristic curve is obtained. Combined with the characteristic parameters of the exciter, the relationship curve between the excitation frequency and the excitation force under the influence of the natural frequency of the device is obtained. The comprehensive of finite element method and calculation results show that the designed dynamic force generating device can meet the requirements of use. According to the analysis results, it is found that the excitation capability can be maintained at a low frequency state to achieve high amplitude fixed frequency excitation, and low amplitude fixed frequency excitation is realized at a high frequency state.

1. Introduction

With the rapid development of high-precision test instruments, the methods of static force measurement have become more and more perfect, and the measurement accuracy is also constantly improving. For dynamic forces, dynamic force measurement is different from static, and is often used in a "Static calibration to measure dynamic forces" approach. When dynamically calibrating a force sensor, the important thing to address is how to get the standard dynamic force required for dynamic calibration. Dynamic force generating devices are primarily used to generate standard dynamic forces for dynamic calibration of test devices [1]. According to the standard dynamic force generating device output force signal form, the standard force generating devices commonly used at home and abroad can be divided into steady-state sinusoidal excitation type force source.

The research of the German PTB National Laboratory is mainly based on the steady-state sinusoidal[2].Professor Y. Fujii of Japan has been involved in the above three research directions, while domestic research is mostly stepped.The PTB National Laboratory of Germany[3] has developed a dynamic calibration device for the PTB steady-state sinusoidal force sensor.It is a typical steady-state sinusoidal excitation force source with an exciter as the actuator.By changing the frequency and load quality of the sinusoidal signal,the dynamic force output of the variable force value

IOP Conf. Series: Materials Science and Engineering **616** (2019) 012023 doi:10.1088/1757-899X/616/1/012023

and frequency can be realized.Japan's Yusaku Fujii[4] uses a hammer to strike a mass connected to the spring to generate a sinusoidal force source for the force sensor to be calibrated. Wang Hongyan et al[5] developed a trapezoidal pulse force generating device based on a proportional electromagnet.By inputting a certain excitation signal to a proportional electromagnet, the dynamic force required for calibration can be obtained.Wang Weijun[6] proposed a shock wave step force generating device.By connecting the air source to the outside,the air pressure difference from the high pressure section to the low pressure section forms a very ideal pressure step before and after the shock wave,acting on the force sensor installed at the end of the low pressure section, so that the force sensor is subjected to the positive step force.Wang Qingming et al[7] designed and manufactured the drop hammer-pre-stressed rod type minus step force signal generating device, which uses the drop hammer impact load method to generate minus step force.Japan's Yusaku Fujii[8] produced a positive step force source that shears the wire which hangs heavy objects.

In summary,from the dynamic calibration source generation form and the dynamic calibration requirements of a six-dimensional force test platform,a six-dimensional dynamic force generation device based on the exciter is designed.

2. Overall structure of six-dimensional dynamic force generating device

The electric exciter is connected to a specific signal generator, in addition to generating a fixed amplitude sinusoidal force with adjustable frequency, it can also generate various complex dynamic force signals related to time. The generation of dynamic force is different from pulsation mechanism, and the electromagnetic induction force can be realized under the condition of no displacement. Therefore, the experimental device uses an exciter as an excitation unit, and the structural design can apply the excitation force to the excited object, so that the excited object obtains a certain amount and magnitude of vibration, so vibration and strength testing of the objects can be done. It also can calibrate vibration test instruments and sensors. The overall structure of the designed six-dimensional force generating device is shown in Figure 1.

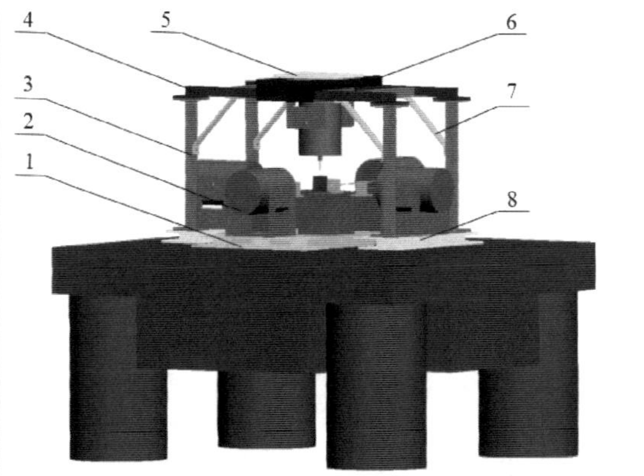

1-Base A 2-Exciter 3-Column 4-Upper beam
5-Guide A 6-Guide B 7-Reinforcement 8-Base B

Figure1 Overall structure of experimental device

3. Modal analysis

The essence of modal analysis is to solve eigenvectors (mode shapes) and eigenvalues (frequencies) of motion equations with finite degrees of freedom and without external loads and dampers, so as to avoid resonance caused by equal external excitation frequencies and natural frequencies of the system itself. As shown in Figure 2, after the constraint is added to the bottom end of the experimental device, the first 6 natural frequencies and natural modes of the device itself are calculated by finite element analysis. Since the 6th-order natural mode has the greatest influence on the experiment, it is important to consider the 6th-order natural frequency affecting the Z direction of 126.9 Hz.

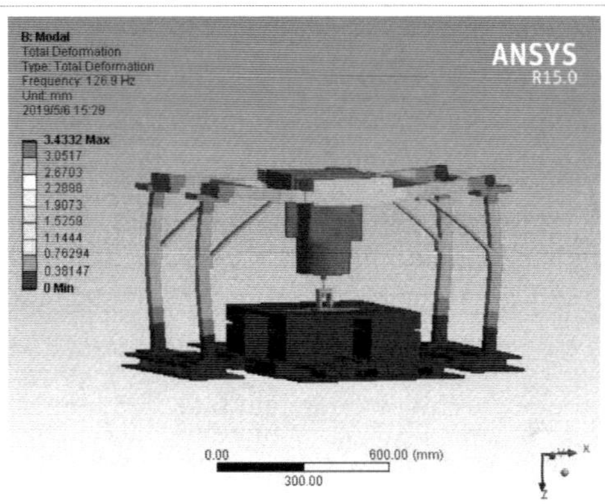

Figure2 6th-order mode vibration cloud

4. Dynamic performance study

4.1. Analysis of amplitude and phase frequency characteristics

The dynamic force generator device accords with the typical 2-order mechanical system, the differential equation of motion for a single degree of freedom damping system is as follows:

$$m\ddot{x} + c\dot{x} + kx = F \tag{1}$$

Therefore, based on the results of finite element analysis, the 2-order system transfer function at the 6-order natural frequency point of the experimental device is established:

$$G(s) = \frac{\omega_n^{\ 2}}{s^2 + 2\zeta\omega_n s + \omega_n^{\ 2}} \tag{2}$$

Among, m is quality, c is damping, k is stiffness, F is an incentive, x is the displacement response. Undamped natural frequency $\omega_n = \sqrt{k/m}$, Damping ratio is $\xi = c/2\sqrt{km}$,Natural frequency is $\omega_d = \sqrt{1-\xi^2}\omega_n$ .According to the MATLAB program, the Bode diagram of the experimental device can be fitted, as shown in Figure 3.

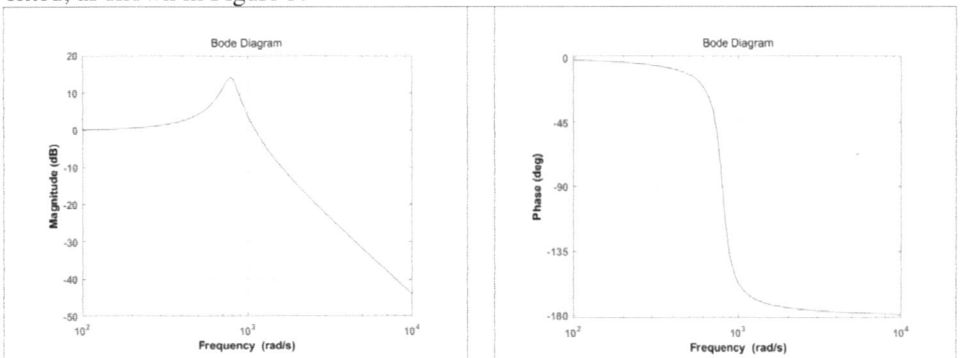

Figure 3 Bode diagram of the experimental device

According to the amplitude- phase frequency characteristic curve, it can be seen that the device is prone to Z-direction resonance at the sixth-order natural frequency. The upper beam of the dynamic force generating device is a monolithic structure, and the upper beam stiffness K is further analyzed according to the finite element is 9.43×10^6N/m.

$$F = Kx \tag{3}$$

$$L(\omega) = 20\lg\left|\frac{A_1}{A_2}\right| \tag{4}$$

Among F is the external force source, K is the upper beam stiffness, x is the displacement, A_1 is the effective displacement, A_2 is the displacement under different force values, A_2 is also the x in (3).The effective displacement of the upper beam is specified to be 0.2mm.When the external force source force value is between 10N and 500N,the frequency is in the range of 5Hz to 200Hz.The effective frequency band is calculated according to formula (2.3) and formula (2.4),as shown in Figure 4.

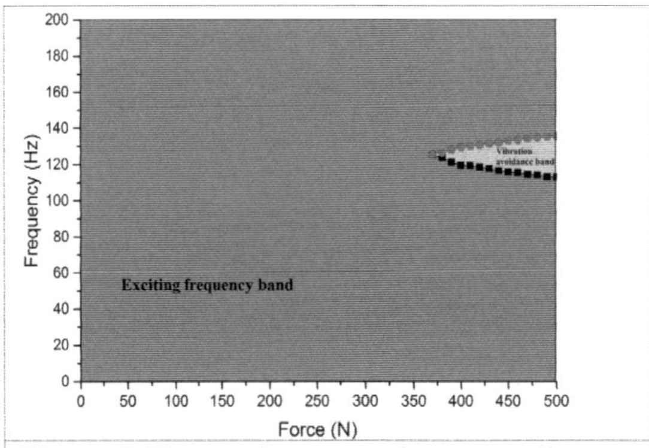

Figure 4 Force value versus frequency diagram

According to the relationship between the force value and the frequency, a preliminary conclusion can be drawn. When the external force is less than 370N, the excitation experiment can be performed in the full frequency band (0Hz to 200Hz). When the force value is more than 370N, in order not to damage the experimental device, the excitation experiment should be performed in the safe frequency band. Based on this, it can provide some theoretical basis for the experimental operators, and also play a certain protective role on the experimental equipment.

4.2. Analysis of excitation capacity characteristics

The six-dimensional dynamic force generating device uses a exciter as a force source for dynamic calibration experiments. According to the external force source requirements set forth above, it can be seen that the dynamic force excitation frequency band is wide and the force value range is wide too. Therefore, in order to be more suitable for dynamic calibration experiments, the excitation force source will be analyzed for the excitation capability characteristics. It is generally believed that the excitation capacity increases simultaneously with the increase of the excitation amplitude at the same excitation frequency. In addition, in dynamic calibration experiments, it is also important to determine the relationship between excitation force and frequency.

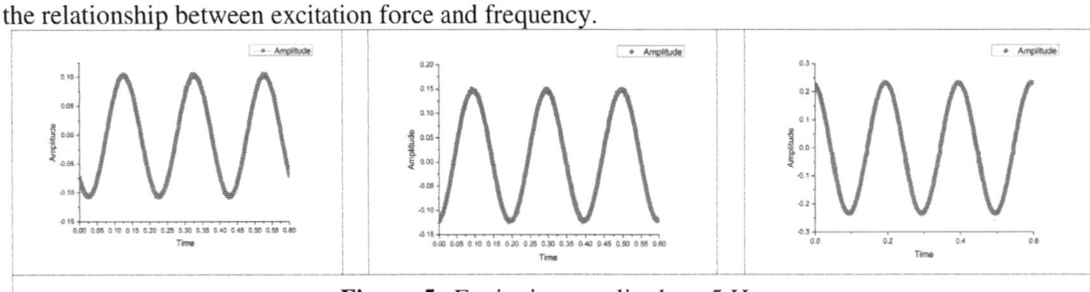

Figure 5 Excitation amplitude at 5 Hz

It can be seen from Figure 5 that, under the same frequency, the excitation capacity increases with the increase of the excitation amplitude. Based on this, the theoretical basis for the dynamic experiment of the excitation force can be provided. Next,5Hz,50Hz and 100Hz are selected for the excitation test, and the experimental data shown in Fig. 6 is obtained by multiple experiments.

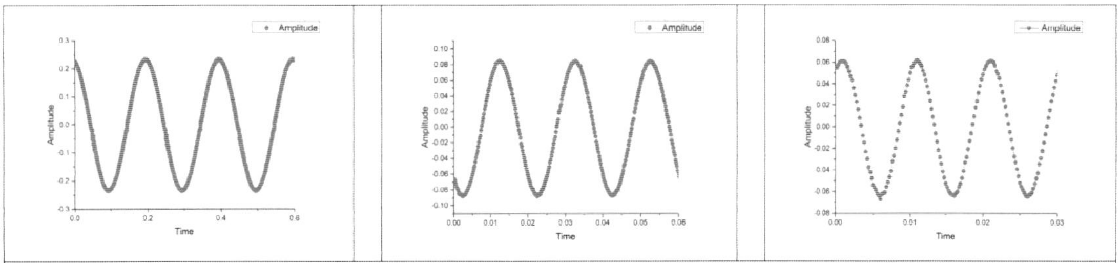

Figure 6 Excitation amplitude at different frequencies

It can be seen from Figure 6 that the relationship between the excitation frequency and the excitation amplitude is obvious. When the excitation frequency is low, the excitation amplitude is high, and when the excitation frequency is high, the excitation amplitude is correspondingly low. According to the above analysis, it can be concluded that the overall characteristics of the exciter's excitation capability is that high-amplitude fixed-frequency excitation force value can be achieved in the low-frequency input state, while low-amplitude fixed-frequency excitation force value can only be realized in the high-frequency input state. Therefore, in the background of wide excitation frequency and wide range of force values, dynamic calibration experiments can be carried out with reference to the characteristics of excitation capability.

5. Conclusion

In this paper, the six-dimensional dynamic force generating device is modeled by Pro/E software, and the modal analysis of the six-dimensional dynamic force generating device is solved by ANSYS Workbench finite element software, and the modal shape of the six-dimensional dynamic force generating device is obtained. From the analysis of the amplitude-phase frequency characteristic curve, it is concluded that the excitation experiment should be performed in the safe frequency band to prevent the device from generating resonance. In addition, according to the analysis of the excitation capacity, it is found that the overall characteristic is that high-amplitude fixed-frequency excitation can be realized in the low-frequency state, and low- amplitude fixed-frequency excitation can only be realized in the high-frequency state. In summary, the initially concluded is that the dynamic force experimental device can generate six-dimensional dynamic force and meet the test requirements.

References

[1] He Wen. Research Status of Standard Dynamic Force Generators at Home and Abroad[J]. Mechanical and Electrical Engineering, 1999 (02): 51-53+4.

[2] Schlegel C, Kieckenap G, Glöckner B, et al. Traceable periodic force calibration[J]. Metrologia,2012,49(3):224-235.

[3] Bruns T, Kumme R, Kobusch M, et al. From oscillation to impact: the design of a new force calibration device at PTB[J]. Measurement,2002,32(1):85-92.

[4] Fujii, Yusaku. A method for calibrating force transducers against oscillation force[J]. Measurement Science and Technology,2003,14(8):1259-1264.

[5] Wang Hongyan, Shen Jianwei, Dai Xuefeng. Application Research of Small Force Value Generator Based on Proportional Electromagnet[J]. Hydraulic & Pneumatics, 2006(12):49-51.

[6] Wang Weiwei. Principles and methods of dynamic force measurement[M]. China Metrology Press, 1986: 100-101.

[7] Wang Qingming, He Wen, Jia Shushi, et al. A new negative-threshold signal generating device [J]. Practical Testing Technology, 1996 (01): 46-47.

[8] Fujii, Yusaku. Proposal for a step response evaluation method for force transducers[J]. Measurement Science and Technology,2003,14(10):1741-1746.

IOP Conf. Series: Materials Science and Engineering **616** (2019) 012024 doi:10.1088/1757-899X/616/1/012024

Influence of Sampling Frequency in DCPD Crack Monitoring

Liang Zhang, He Xue, Chenqiang Ni and Siyu Gou

School of Mechanical Engineering, Xi'an University of Science and Technology,710000, China

Email: 13022863497@163.com

Abstract. Due to the particularity of autoclave environment, it is difficulty to real-time monitor the crack growth state in the autoclave, and Direct Current Potential Drop method (DCPD) is widely in the crack propagate monitor in this kind experiment. The signal characterized by DCPD method is usually submerged by strong noise. In this paper, the extraction and filtering methods of crack characteristics in high temperature and high pressure environment are studied. The SINC filter is used to obtain crack initial data information and filter out the pulse interference signal, and the FIR filter is used to filter out high frequency interference. Finally, crack monitoring experiments at different sampling frequencies were studied. The experimental results show that the sampling frequency has a great influence on the stability and anti-interference ability of the data.

1. Introduction

The key issue that must be overcome in nuclear power plant life extension is that environmental cracking represented by Stress Corrosion Cracking (SCC), it reduces the design life and safety of nuclear power equipment[1]. It is one of the important experiment to obtain the stress corrosion cracking growth rate in an autoclave simulated the high temperature and high pressure water environment of the nuclear power equipment in the nuclear structural materials selection and safety evaluation. Due to the particularity of autoclave environment, it is very difficult to detect and monitor cracks. At present, the potential drop method(DCPD)[2] is the most effective indirect measurement method for crack length applied in high temperature and high pressure environments[3-4].Compared with the AC potential drop method (ACPD)[5], the DC potential drop method (DCPD) has no skin effect. The existence and size of internal defects of the material are inferred by detecting the potential changes on both sides of the crack[6]. This technology was applied to nuclear power materials by P.L.Andresen et al.[7-8],they simulated the actual environment and studied the crack propagation of the material. The DC potential drop method was used by I.Černý[9] for experimental monitoring of some large pipeline components under high temperature conditions, he studied the initiation and propagation of crack defects in pipeline components, and took large pipeline specimens as an example. It was proposed to carry out certain temperature compensation and irregular area correction during the test, and the influence of crack closure should be considered.

The DCPD method monitors the crack signal is weak, it is difficult to extract and filter the signal in an autoclave environment,because noise is superimposed on weak signals.The method of FIR filtering to eliminate observation noise in GPS dynamic monitoring was proposed by Fengbo Wu[10], and he proposed a method based on FIR filtering for frequency shift filtering of high frequency signals.SINC3 filter designed by Xiaohang Wang[11], the method effectively filters out high frequency noise and impulse noise. In this paper, the crack monitoring of 304 steel is studied, and the FIR filter algorithm is used to filter out the high frequency interference signals in the crack signal, and SINC filtering is used to eliminate spikes. We use the test instrument developed by ADS1235 chip as the terminal for data acquisition. We choose different sampling frequency, and find out the best filtering algorithm, and improve the accuracy and stability of the instrument.

2. Potential drop method for monitoring weak signal principle

The DC potential drop method is based on the detection method of the conductive properties of metal materials. When the CT sample is supplied with a constant current,as the crack length increases, the potential at both sides of the crack increases,but its potential changes have nonlinear characteristics. The Johnson formula gives a function to characterize the relationship between the crack size and the potential drop, this method is used to describe the length of the crack. Figure 1 shows a schematic diagram of the DCPD measurement principle of a crack-containing sample. The solution of the potential drop is shown in Equation 1:

$$U = U_1 - U_2 \tag{1}$$

In the formula:U is the potential drop, U1 and U2 are the two measuring point potentials.
For the infinitely long finite-width plate center crack specimen, the relationship between potential drop and crack length is given by Johnson's formula:

$$U = \frac{4\rho l}{\pi B} \bullet \cosh^{-1}\left[\cosh\left(\frac{\pi y}{2W}\right) / \cos(\pi a / 2W)\right] \tag{2}$$

In the formula:a is the length of the crack half length, y is the half length of the distance between the two sides, W is the half width of the sample, B is the width of the sample, U is the potential difference between the two points, and I is the current intensity.

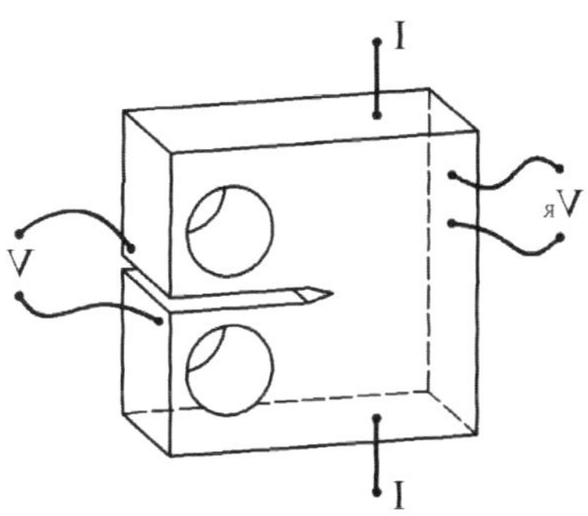

Figure 1. sample wiring layout scheme

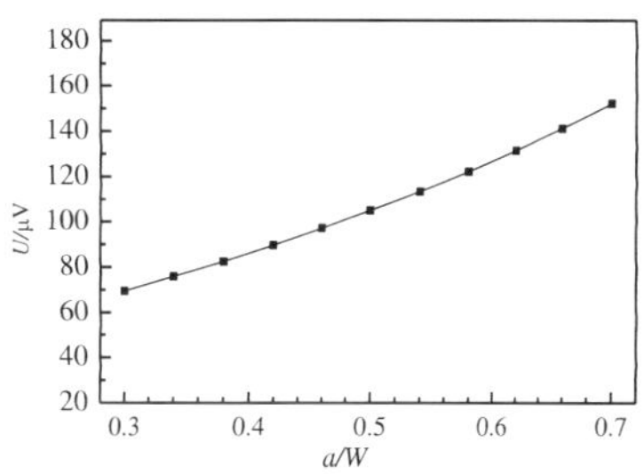

Figure 2. Relationship between crack length and potential drop

For compact tensile specimens (CT),The applicability of the Johnson formula in CT samples was analyzed by Chenqiang Ni[12], which pointed out that in terms of boundary conditions, the position of the input current point is no longer equivalent to the "distal" to the crack tip.At the same time, the position change of the current input point will have a considerable influence on the potential field distribution and the corresponding crack potential calibration function, and propose the most reasonable wiring point layout scheme[13]. As shown in Figure 1, The nuclear welding material of 304 steel was used as the research object, and a constant 1A current was passed at both ends. Monitor potential changes at both sides of the crack, the physical object is shown in Figure 3. Using a CT specimen with a pre-crack length of 1 mm, the voltage signals on both sides of the crack were measured, and the effects of the instrument on the weak signal acquisition were compared at different sampling rates.

As shown in Figure 2, according to the results of finite element analysis[14], it can be seen that the crack length is from 0.3m to 0.7m, and the potential change is 70uV. The results show that the relationship between the crack length and the potential drop is nonlinear, which verifies the consistency and accuracy of the finite element analysis and the Johnson formula.

3. Small signal detection algorithm

The small signal detection algorithm has multiple filtering measures. Firstly, the FIR signal is used to separate the high frequency interference from the low frequency crack characteristic signal. Then the SINC filter is used to filter the pulse interference, and the potential signal characterizing the crack length is obtained. Finally, the algorithm is used. The error analysis shows the sampling frequency with the best stability of DCPD crack monitoring. The algorithm flow is shown in Figure 3.
The text of your paper should be formatted as follows:

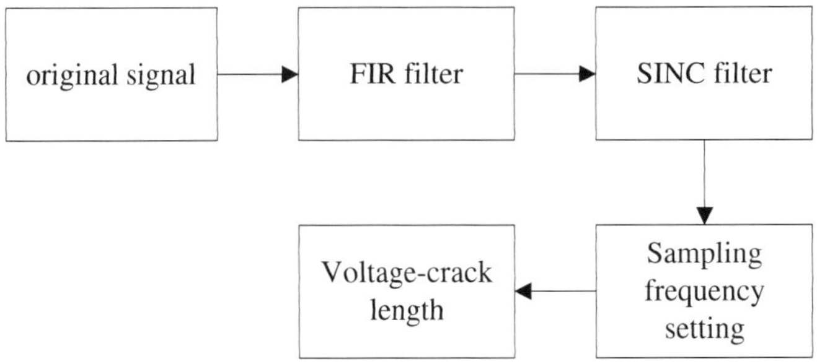

Figure 3. algorithm flow

3.1. FIR Adaptive Filtering Algorithm

According to the characteristics of DC potential drop method, the useful signals are all in the DC component. Most of the interference comes from 50Hz/60Hz power frequency interference and random noise. These interferences are high frequency interference. The FIR filter works in low-pass mode, and it needs to filter out high-frequency interference to meet the design criteria of the filter circuit. There are many FIR filter algorithms, The minimum mean square error criterion adaptive algorithm is small in computation and easy to design and implement. Used in all kinds of filtering occasions. Due to the weak signal, differential circuits are used to acquire signals, according to the basic formula of FIR (3):

$$Y(i) = \sum_{k=0}^{n-1} W(k) \cdot X(k-i), i = 0, \cdots, M \tag{3}$$

$W(k)$ is the filter weight coefficient, the best design principle is that the sum of the squares of the error sequence is the smallest:

$$\varphi = \sum_{i=0}^{M} e^2(i) = \sum_{i=0}^{M} d^2(i) - 2\sum_{k=0}^{N-1} w(k) r_{dx}(k) + \sum_{k=0}^{N-1}\sum_{j=0}^{N-1} w(k) w(j) r_{xx}(k-j) \tag{4}$$

In the formula: r_{dx} the relationship between the expected signal and the input signal Describe the formula r_{xx} is the autocorrelation coefficient of the input signal.

$$\frac{\partial \varphi}{\partial w(k)} = 0, 0 \le K \le N-1 \tag{5}$$

Minimum principle based on squared sum of error sequences

$$w_{i+1} = w_i(k) + 2\mu e(n)x_i(i+k), i = 0,2,\cdots,M \tag{6}$$

The main determinants of the convergence speed of the iterative step size and the filter order. It can theoretically be any value. In the actual calculation, the initial value of the weighting value is zero. For the filtering order, the general filtering order is higher. The better the convergence range of the filter is, the stricter the convergence condition is. The optimal length of the different signals is different. Generally, the selection is based on experience. Generally, the length is set to be larger.

3.2. SINC filtering algorithm

The potential signal characterizing the crack length is filtered by FIR, and the high frequency signal interference is filtered out. Due to the special environment in the autoclave, the experimental results show that there is a random pulse signal as shown in Figure 4. In order to filter out the pulse interference, SINC filter is used, the SINC filter has no excessive bandwidth, which limits the signal content that the filter can measure. According to the technical characteristics of the DC potential drop method, the useful signal is a DC component. It is just right to solve the filtering of such signals.As shown in Figure 4, the test result is filtered after SINC filter.

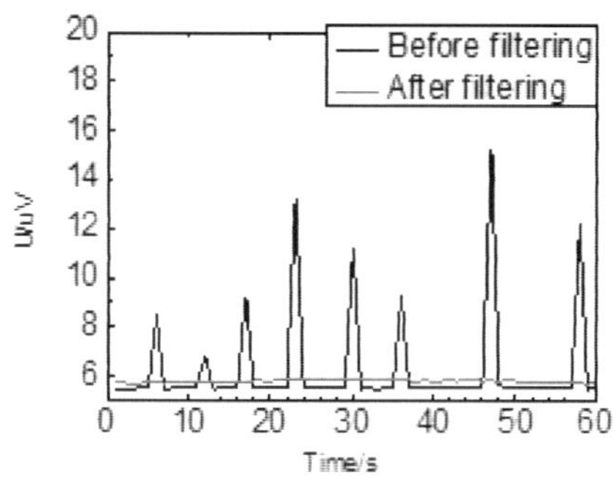

Figure 4. Comparison before and after filtering

4. Experiment and analysis

4.1. Effect of sampling frequency

In this experiment, the 34420A nanovoltmeter of Keysight Technology was used as the experimental comparison. The design instrument based on ads1235 chip was used as the acquisition equipment, and the acquisition rate was adjusted from 2.5SPS to 7200SPS. For the same CT sample measured for 60s, after a lot of experiments, The sampling frequency has a great influence on the instrument test results, and the data fluctuates between 5uV and 15uV. The result is shown in Figure 5.

The stability of the instrument is related to the sampling frequency. The test results of the instruments at different sampling frequencies have large differences.Although the FIR filter filters out high-frequency interference, the SINC filter filters out the pulse signal, and there are still 11 kinds of sampling frequencies with data fluctuations around 15uV.The data fluctuation is around 0.2uV and the sampling frequency is 400SPS and 5SPS. According to the theoretical calculation, the CT signal with a crack length of 1 mm has a crack potential signal of 6 uV. As can be seen from Figure 6, the different sampling frequencies make a large change in the accuracy and stability of the instrument acquisition results. The weak signal and the interference signal are superimposed together, and the peak-to-peak voltage of the collected voltage is about 15uV, which greatly affects the crack monitoring accuracy. According to the theoretical calculation, the crack growth 1mm potential signal changes to 3uV, therefore, the instrument signal fluctuation is 1uV. The best resolution of the instrument.

As shown in Figure 6, the sampling frequency is less fluctuating at 400 SPS and 5 SPS, and the fluctuation range is 1 uV, which can meet the acquisition accuracy of the crack monitoring instrument.

Both sampling frequencies can meet the accuracy requirements of the instrument test.The stability of the data is an important indicator of the constant instrument[15]. The stability of the variance analysis is used for the test data. The sampling frequency is 400SPS and the 5SPS variance results are $\sigma^2 = 0.001540321$ and $\sigma^2 = 0.006835235$ respectively. The results show that the stability is best when the sampling frequency is 400SPS in the short term.

4.2. Crack dynamic simulation experiment

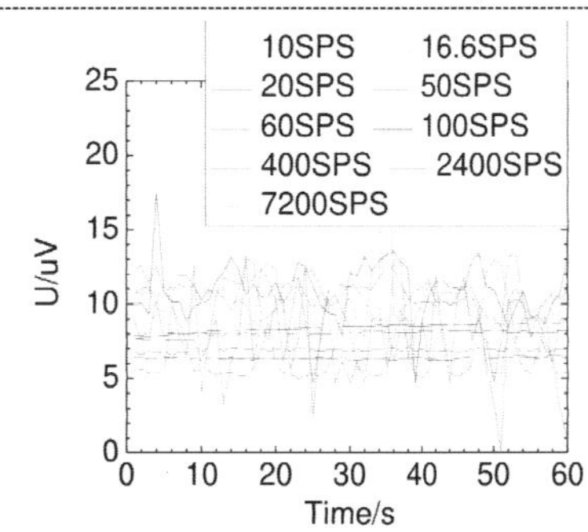

Figure 5. different sampling frequency test chart

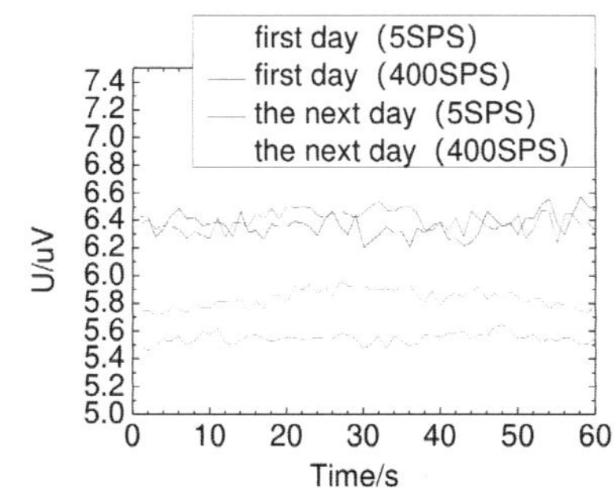

Figure 6. Samples of different days

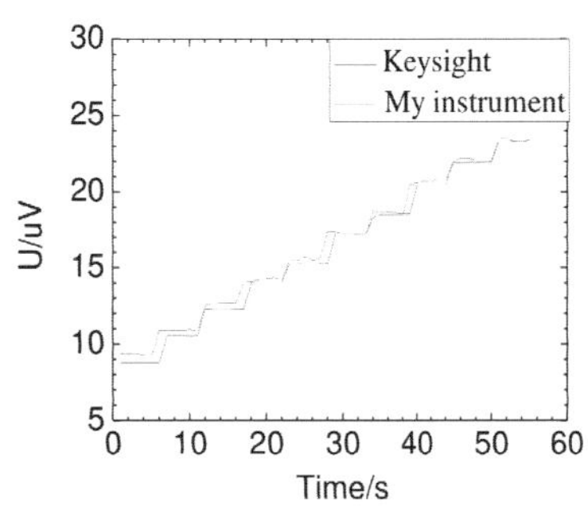

(a) Comparison with 34420A

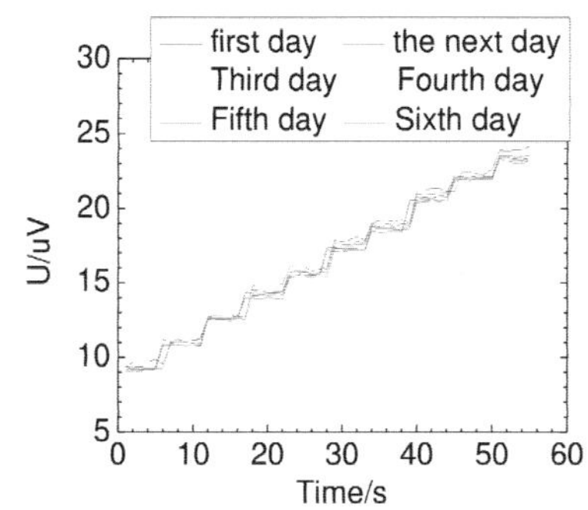

(b) Instrument test results at different times

Figure 7. crack simulation test chart

The crack growth period is long, and the CT specimen cannot exhibit the crack propagation in a short time. In order to verify the instrument, a total of 10 CT samples with a crack length of 1 mm to 5 mm were prefabricated, and the dynamic growth of the crack was simulated by the contact relay switching. As shown in Fig. 7(a), the crack length is increased by 0.5 mm in turn, and a significant step signal can be seen on the graph. As with the finite element analysis, the voltage change of 4uV per 1mm crack is also consistent with the expected results.Figure 7(a) shows the comparison between Keysight's products and this product. The test results of foreign related instruments and the instrument are basically the same, and the data collected by the instrument is reliable.Figure 7(b) shows the sampling results of different dates. According to the analysis of the test results of six days, it is found that the stability of the instrument for a long time is high, but there are still fluctuations, and we will further improve the stability of the instrument.

5. Conclusion and future work

In order to test the environmental cracking crack growth rate of nuclear power structural materials in an autoclave simulating nuclear power high temperature and high pressure water environment, a crack monitoring instrument based on ADS1235 chip was developed, and FIR filtering and Sinc filtering are used in the acquisition of weak signals. The experimental results show that the proposed method not only preserves the original signal characteristics, but also improves the accuracy and resolution of the monitoring system. According to the experiment, the following conclusions are obtained:

- The FIR filtering algorithm used in the microvolt signal monitoring process has a good suppression effect on the high frequency signal, and the SINC filtering algorithm has a good filtering effect on the pulse signal.
- Microvolt-level signal acquisition, the change of sampling frequency has a greater impact on the stability and accuracy of the measured data.

ACKNOWLEDGMENT
This work was supported by the National Natural ScienceFoundation of China (No. 51475362), and the Key Research and Development Plan Funded by Shaanxi Provincial Science and Technology Department (No. 2017GY-034).

REFERENCES

[1] P L Andresen, K Gott, J L Nelson. Stress corrosion cracking of sensitized type 304 stainless steel in 288°C water: a five laboratory round bobbin, *Proceedings of the 9th International Symposium on Environmental Degradation of Material in Nuclear Power Systems-Water Reactors*, Newport Beach, **1999, 38(17)**:423-433.
[2] Hongfan Zhou. A new technique for measuring crack propagation, a flip DC potential method. *Testing machine and material test*, **1983, 22(6)**: 45-68.
[3] Wei Yuan, Qing Liu. Experimental study on the determination of surface crack depth by DC potential method. *Pressure vessel*, **1986, 4(6)**: 49-57.
[4] Donghai Du, Xuan Yu, Kai Chen, Lefu Zhang, Xiuqiang Shi. On-line measurement test system for DC voltage drop method stress corrosion crack growth rate. *Physical and Chemical Testing (Physical Section)*, **2014, 17(01)**: 25-30.
[5] Zhicheng Fang, Lihua Lu, Junqi Cheng. AC potential method crack depth measuring instrument. *Journal of Huaqiao University: Natural Science Edition*, **1982, 62(2)**: 61-70.
[6] Jun Xu. Environment-promoted crack propagation behavior and crack monitoring techniques for two structural steels. *General Institute of Mechanical Science*, **2010, 24(26)**: 70-78.
[7] P L Andresen. Effect of MnS and Corrosion Potential on Environmentally Assisted Cracking of Low Alloy Steel. *Minutes of the Mechanisms Sub-Committee International CooPerative GrouP on Environmentally Assisted Cracking*, Malelane, South Afriea, **1993, 30(23)**: 24-26.
[8] P L Andresen. M M Morra.PWSCC of Alloys 690, 52 and 152.*13th International Conference on Environmental Degradation of Materials in Nuclear Power Systems*, Whistler, British Columbia, **2007, 18(14)**: 19-23.
[9] I Čern. Measurement of subcritical growth of defects in large components of nuclear Power Plants elevated temperatures.*International Journal of Pressure Vessels and Pipings*, **2001, 78(12)**: 893-902.
[10] Fengbo Wu. Research on GPS Dynamic Deformation Multipath Effect Extraction Based on FIR Filter[J]. *Urban Survey*, **2019, 12(01)**: 91-94.
[11] Yihang Wang. FPGA-based Sigma-Delta ADC Digital Decimation Filter Sinc3 Design[J]. *Microelectronics*, **2012, 42(5)**: 655-658.
[12] Chenqiang Ni, He Xue, Xiurong Fang, et al. Study on Applicability of Johnson Formula in Crack Propagation Test[J]. *Foundry Technology*, **2017, 38(12)**: 2877-80.
[13] Chenqiang NI, He Xue, Yinghao Cui. Analysis of Junction Point Position in DC Potential Drop Crack Monitoring Test[J]. *Thermal Process Technology*, **2016, 45(24)**: 122-125.
[14] Zhijun Li, He Xue. Numerical Calibration of DC Potential Drop Crack Depth Sounder[J].*Journal of Xi\'an University of Science and Technology*,**2012, 32(01)**: 116-120.
[15] Junqiang Dong, Lihuan Liu, Hongxiang Han, et al. Allan variance random error identification sampling frequency impact analysis[C],*2018 China Aviation Measurement and Control Technology Annual Meeting*, **2018,37(2)**: 331-334.

IOP Conf. Series: Materials Science and Engineering **616** (2019) 012025 doi:10.1088/1757-899X/616/1/012025

Research on Intelligent Grasping System of Monocular Vision Guided Manipulator

Han Jiabu[1], Chen Saixuan[1], Cui Guohua[1], Cheng Zhenyi[1] and Li Shipei[1]

[1]School of Mechanical and Automotive Engineering, Shanghai University Of Engineering Science, Shanghai, China

E-mail: 2022653432@qq.com

Abstract. In order to solve the problem of how a cooperative robot can quickly and accurately locate and grasp the specified target work piece and avoid obstacles autonomously under the condition of man-machine cooperation and the variable position of the object to be grasped, a new method of robot arm grasping under the visual guidance of ROS system was proposed. Using ROS framework and tools, the color threshold algorithm in ROS-Opencv is used to realize the rapid recognition and positioning of grasping targets. Use the ROS-Moveit! The kinematics calculation, motion planning and obstacle avoidance of the manipulator are completed the relationship between the image coordinate system and the robot coordinate system is established by hand eye calibration technology. In addition, ROS tools can be used to obtain information such as the position and speed of each joint during the movement of the mechanical arm. Through experimental testing and data analysis, it is shown that the hand-eye system is accurate in calibration, object positioning, and the robot arm can accurately grasp the target object, which has certain practical application value.

1. Introduction

In the current industrial production process, the vast majority of industrial robots rely on the teaching reproduction method to capture and place the workpiece in a fixed position. If the external environment or the state of the workpiece changes, the capture task will be interrupted or fail. In order to solve this problem and improve the flexible production of the factory, the technology of vision-guided robot was produced. The robot uses the visual system to identify and locate the objects in the working space, avoid obstacles autonomously, and at the same time, work in cooperation with human safety. Zhang chi makes use of the VS software and the grasping platform developed by ABB robot[1]. This platform has a long development cycle and great difficulty in development. Xu bofan used the grab system platform of visually guided FANUC Robot Interface developed by FANUC Robot Interface[2]. This platform has poor adaptability and is only suitable for FANUC robots. To solve the above problems, ROS system framework and tools are adopted in this paper to build a gripper system based on UR5 cooperative robot and flexible gripper. Images are collected by the monocular camera, ROS image processing nodes get the image features of the workpiece, and the feature information is converted into positioning information and sent to ros-moveit! The motion planning parameters are sent to the manipulator controller to realize flexible grasping under visual guidance. Improve the code reuse rate of robot development and research, reduce the robot project research and development cycle. And greatly improve the system integration, portability[3,4].

IOP Conf. Series: Materials Science and Engineering **616** (2019) 012025 doi:10.1088/1757-899X/616/1/012025

2. System composition

2.1. Hardware system

The hardware of the platform is composed of UR5 cooperative robot, USB camera, flexible gripper, upper computer and lower computer. Hardware composition of the platform is shown in Figure.1.

The camera is installed above the experimental platform and connected to the upper PC through a USB interface. The lower PC communicates with the upper PC through a network port based on the Ethernet protocol. The driving and image processing of the camera and the control of the robot are based on the ROS system in PC. Hardware communication architecture is shown in Figure.2.

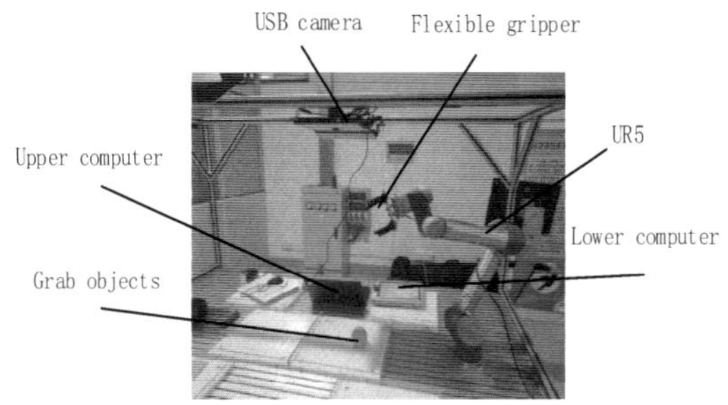

Figure 1. Hardware composition of the platform

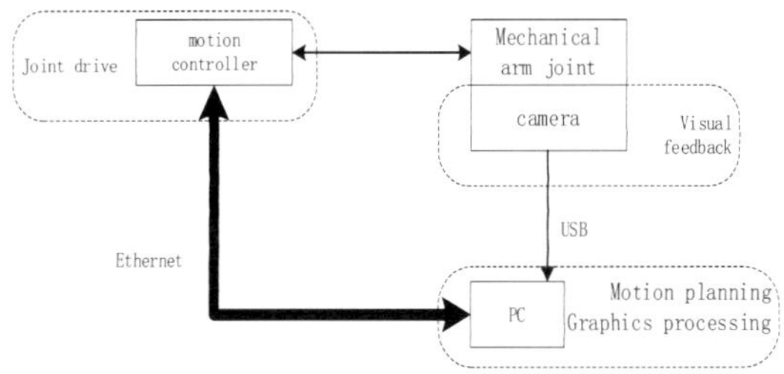

Figure 2. Hardware communication architecture

2.2.Software system

The whole software system is based on the ROS system[5]. ROS system adopts a distributed computing and communication architecture, providing functions such as hardware abstract description, underlying driver management, inter-program messaging and program distribution package management. This software system mainly includes three modules: camera calibration and hand-eye calibration module, object recognition and positioning module based on ros-opencv and ros-moveit! of the motion planning module. Object feature recognition was developed by ros-opencv computer vision library using C++.After the target object location was obtained, ros-moveit! was used the trajectory planning of the end of the manipulator from the starting point to the target point is completed. The working process of the system is shown in Figure.3.

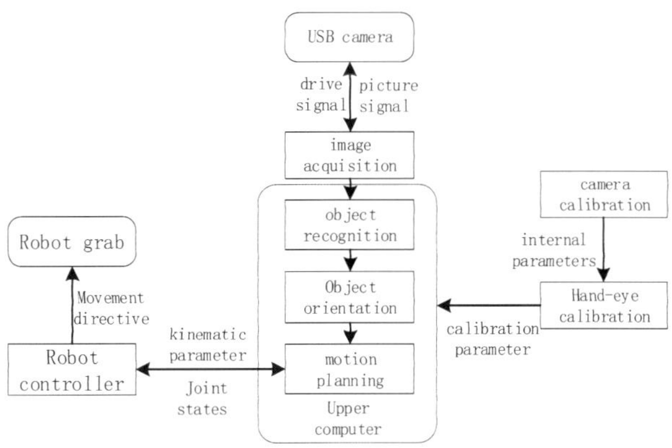

Figure 3. The working process of the system

Camera calibration is carried out to obtain the internal and external parame-ters of the camera, and the internal and external parameters are imported into the camera driver file and called in the parameter server. All information obtained by the camera are described in the camera coordinate system, through hand-eye calibration technology of the workpiece under the camera coordinate system transformation to the robot pose information base coordinates, this experiment adopts the eye-to-hand camera installation, determine the camera coordinate system and robot hand-eye calibration relative relationship between the base coordinate system. After a series of image processing tasks such as filtering, color segmentation and feature extraction, the image acquired by the camera can determine the target object contour to be recognized. Then, according to the central point of the target object contour and the internal and external parameter matrix of the camera, the plane coordinate information of the target object in the camera coordinate system can be obtained. Then, according to the conversion relationship between the camera coordinate system calibrated by hand-eye and the manipulator coordinate system, the spatial position information of the target object in the manipulator coordinate system can be obtained, and the object information can be transmitted to the ROS system through the ROS communication mechanism for task planning. Ros-moveit! is used according to the planning task. The kinematics library, the kinematics planning library and the obstacle avoidance algorithm are respectively used for inverse kinematics solution, trajectory planning and obstacle avoidance movement. Finally the motion planning of robotic arm motion planning information queue through Ethernet communication mode is passed to the next bit machine and lower machine receives the mechanical arm movement and parses the message queue, at the same time to drive the mechanical arm according to the planning of the path of the movement and scraping, and transfer the real-time position and posture information of the robotic arm to PC, the realization of ultimate target accurately grasping.

3. Principle of visual localization

3.1.Monocular localization model

Inside and outside the camera parameters（f u_0 v_0 α_x α_y R T） were obtained through the camera calibration experiment[6], and the monocular localization model was established according to the pinhole model[7]. Monocular localization model is shown in Figure 4.

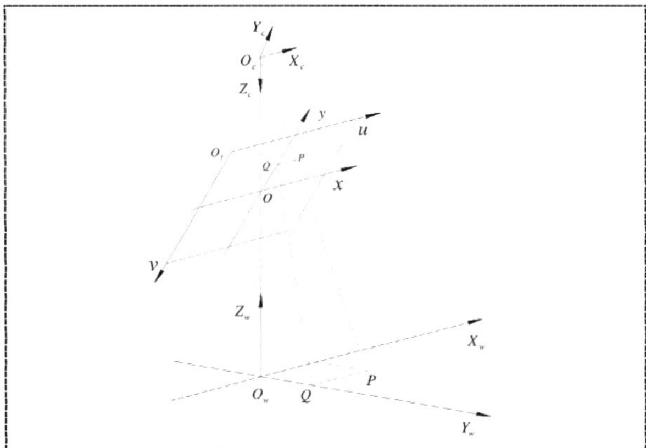

Figure 4. Monocular localization model

The origin o_c of camera coordinate system $o_c x_c y_c z_c$ is the optical center of the camera, the height of the camera optical center from the table plane xoy is fixed, the projection of camera optical center on the table surface coincides with the origin o, so the world coordinate system $o_w x_w y_w z_w$ is established with as the origin o ,the optical axis intersects the equivalent image plane oxy at the coordinate value (u_0, v_0) of point o, the distance from o to o_c is the focal length f, point P is any point on the plane of the carrying object, and its world coordinate value is expressed as (x_w, y_w, z_w), P' is its projection position on the equivalent imaging plane, Q , Q' is the projection of P and P' on the vertical axis. Frame $o_f uv$ is the image pixel coordinate system. u and v re-spectively represent the columns and rows of the image matrix. α_x and α_y are the coefficients along the axis x and y respectively.

3.1.1. Longitudinal positioning. The longitudinal coordinate value y_w ,namely the length of O_wQ , of point P is derived in plane $O_cO_wY_w$

$$y_w = h\tan\angle O_wO_c = h\tan\left[\arctan\left(\frac{|OQ'|}{f}\right)\right] \tag{1}$$

In the pixel coordinate system of the image

$$|OQ'| = (v_o - v_{Q'})\alpha_y \tag{2}$$

The formula can be written as

$$y_w = h\tan\angle O_wO_c = h\tan\left[\arctan\left(\frac{(v_o - v_{Q'})\alpha_y}{f}\right)\right] \tag{3}$$

3.1.2 Lateral positioning. Determine the transverse coordinate of points in plane, and get from the proportion relation

$$x_w = |PQ| = |O_cQ|\tan\angle PO_cQ \tag{4}$$

$$|O_cQ| = \frac{h}{\cos\angle O_wO_cQ} = h/\cos\left[\arctan\left(\frac{|OQ'|}{f}\right)\right] \tag{5}$$

$$\tan\angle PO_cQ = \frac{|P'Q'|}{|O_cQ'|} = \frac{|P'Q'|}{\sqrt{f^2 + |OQ'|^2}} \tag{6}$$

Represent $|OQ'|$ and $|P'Q'|$ in the image pixel coordinate system

$$x_w = \frac{h}{\cos\left\{\arctan\left[\left(\frac{(v_c - v_{Q'})\alpha_y}{f}\right)\right]\right\}} \bullet \frac{\left[u(P') - u_o\right]\alpha_x}{\sqrt{f^2 + \left[(v_o - v_{Q'})\right]}} \tag{7}$$

Through the above longitudinal positioning y_w and lateral positioning x_w calculation, the coordinates of any point on the plane within the scope of camera vision in the world coordinate system can be obtained, so as to determine the position of the target on the experimental platform.

3.2 Hand-eye calibration

According to the application scenarios of the experimental platform[8], the camera of the experimental platform is installed in an eye-to-hand way. The eye-to-hand calibration model is shown in Figure 5.

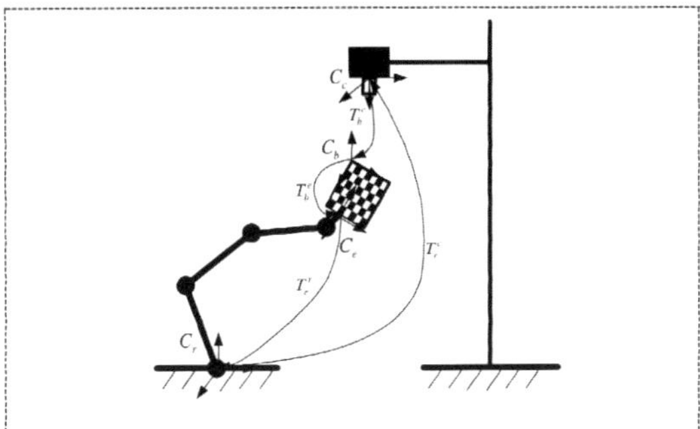

Figure 5. The eye-to-hand calibration model

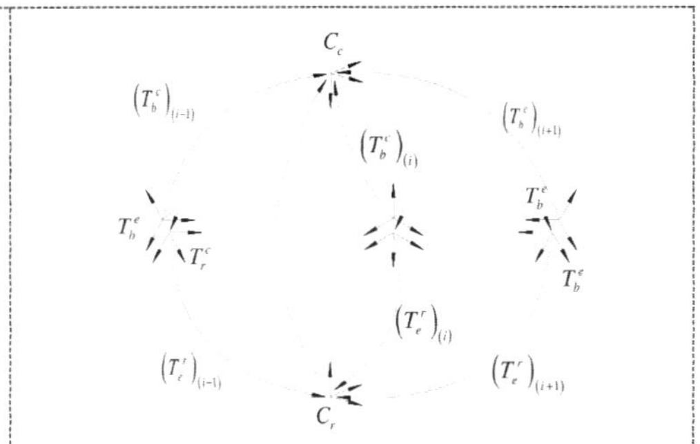

Figure 6. The coordinate transformation relationship model

$T=\begin{bmatrix} R & t; & 0 & 1 \end{bmatrix}$ represents the rotational translation matrix in homogeneous coordinates, R represents the rotation matrix, t represents the translation vector, T_b^c represents the transformation relationship between the calibration plate coordinate system and the camera coordinate system, T_b^e represents the transformation relationship between the calibration plate coordinate system and the end coordinate system of the mechanical arm, T_e^r represents the transformation relationship between the end coordinate system of the mechanical arm and the coordinate system of the robot base. T_r^c represents the transformation relationship between the coordinate system of the robot base and the camera coordinate system, namely the hand-eye transformation matrix to be solved.

According to the transformation relationship between the camera, calibration plate, the end of the mechanical arm and the coordinate system of the robot base in the motion state transition as shown in the figure6, the following state transition equation is established:

$$\begin{aligned}
\left(T_b^c\right)_{(1)} &= T_r^c \cdot \left(T_e^r\right)_{(1)} \cdot T_b^e \\
\left(T_b^c\right)_{(2)} &= T_r^c \cdot \left(T_e^r\right)_{(2)} \cdot T_b^e \\
&\vdots \\
\left(T_b^c\right)_{(i)} &= T_r^c \cdot \left(T_e^r\right)_{(i)} \cdot T_b^e \\
\left(T_b^c\right)_{(i+1)} &= T_r^c \cdot \left(T_e^r\right)_{(i+1)} \cdot T_b^e \\
&\vdots \\
\left(T_b^c\right)_{(n)} &= T_r^c \cdot \left(T_e^r\right)_{(n)} \cdot T_b^e
\end{aligned} \tag{8}$$

Then get rid of the T_b^e

$$\left(T_b^c\right)_{(i+1)} \cdot \left(T_b^c\right)_{(i)}^{-1} \cdot T_r^c = T_r^c \cdot \left(T_e^r\right)_{(i+1)} \cdot \left(T_e^r\right)_{(i)}^{-1} \tag{9}$$

If

$$\begin{aligned}
A &= \left(T_b^c\right)_{(i+1)} \cdot \left(T_b^c\right)_{(i)}^{-1} \\
B &= \left(T_e^r\right)_{(i+1)} \cdot \left(T_e^r\right)_{(i)}^{-1} \\
X &= T_r^c
\end{aligned}$$

A represents the motion relation of the calibration plate coordinate system an B represents the motion relation of the end coordinate system of the mechanical arm, so solving for is the same thing as solving for T_r^c the unknown X in equation $AX = XB$.

According to the formula, to solve the matrix X at least two relative position trans-formations of the robot are required, and the rotation axes of these two position changes are not parallel. In order to improve the calibration accuracy and offset the accidental error, this experiment collected the relative pose relationship of 17 groups of mechanical arms. Hand-eye calibration pose transformation of drawing is shown in Figure 7.

IOP Conf. Series: Materials Science and Engineering **616** (2019) 012025 doi:10.1088/1757-899X/616/1/012025

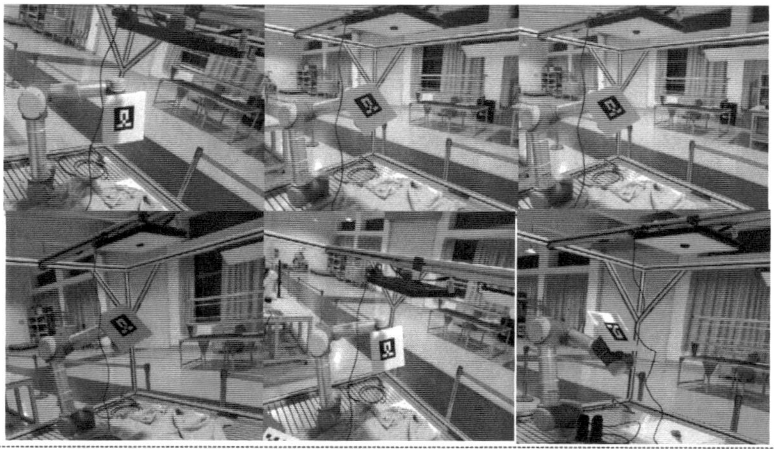

Figure 7. Hand-eye calibration pose transformation of drawing

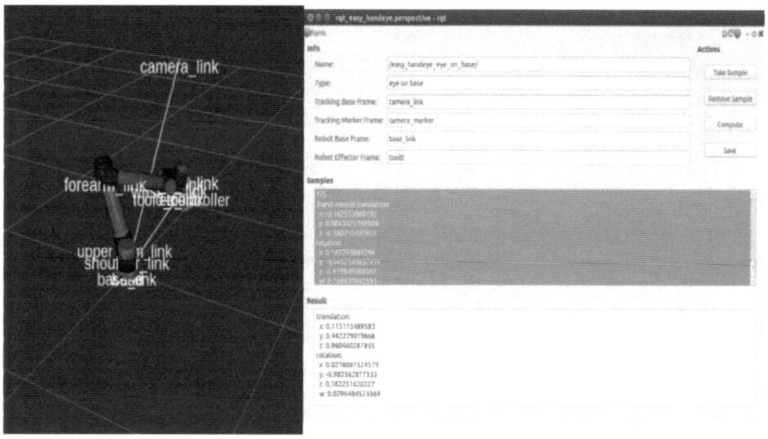

Figure 8. Hand-eye calibration pose transformation results

In the ROS system, the ROS calibration function package easy_handeye and Opencv are used to write corresponding programs to complete the corner feature extraction of the calibration board and the pose calculation of the camera coordinate system relative to the end-effector coordinate system.

The usb_cam node in the ROS system is used to start the camera and send image messages. Then, the ros-opencv image processing node is used to extract and describe image features. Ros-moveit! is used multiple different positions of the calibration plate at the end of the mechanical arm were planned, and then parameters were calculated based on image feature points through the easy_handeye function package. The position and attitude of the calibration plate relative to the camera coordinate system were obtained, and the hand-eye transformation matrix was obtained to complete the calibration. Hand-eye calibration pose transformation results is shown in Figure 8.

4. Robot motion control

4.1 Target identification and location

We driven camera and get ROS of image information by ROS, using the cv_bridge class into ROS-Opencv image, then Opencv vision library preprocessing, image processing algorithms to use filtering algorithm of median filter was carried out on the original image denoising and according to the experimental platform for fetching in target based on color threshold method for image segmentation

IOP Conf. Series: Materials Science and Engineering **616** (2019) 012025 doi:10.1088/1757-899X/616/1/012025

and edge features based on target for the workpiece position measurement and calculation[9,10]. Object orientation is shown in Figure.9. Finally, the image is converted into a ROS message through the cv_bridge class and the image data is sent to the corresponding ROS topic. Target object contour recognition flow chart is shown in Figure.10.

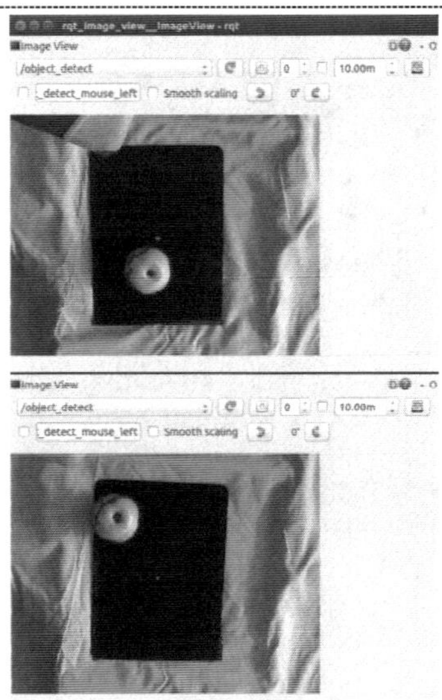

Figure 9. Object orientation

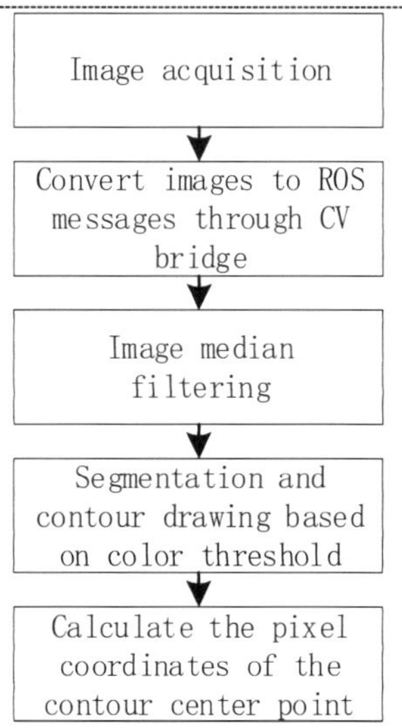

Figure 10. Target object contour recognition flow chart

4.2 Robot grasp control

The UR5 robot motion plan using ROS-Moveit! module[11]. Through Moveit! configuration Assistant (Moveit Setup Assistant Tool) created call URDF mechanical arm model file, and then in turn to collision detection configuration, virtual joint configuration, mechanical arm planning initial position mechanical arm joint group configuration, configuration, mechanical arm end actuator configuration, passive joint configuration, generate Semantic Robot Description. SRDF file, the kinematics and motion planning library configuration file configuration file and the corresponding startup files, in the mission planning to load the corresponding configuration file, the motion planning, visualization and simulation of the robot are completed. Schematic diagram of UR5 grasping results is shown in Figure.11.

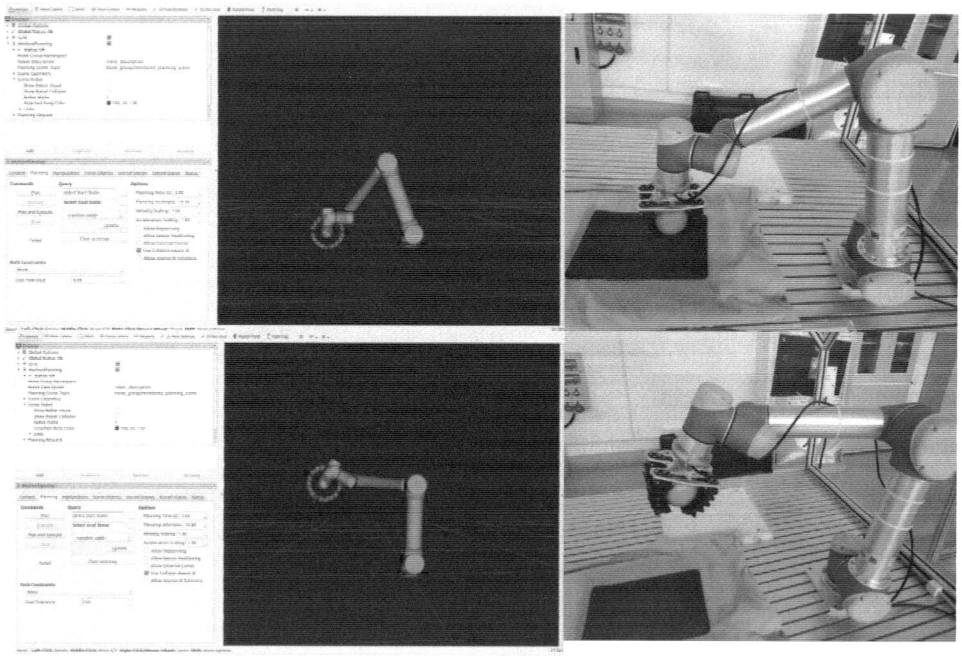

Figure 11. Schematic diagram of UR5 grasping results

According to the working condition information obtained by the camera and the state of the robot arm, the robot arm grabbed the task planning, and loaded the corresponding Moveit! in the process of executing the task configuration assistant to generate configuration files, the use of open source kinematics, motion planning complete library of kinematics calculation and motion planning, and planning of the mechanical arm movement information released to joint server, PC through the Socket communication will further movement information sent to the robot controller and motion information analysis, the driving mechanical arm in accordance with the plan action execution fetching. While executing the planning action, the position sensor on the mechanical arm sends the actual pose of the mechanical arm to the upper computer through the socket.

5. The experimental example

The image pixel coordinates of the center of the upper surface of the object to be captured are obtained through feature extraction of the image of the object to be captured. The pixel coordinates of the center image of the object to be captured are converted into the coordinates under the coordinate system of the robot base by camera calibration and hand-eye calibration technology to complete the object positioning. Ros-Moveit! is used according to the starting point of the end of the robot arm and the coordinate position of the target.

The motion planner is used to obtain the trajectory parameters, which are sent to the robot controller through socket communication and analyzed, and the mechanical arm is driven to complete grasping according to the predetermined action.During the experiment, the object to be grasped was placed in different positions in the photographing area, and the grasping system could successfully complete the visually guided grasping experiment. Visual localization and motion position results is shown in Table1.

Table1 Visual localization and motion position results unit:m

Serial number	Visual location （x, y）	Motion position（x, y）
1	（-0.335, -0.483）	（-0.328, -0.480）
2	（-0.326, -0.451）	（-0.315, -0.446）

3	(-0.340, -0.455)	(-0.335, -0.461)
4	(-0.339, -0.457)	(-0.330, -0.451)
5	(-0.330, -0.465)	(-0.335, -0.470)

By analyzing the experimental data, it can be concluded that the measurement error of the target position grabbed by the experimental platform is less than 1cm, and the positioning accuracy of the cooperative robot for the target is different in different actuators and working scenes. The flexible gripper grasping mechanism adopted in this experimental platform can complete the grasping action with an error value with-in 1cm.The errors mainly come from camera calibration, hand-eye calibration and repeated positioning of the robot. In the process of camera calibration and hand-eye calibration, there are errors in measurement and installation position of calibration plate. At the same time, the horizontal plane of the object to be grabbed and the base plane of the robot will also affect the positioning accuracy, but the positioning accuracy can meet the application requirements of most intelligent grasping platforms.

6.Conclusion

This paper builds a set of intelligent grasping platform based on ROS system, and using color machine vision algorithm in ROS-Opencv to achieve rapid target recogni-tion. The coordinate transformation of the target is completed through camera calibration and hand-eye calibration, and the precise coordinates of the target in the coordinate system of the robot base are obtained. Using the ROS-Moveit! The trajectory planning of the robot from the initial position to the target position is completed to achieve the accurate capture of the target. This platform has certain portability, only a few configuration files need to be modified, which can be quickly transplanted to different types of ROS supported robots. This experimental platform can accurately locate and grasp the shape rule workpiece, which has certain application value in the actual work scene.

References

[1] Zhang chi, Liao huali. Intelligent grasping system for industrial robot based on monocular vision [J]. Journal of Mechanical & Electrical Engineering.2018,35(3):283-287.
[2] Xu bofan, zhaohuadong. Research on grasping system of FANUC robot based on visual guidance[J]. Modular Machine Tool & Automatic Manufacturing Technique,2018(7):111-114.
[3] ROS/technical overview. http://wiki.ros.org/.
[4] Liu qifan, Xie ming. Industrial robot motion planning based on ROS system[J]. Modular Machine Tool & Automatic Manufacturing Technique,2017(5):36-39.
[5] Lei shi.Vision Guiding System for 6-DoF Robotic Manipulator[D].Tallinn University of Technology,2017.
[6] Xu de, Tan min,Li yuan. Robot vision measurement and control[M]. Beijing: National Defence Industry Press. 2011.
[7] Wang tianqi, Chen ming. Research on target recognition and location based on monocular vision[J]. Mechanotronics mechanical-electrical integration,2015(11).
[8] Cheng yuli. Hand-eye calibration and object localization for industrial robotic application[D]. Zhejiang University,2016.
[9] ROS/opencv. https://opencv.org/tutorials
[10] Liu xiangde. Research on motion control of Baxter robot based on visual guidance[J]. Journal of Chongqing University of Posts and Telecommunications,2018,30(4):552-557.
[11] Research on robot grasping planning based on RGB-D vision recognition[D]. Harbin Institute of Technolog

IOP Conf. Series: Materials Science and Engineering **616** (2019) 012026 doi:10.1088/1757-899X/616/1/012026

Numerical Investigation of a New Single Vortex Generation Technique

S Hazel, Y Huang* and M Ait amirat

Collaborative Innovation Centre of Advanced Aero Engines, National Key Laboratory of Science and Technology on Aero Engines, School of Energy and Power Engineering, Beihang University, Beijing 100191, People's Republic of China

*Corresponding author e-mail: yhuang@buaa.edu.cn

Abstract. In this work a new single vortex generation method based on the flow past a rotating cylinder is proposed. Two dimensional numerical simulation for a laminar flow at Re = 100 is conducted to investigate this method. An impulsive drop of the rotation speed appeared to trigger a vortex roll up process; one single vortex is growing behind the cylinder and detached from it. The process is studied for different rotation drop rates and several final rotation rates. Vortex characteristics are computed and compared for different cases. Increasing rotation drop rate generates a more rapidly growing vortex. Its trajectory is closer to the geometric centreline and its size and intensity are higher. This new vortex generation method allows the investigation of new experimental configurations for studying vortex flame interaction.

1. Introduction

Flame vortex interaction plays an important role in understanding basic mechanisms in turbulent combustion and combustion instability. The simple configuration of one vortex or a pair of vortices interacting with a premixed or diffused flame helps understanding characteristics like flame structure, flame instabilities, mixing, ignition and extinction of more complex combustion configurations observed in practical applications such IC engines and turbine combustors.

Flame vortex interaction has been extensively studied in the past years [1]. Experimental studies have provided a considerable amount of data in a wide variety of geometries using several dynamic methods. For all these experimental studies either vortex rings [2] [3] [4], vortex dipoles, Van Karman vortex street [5] or jet flows [6] [7] were used. These configurations gave a very important advancement in understanding flame/vortex dynamics; however some limitations exist rising the need to more improvements. In fact, all those experiments were limited by flame stability phenomena, reproducibility of the interaction, optical access, and quality of the vortical structure. For the last point, whereas the ideal and simple vortical structure seems to be a single coherent large scale two-dimensional span wise vortex with a well defined size and intensity, none of the configurations presented reached that ideal pattern. Other generation techniques could be investigated to reach this ideal structure with less considerations and parasitic phenomena.

One idea proposed in this paper is to modify the Van Karman Street shedded from a cylinder to generate a single two-dimensional vortex. The modification of the Van Karman Street by rotating the cylinder and suddenly decrease its rotation speed seems to be a potential configuration to reach that goal.

Flow past rotating cylinder attracted the attention of scientists few centuries ago and it was extensively studied since then. The rotation rate α defined as the ratio of tangential cylinder velocity to the uniform flow velocity adds another control parameter to the flow, which, in the case of a Van Karman street, is only governed by Reynolds number. The first relevant works concerning flow past rotating cylinder were conducted principally by Prandtl [8], several authors pursued his path [9] [10 [11] [12] and agreed on some very attractive features of this flow: Flow topology is affected by the cylinder's rotation. The wake behind the cylinder is modified and deflected towards the rotation direction; it shrinks gradually with rotation rate been increased. The strengths of the upstream and downstream vortices are not equal and The Von Karman Street is suppressed for a critical value that depends on the Reynolds number. For low Reynolds numbers, not only one steady state exists but actually two, with a narrow unsteady region between them. This unsteady region is featured by a one side single vortex shedding with a remarkably smaller frequency [13] and [14]. The exact point of onset being dependent on Reynolds number: with increasing Re the second shedding mode appeared at progressively smaller values of α but will have a narrower region between the two steady states [15]. The two steady state regions present an interesting characteristic for our study. Because we are focusing on generating a single vortex, selected rotation rates must be inside these two regions. If not, the vortex shedding caused by the drop of the rotation speed (if any) will be accompanied by the vortex shedding of the unsteady state.

In this paper will numerically investigate the applicability of our idea to generate a single spanwise vortex in the laminar regime. The results will be the starting point of a set of experimental tests planned in the future.

2. Methodology

We will numerically study the effect of a sudden drop on the vortex generation process for a laminar flow. Unsteady two-dimensional laminar flow of a viscous incompressible fluid past a rotating cylinder is considered.

2.1. Governing equations

The flow is governed by the Navier-Stokes equations. The non-dimensional equations are given below:

$$\frac{\partial U}{\partial \tau} + (U.\nabla).U = -\nabla P + \frac{1}{Re}\nabla^2.V \qquad (1)$$

$$\nabla.U = 0 \qquad (2)$$

Where U is the non-dimensional flow velocity vector in the Cartesian coordinate system. τ is the non-dimensional time defined as $\tau = tu_\infty/D$, P is the non-dimensional static pressure and Re is the Reynolds number of the flow given by $Re = Du_\infty/\gamma$ where D is the diameter of the cylinder, u_∞ is the free stream inlet velocity and γ is the kinematic viscosity of the fluid. Non-dimensional rotational rate α is defined as $\alpha = \omega D/u_\infty$, where ω is the angular velocity of the rotating cylinder.

The commercial computational fluid dynamics (CFD) code ANSYS Fluent (version 15.0) is used to simulate the flow. A control-volume method is applied by the software to integrate the equations of motion. A second order upwind scheme is used to discretize the convective term in the momentum equations. SIMPLEC (Semi-implicit method for pressure-linked equations Consistent algorithm is selected as the pressure-velocity coupling scheme. Finally, the time integration of the unsteady momentum equations is performed using a second-order approximation. The simulation is run with the initial rotation rate until a steady state is reached. The rotation rate is then impulsively reduced and the flow behavior is observed. To monitor the simulation evolution, Lift and drag coefficients along with the residuals are observed. A convergence criterion of 10^{-6} is used for continuity and x-y components of momentum equations. The Strouhal number is a non-dimensional parameter computed using the vortex shedding frequency f_s obtained by fast Fourier transform (FFT) of the lift fluctuating-time history. These parameters are used for validation and can give indications of flow characteristics including steadiness and vortex shedding. They are defined, respectively, as follows:

$$C_D = \frac{2\,F_D}{\rho u_\infty^2 D} \tag{3}$$

$$C_L = \frac{2\,F_L}{\rho u_\infty^2 D} \tag{4}$$

$$S_t = \frac{f_s D}{u_\infty} \tag{5}$$

where ρ is the fluid density; F_D and F_L are the total drag force and total lift force, respectively. $\overline{C_D}$ and $C_L{}'$ will be used for results validation, they represent the mean drag coefficient and the r.m.s. lift coefficient respectively.

2.2. *Computational domain and boundary conditions*

The geometry to be considered is a cylinder with diameter D submerged in a uniform fluid flow. The cylinder rotates about its axis with angular velocity ω. The computational domain used for the simulations is set at 24D ×36D in x, y directions of a fixed Cartesian coordinate system. The center of the cylinder is located at a distance of 12D from the inlet boundary; 24D from the outlet boundary and an equidistance of 12D from top and bottom boundaries.

Uniform velocity is prescribed at the inlet; while at the outlet, Neumann type boundary condition is imposed for velocities. Pressure boundary condition is given as atmospheric at the outlet. The end wall effects are neglected in present work, no-shear boundary condition that corresponds to slip at top and bottom walls is prescribed. The cylinder rotates in the counter-clockwise direction and a no-slip boundary condition with rotating speed is imposed at the cylinder surface.

2.3. *Grid independency and validation*

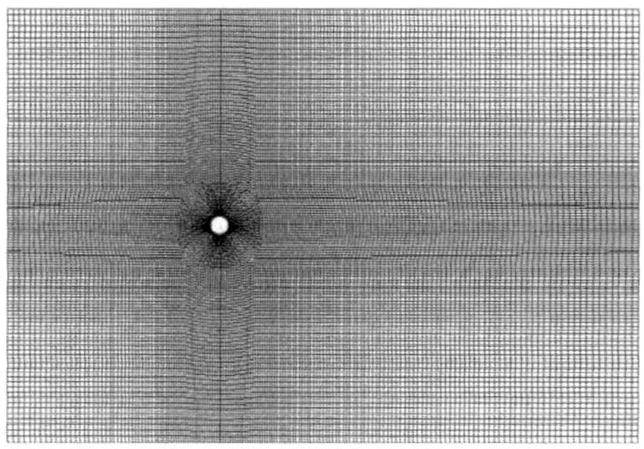

Figure 1. Computational domain and mesh distribution

Figure 1 shows the detailed diagram of the grid system. An O-type grid structure with a non-uniform distribution was used; a finer grid was generated near the cylinder and gradually became coarser in the wake and the far field. The grid was clustered near the cylinder with an expansion rate of 1.2 in the radial direction from the cylinder surface. The distance between the first grid and the cylinder surface was 0.01D. The dimensionless time steps $\tau = \frac{tu_\infty}{D} = 0.0125$ and 0.025 were used.

Table 1. Comparison with published results for circular cylinder

		S_t	$\overline{C_D}$	$C_L{}'$
Williamson (1996) [16]	Exp	0.164		
Park, Kwon and Choi (1998) [17]	Num	0.165	1.33	0.23
Sharman et al. (2005) [18]	Num	0.164	1.33	0.23
Present study	Num	0.168	1.36	0.238

Two dimensional flow around a fixed circular cylinder at Re=100 was computed for mesh independency and validation. With different grid numbers, grid independence tests were satisfied if the variation in the values of $S_t, \overline{C_D}$ and $C_L{}'$were very small. Moreover, the two different time steps had negligible effect on the values above. As shown in table 1, the values of $S_t, \overline{C_D}$ and $C_L{}'$of the present simulations case are all in good agreement with other published results of a circular cylinder. Based on these analyses, the mesh model with cell numbers of 22800 was adopted for capturing detailed vortex structures and concurrently saving the computational time. The zone close to the cylinder surface was meshed with 120 grids (Nc =120) uniformly distributed along the circumferential direction.

2.4. Vortex identification and tracking

To objectively and accurately determine the characteristics and dynamics of the vortical structures generated by the proposed method, a vortex core Σ is identified Based on the Q-criterion [19]. To complete this description, specific parameters such as size, core position and intensity evolution are used. These relevant parameters are computed as follows [20]:

$$\Gamma = \iint_{\Sigma} \omega \, dx \, dy \tag{6}$$

Where ω refers to the vorticity field distribution in the core and Γ is the vortex circulation.

The position (x_c, y_c) of the vortex centroid is then given by:

$$x_c = \frac{1}{\Gamma} \iint_{\Sigma} x\omega \, dx \, dy \tag{7}$$

$$y_c = \frac{1}{\Gamma} \iint_{\Sigma} y\omega \, dx \, dy \tag{8}$$

Size and intensity of a vortex are difficult to define as a qualitative measurement. For a meaningful comparison, a vortex can be visualized as a rotating disc with a definite radius. Two quantities can be defined using this approximation: equivalent radius R and tangential velocity V_θ.

R is defined based on polar moments of vorticity:

$$R = \left[\frac{1}{2\Gamma} \iint_{\Sigma} [(x - x_c)^2 + (y - y_c)^2]\omega \, dx \, dy\right]^{1/2} \tag{9}$$

V_θ is the tangential velocity at the periphery of the vortex core defined as:

$$V_\theta = \frac{\Gamma}{2\pi R} \tag{10}$$

All parameters previously defined will be normalized with respect to D and u_∞. Normalization will be indicated by superscript *.

3. Results and discussion:

3.1. Qualitative analysis

A qualitative description of the flow topology is conducted using instantaneous normalized vorticity field visualizations for Re=100 and an anticlockwise rotation rate drop from $\alpha_1 = 6$ to $\alpha_2 = 3$. Some selected instants are presented in figure 2.

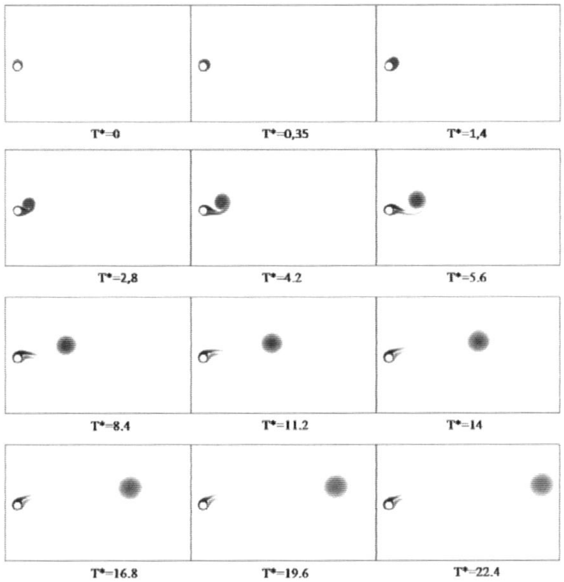

Figure 2. Normalized vorticity time evolution

The main observation is the generation of a single anticlockwise vortex from the cylinder. Once the rotation rate is reduced, positive and negative vorticity field growth is observed. The positive vorticity is gradually growing untill the vortex is generated, shedded from the cylinder and advected away with the flow stream. No other vortices are observed; the flow is gradually stabilized and finally reaches the first steady state.

A vortex is rolling up and growing in size and strength as long as there is a circulation flux into it. This flux was possible because of the impulsive drop of the cylinder rotation speed. The initial high vorticity caused by the first rotation rate is now released in a lower speed region instead of being flushed by the highly rotated cylinder. The upper opposite circulation is also increasing and pushing away the generated vortex until it is totally shedded from the cylinder.

3.2. Quantitave analysis

In the previous section, the possibility of generating a single vortex by suddenly dropping the rotation speed was put in evidence. More quantitative analysis is now needed to understand the effect of reducing the rotation speed on the generating process and the vortex characteristics. To get the quantitative description of the generated vortex, equations (3) to (7) are used.

A non dimensional parameter called the rotation drop rate is defined us:

$$\Delta\alpha = \frac{\alpha_1 - \alpha_2}{\alpha_1} \tag{11}$$

Starting from an initial rotation rate of $\alpha_1 = 6$, rotation speed is impulsively decreased to five different final values with a rotation drop rate of $\Delta\alpha_1 = 0.4$, $\Delta\alpha_2 = 0.5$, $\Delta\alpha_3 = 0.6$, $\Delta\alpha_4 = 0.7$ and $\Delta\alpha_5 = 0.8$ respectively. Results for different vortex tracking parameters are presented in figure 3.

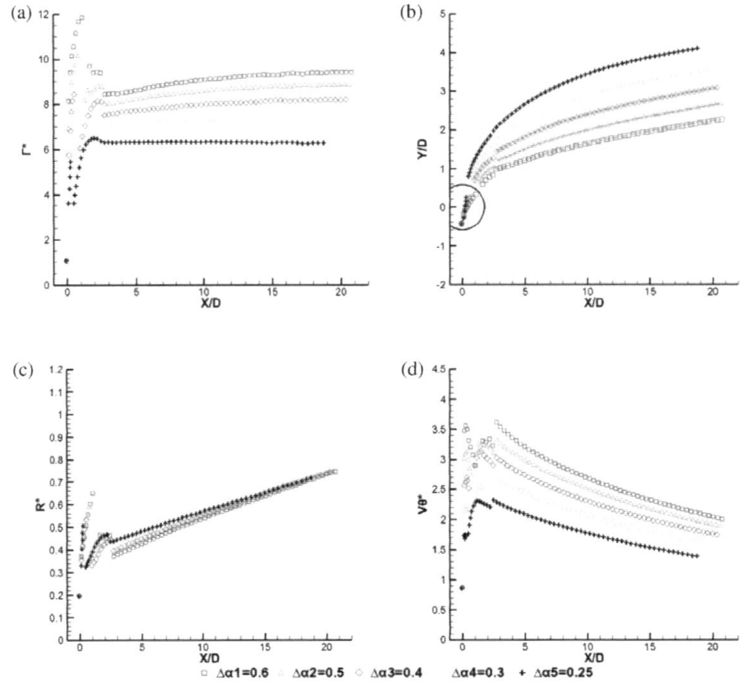

Figure 3. Vortex identification and tracking, (a) normalized circulation, (b) vortex trajectory, (c) vortex normalized radius, (d) vortex normalized tangential velocity

In figure 3(a), three regions separated by two spatio-temporal discontinuities are identified. These regions are respectively characterizing the vortices's formation, shedding and far-wake advection. The first region reflects the intensity of the circulation feeding process allowing the formation of the vortex. For the cases $\Delta\alpha_1$ and $\Delta\alpha_2$, the vortex shedding process is well defined; the circulation is observed to increase in this region. The size of the vortex is also increased causing the tangential velocity to drop. However, for other cases, a strange behavior is observed. Although the discontinuity reflects a vortex separation, the circulation is kept increasing meaning that the feeding process is still present. The circulation is appeared to be underestimated by the identification method presented in section 2. This behavior is due to the limitation of the velocity gradiant based methods to identify the vortex region. These methods tend to eliminate patches from the vortex core when a high radial or axial stretching is present. The reader is referred to [21], and [22] for more details.

After the first discontinuity, the vortex is pinched off and started to move away from the cylinder. The second discontinuity is distinguished by a vortex-tail breakup. The vortex tail is elongated, pronouncing stretching and finally separating from the vortex core. This leads to a sudden drop of the vortex radius and an increase of the vortex core tangential velocity. The process is better observed for higher rotation drop rates and seems to totally disappear for the lowest drop rate case. The normalized radius is observed to converge to almost identical values but the vortex intensity has higher values for larger drop rates.

The vortex trajectory is also affected by the rotation drop rate. More the drop rate is significant, less the vortex core is deviated from the cylinder centerline.

4. Conclusion

In this work we numerically evaluate the possibility of generating single two dimensional vortex using flow past a rotating cylinder. The impulsive drop of the rotating speed is accompanied with a single vortex shedding. A vortex identification approach based on the Q criterion method was used to track this vortex. This method appeared to ignore some vortex regions for cases where a strong radial stretch

IOP Conf. Series: Materials Science and Engineering **616** (2019) 012026 doi:10.1088/1757-899X/616/1/012026

was present. In general, the vortex generation process is divided into three distinct regions separated by two discontinuities. The discontinuities reflect both the vortex shedding and the vortex-tail breakup processes. For the same initial rotation speed, the vortex is stronger for larger rotation drop rates. It is also shedded closer to the cylinder with a trajectory more adjacent to the cylinder centerline. These results give an interesting opportunity for new configurations to study the flame vortex interaction that could be experimentally tested in the future.

References

[1] Renard P H, Thévenin D, Rolon J C, Candel S 2000 *Prog. Energy Combust. Sci* **26** 225–282.

[2] Ishizuka S, Yamashita T and Shimokuri D *Proc. Combust. Inst.* **34** 745–753.

[3] Park J and Shin HD 1997 *Combust.Flame* **110** 67–77.

[4] Rolon JC, Aguerre F and Candel S 1995 *Combust. Flame.* **100** 422–9.

[5] Lee J G, Lee T W, Nye D A and Santavicca D A 1995 *Combust. Flame.* **100** 161–168

[6] Kim M , Choi Y, Oh J and Yoon Y 2009 *Combust. Flame.* ***156*** 2252–2263.

[7] Ahn K and Yu K H 2012 *Combust. Flame.* **159** 686–696.

[8] Prandtl L 1925 *Die Naturwissenschaft.* **13** 93–108

[9] Badr H M, Coutanceau M, Dennis S C R and Ménard C. 1990 *J. Fluid. Mech.* **220** 459–484

[10] Kang S, Choi H and Lee S 1999 *Phys.Fluids.* **11** 3312–3321

[11] Lam K M 2009 *J.Fluids. Struct.* **25** 245–262

[12] Karabelas S J 2010 *Int. J. Heat. Fluid. Flow* **31** 518–527

[13] Stojkovic D, Schön P, Breuer M and Durst F 2003 *Phys. Fluids.* **8** 1683–1685

[14] Mittal S and Kumar B 2003 *J. Fluid. Mech.* **476** 303– 334

[15] El Akoury R, Martinat G, Braza M, Perrin R, Hoarau Y, Harran G, and Ruiz D 2009 *IUTAM Symp.* Unsteady Separated Flows their Control **14** 187–200

[16] Williamson C H K 1996 *Annu. Rev. Fluid Mech.* **28** 477-539

[17] Park J, Kwon K and Choi H 1998 *KSME Intl. J.* 12, 1200–1205.

[18] Sharman B, Lien F S, Davidson L and Norberg C 2005 *Intl J. Numer. Meth. Fluids* **47** 423–447.

[19] Jeong J, & Hussain F 1995 On the identification of a vortex *J. Fluid. Mech.* **285** 69-94.

[20] Jardin T, Bury Y 2012. *J. Fluid. Mech.* **696** 285–300.

[21] Kolář V 2007 *Int. J. Heat. Fluid. Flow.* **28** 638-652

[22] Wu J Z, Xiong A K, and Yang Y T 2005 *Phys. Fluids.* **17** 038108

IOP Conf. Series: Materials Science and Engineering **616** (2019) 012027 doi:10.1088/1757-899X/616/1/012027

Comparative Study on Two Solutions of 3-RPS Parallel Mechanism Pose

Jin-yi ZHU[a], Chun-yan ZHANG[b] and Yu WAN[c]
Shanghai University of Engineering Science .Shanghai P.R. China

[a] 2549541565@qq.com; [b] cyzhang@sues.edu.cn; [c] 976809344@qq.com

Abstract. In this paper, the performance of the different methods for solving the pose of parallel mechanism is different in the amount and precision of calculation. Taking typical 3-RPS parallel mechanism as the analysis object, kinematics models of the mechanism were established with quaternion and Rodriguez parameter method respectively, and the bar length of the maximum reachable space was solved by genetic algorithm, so as to compare the advantages and disadvantages of the solution of the model. Finally, numerical simulation is used to compare the amplitude and fluctuation degree of the curve according to the smooth spline interpolation in both calculation and precision. The results show that the Rodriguez parameter method is superior to the quaternion method in the accuracy of pose resolution of 3-RPS parallel mechanism, while the quaternion method is superior in the real-time performance of the solution. This result provides a theoretical basis for the pose control of parallel mechanism.

1. Introduction

Parallel mechanism is widely used in industrial fields due to its high rigidity, strong bearing capacity and lack of input error, and has become a hotspot in robotics and mechanism research [1]. The pose description method of the parallel mechanism dynamic platform is of great significance for the kinematics research of the parallel mechanism. The commonly used pose description methods include Rodriguez parameter method [2], Euler angle method [3], quaternion method [4], etc. Over the years, many scholars have used these methods to conduct a large number of researches and have achieved fruitful results. Wu Yuyao uses the Rodriguez parameter method to establish a spacecraft model and perform pose control so that the closed-loop system can complete the "stationary to stationary" attitude maneuver [5]. Tian Xin, Ge Xinsheng et al. used the Rodriguez parameter method to estimate the pose dynamics of the 3D rigid body pendulum and constructed the Lyapunov function of the system [6]. Wang Shun used the quaternion method to interpolate the pose space and solved the continuous multi-position interpolation problem of the angular velocity of the 6-3 platform parallel machine tool [7]. Yukio- Takeda, Daisuke Matsuura et al. used the Euler angle to analyze the kinematics of the 3-RPSR parallel mechanism, clarified the relationship between design parameters and orientation capability, and designed a parallel mechanism with high orientation capability [8]. However, Euler's law has singularity in mathematics [9], so this paper only describes the pose of Rodriguez parameter method and quaternion method.

What differences exist between these two kinds of pose descriptions in attitude description of parallel mechanism and what differences exist in their solutions? Based on this, this paper uses the Rodriguez parameter method and the quaternion method to analyze the difference of the position of the 3-RPS parallel mechanism by using the typical 3-DOS parallel mechanism with less degrees of freedom.

2. Organizational structure description

Figure 1 shows the structure of the 3-RPS parallel mechanism. The moving platforms are all equilateral triangles, which are hinged by three spatially symmetric RPS branches. The lengths of the regular triangles on the dynamic platform are defined as Lm and L respectively, and the introduction of the scale factor μ is $L=\mu L_m$. The structural parameters of the moving platform can be determined by Lm. The coordinate system $\{O_1\}$ and $\{O_2\}$ of the moving platform are established, and the coordinate parameters under the two coordinate systems of each apex of the mechanism are shown in Table 1. In this organization, the three-branched mobile pair is the active pair and is described and adopted as $L_{A_1A_2}, L_{B_1B_2}, L_{C_1C_2}$.

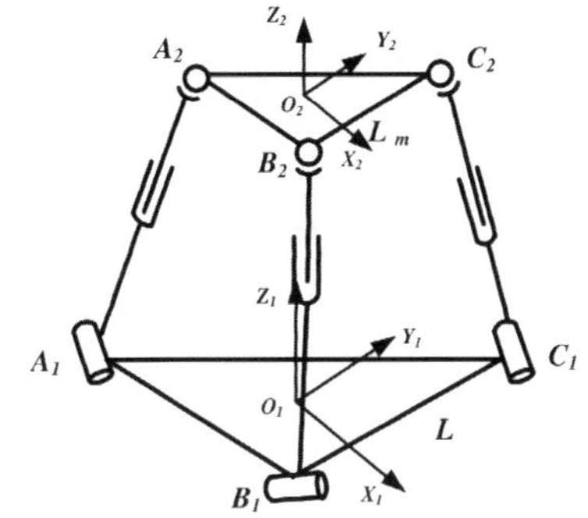

Figure 1. 3-RPS structure diagram

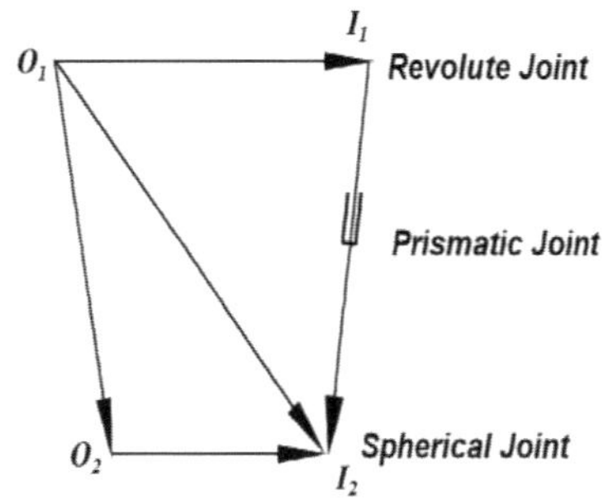

Figure 2. branch vector

Table 1. Coordinate parameters of fixed and fixed platforms

	$A_{1,2}$	$B_{1,2}$	$C_{1,2}$
Relative to $\{O_1\}$	$(0,\frac{\sqrt{3}}{3}\mu L_m,0)$	$(\frac{\mu L_m}{2},-\frac{\sqrt{3}}{6}\mu L_m,0)$	$(-\frac{\mu L_m}{2},-\frac{\sqrt{3}}{6}\mu L_m,0)$
Relative to $\{O_2\}$	$(O,\frac{L_m}{\sqrt{3}},O)$	$(\frac{L_m}{2},-\frac{L_m}{2\sqrt{3}},0)$	$(-\frac{L_m}{2},-\frac{L_m}{2\sqrt{3}},0)$

3. Institutional pose inverse model Rodriguez parameter method modeling

3.1 Basic definition of Rodriguez parameter method

The rigid body having a fixed point O is rotated about the unit vector n. There is a certain point vector r on the rigid body, and the angle θ is rotated around the n-axis to the position of r'. In 1840, the French mathematician Rodriguez proposed [10][11], introducing parameters:

$$\xi = \tan\frac{\theta}{2}n \tag{1}$$

The moving platform coordinate system $Ox_2y_2z_2$ has a reference point position vector $\vec{P}$, $Ox_2y_2z_2$ coincides with the ground coordinate system $Ox_1y_1z_1$, and the component of $\vec{P}$ on $Ox_2y_2z_2$ is denoted as $r = [x_2\ y_2\ z_2]^T$, when $Ox_2y_2z_2$ is equivalent to Ox1y1z1 rotation When it reaches a new position, the component of $\vec{P}$ is unchanged in $Ox_2y_2z_2$, and its component on $Ox_1y_1z_1$ is $r' = [x_1\ y_1\ z_1]^T$, and the triple-depletion formula is applied according to formula [12] can be re-displayed as:

$$r' = r + \frac{2}{1+\emptyset^2}[\xi \times r + \xi \times (\xi \times r)] = Ar \tag{2}$$

The transformation matrix A can be represented by the Rodriguez parameter:

$$A = \frac{1}{1+\xi_1^2+\xi_2^2+\xi_3^2}\begin{bmatrix} 1+\xi_1^2-\xi_2^2-\xi_3^2 & 2(\xi_1\xi_2-\xi_3) & 2(\xi_1\xi_3+\xi_2) \\ 2(\xi_1\xi_2+\xi_3) & 1-\xi_1^2+\xi_2^2-\xi_3^2 & 2(\xi_1\xi_2+\xi_3) \\ 2(\xi_1\xi_2+\xi_3) & 2(\xi_2\xi_3+\xi_1) & 1-\xi_1^2-\xi_2^2+\xi_3^2 \end{bmatrix} \tag{3}$$

IOP Conf. Series: Materials Science and Engineering **616** (2019) 012027 doi:10.1088/1757-899X/616/1/012027

3.2 Rodriguez parameter inverse solution model

With the Rodriguez parameter transformation matrix A1 and the slave branch vector Figure 2, the vector of the dynamic coordinate system is transformed into a fixed coordinate system by coordinate transformation, and the coordinate parameters of each point in Table 1 are used to obtain three driver rod length vectors at this time. $L_i(i=1,2,3)$ can be expressed in a fixed coordinate system as:

$$L_i = b_i - B_i \quad i = 1,2,3 \tag{4}$$

Solve to get three rod length expressions:

$$L^2_{A_1A_2} = (\frac{\sqrt{3}L_m}{3\lambda_0^2}(1-2\xi_1^2+2\xi_2^2) - \frac{\sqrt{3}}{3}\mu L_m)^2 + (\frac{2\sqrt{3}L_m\xi_1}{3\lambda_0^2} + z_2)^2 \tag{5}$$

$$L^2_{B_1B_2} = (\frac{L_m}{\lambda_0^2}(\frac{(1+\xi_1^2-\xi_2^2)}{2} - \sqrt{3}\xi_1\xi_2) - \frac{\mu L_m}{2})^2 + (\frac{L_m}{\lambda_0^2}(\xi_1\xi_2\frac{\sqrt{3}(1+\xi_1^2-\xi_2^2)}{6}) + \frac{\sqrt{3}}{6}\mu L_m)^2 + (\frac{L_m}{\lambda_0^2}(-\xi_2 - \frac{\xi_1}{\sqrt{3}}) + z_2)^2 \tag{6}$$

$$L^2_{C_1C_2} = (\frac{L_m}{\lambda_0^2}(-\frac{(1+\xi_1^2-\xi_2^2)}{2} - \sqrt{3}\xi_1\xi_2) + \frac{\mu L_m}{2})^2 + (\frac{\mu L_m}{\lambda_0^2}(-\xi_1\xi_2\frac{\sqrt{3}(1+\xi_1^2-\xi_2^2)}{6}) + \frac{\sqrt{3}}{6}\mu L_m)^2 + (\frac{L_m}{\lambda_0^2}(\xi_2 - \frac{\xi_1}{\sqrt{3}}) + z_2)^2 \tag{7}$$

among them: $\lambda_0^2 = 1+\xi_1^2+\xi_2^2+\xi_3^2$

The above three equations are the inverse pose model of the kinematic Rodriguez parameter of the 3-RPS symmetric parallel mechanism.

3.3 Quaternion Modeling

$Q = \frac{\cos\theta}{2} + usin\frac{\theta}{2} = [q_0\ q_1\ q_2\ q_3]^T = \left[\frac{\cos\theta}{2}\ lsin\frac{\theta}{2}\ msin\frac{\theta}{2}\ nsin\frac{\theta}{2}\right]^T$ describes the direction of the equivalent rotation of the rigid body and the angle of the rotation. u is a unit vector of the rotation direction, θ is a rotation angle, and l, m, and n are direction cosines of u. Thus, the quaternion $q_0^2 + q_1^2 + q_2^2 + q_3^2 = 1$ constitutes a unit four-dimensional supersphere[13].

By this principle, the pose transformation matrix of the moving platform can be expressed as B:

$$B = \begin{bmatrix} q_0^2+q_1^2-q_2^2-q_3^2 & 2q_1q_2-2q_0q_3 & 2q_1q_3+2q_0q_2 \\ 2q_1q_2+2q_0q_3 & q_0^2-q_1^2+q_2^2-q_3^2 & 2q_2q_3-2q_0q_1 \\ 2q_1q_3-2q_0q_2 & 2q_2q_3+2q_0q_1 & q_0^2-q_1^2-q_2^2+q_3^2 \end{bmatrix} \tag{8}$$

Combining the structural size parameters of Table 1 with Equation 3, the closed-vector method is used to obtain the quaternion length expression of the mechanism[14]:

$$L^2_{A_1A_2} = (\frac{\sqrt{3}L_m}{3}(q_0^2-2q_1^2+2q_2^2) - \frac{\sqrt{3}}{3}\mu L_m)^2 + (\frac{2\sqrt{3}L_mq_1}{3} + z_2)^2 \tag{9}$$

$$L^2_{B_1B_2} = (L_m(\frac{(q_0^2+q_1^2-q_2^2)}{2} - \sqrt{3}q_1q_2) - \sqrt{3}q_1q_2) - \frac{\mu L_m}{2})^2 + (L_m(q_1q_2\frac{\sqrt{3}(q_0^2+q_1^2-q_2^2)}{6}) + \frac{\sqrt{3}}{6}\mu L_m)^2 + (L_m(-q_0q_2 - \frac{q_0q_1}{\sqrt{3}}) + z_2)^2 \tag{10}$$

$$L^2_{C_1C_2} = (L_m(-\frac{(q_0^2+q_1^2-q_2^2)}{2} - \sqrt{3}q_1q_2) + \frac{\mu L_m}{2})^2 + (\mu L_m(-q_1q_2\frac{\sqrt{3}(q_0^2+q_1^2-q_2^2)}{6}) + \frac{\sqrt{3}}{6}\mu L_m)^2 + (L_m(q_0q_2 - \frac{q_0q_1}{\sqrt{3}}) + z_2)^2 \tag{11}$$

The above three equations are the inverse pose model of the kinematics represented by the kinematics of the 3-RPS symmetric parallel mechanism.

4. Model comparison and numerical simulation

4.1 Model Comparison

The initial parameters of the 3-RPS parallel mechanism are defined as follows: the distance between the three connection points A2, B2 and C2 of the moving platform is $L_m = 50\sqrt{3}mm$, and the μ is 1.5, that is, the distance $L_B = 75\sqrt{3}mm$ between the three connection points A1, B1 and C1 on the fixed platform Since the 3-RPS mechanism has three output parameters, the moving platform is defined to have an angle of rotation of -90° to 90° around the X-axis, a rotation angle of -90° to 90° around the Y-axis, and a displacement distance of 110 mm to 150 mm in the Z-axis direction.

Solving the maximum reachable space of the 3-RPS mechanism according to the set parameters can be equivalent to the optimal solution of the inverse length of the model. The genetic algorithm is used to solve the problem. The number of iterations is different due to the difference of the mathematical model. As shown in Fig. 3, the model solved by the quaternion method needs to be iterated 85 times to achieve stability and the fitness value is 18, and the result is 174.6652 mm, 104.3462 mm, and 153.7322 mm. The iteration required by the parametric model reached 131 times, and the fitness value was 20, the result was 174.6653mm, 104.3462mm, and 153.7321mm. In the case that the results of the two methods are similar, the model established by the quaternion method is better than the model established by the Rodriguez parameter, and the quaternion method is more efficient.

4.2 Numerical simulation

By setting the angle parameter of the output Y-axis to 90°, the distance moved on the Z-axis is 130mm, and the X-axis is rotated for one week. The two models solve the required calculation of the rod length (see Table 2) and efficiency. Since the result of a single solution operation is too small to show that it is not obvious enough, the solution is obtained multiple times and obtained by linear interpolation. As shown in the Fig.4, the quaternion method requires less time to solve, and the result is also in line with the calculation of the algorithm.

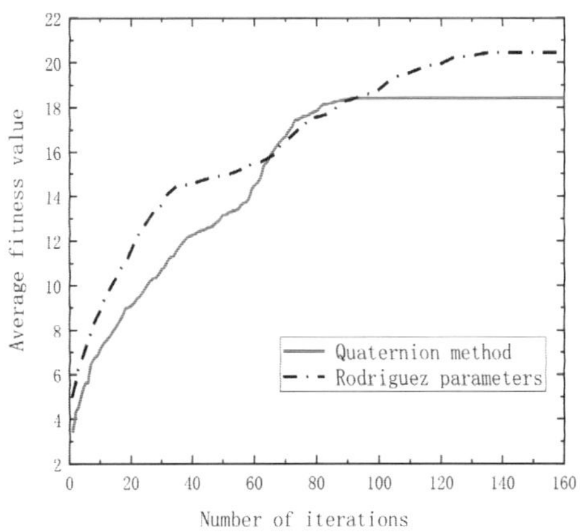

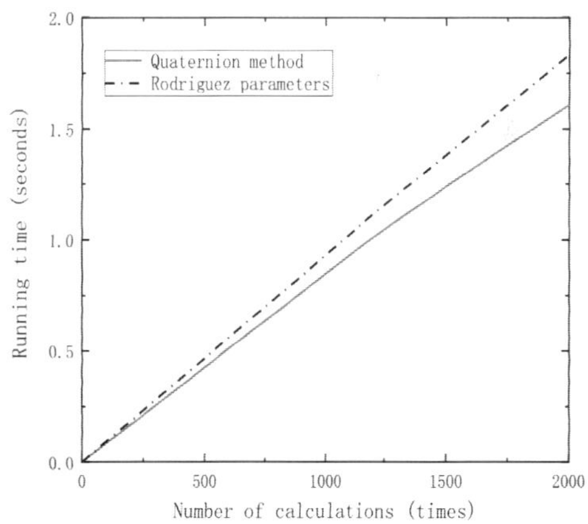

Figure 3. Two model convergence diagram **Figure 4.** Two ways to calculate time

Table 2. Comparison of calculations of two kinds of attitude algorithms

Attitude algorithm	Multiplication and division operation times/times	Addition and subtraction operation times/times	Number of execution operations/times

Quaternion method	40	30	23
Rodriguez parameter method	55	40	26

The rod length parameter calculated by Euler angle is the true value and then the error can be expressed as:

$$v_i = L_o - L_i \quad i= (1,2,3) \tag{12}$$

Where L_o is the true value and L_i is the calculated value.

Table 3. Partial length of the Rodriguez parameter solution

	1	2	3	…	138	139	140
$L_{A_1A_2}$	119.75527 7673313	107.07241 4911061	93.826457 1360416	…	89.434939 0558068	102.68271 9428576	115.62979 2495938
$L_{B_1B_2}$	132.84905 6736857	125.93441 0356644	119.20385 7948781	…	95.152832 0357940	93.458492 6866975	91.482238 1609401
$L_{C_1C_2}$	192.68741 2098202	194.66050 9839215	195.69223 8704550	…	143.16581 0254032	138.09276 5224305	134.15623 5816980

Table 4. Partial length of the quaternion solution

	1	2	3	…	138	139	140
$L_{A_1A_2}$	119.76234 2421414	107.07082 3412327	93.832142 9533476	…	89.421736 782165	102.68431 4124000	115.63123 2187432
$L_{B_1B_2}$	132.85236 7234637	125.92792 3128745	119.21348 7071625	…	95.162873 4124647	93.447234 8985432	91.502138 5726234
$L_{C_1C_2}$	192.69245 6521947	194.67992 3214789	195.70137 6471367	…	143.17326 9178400	138.10312 6546812	134.15482 1784986

According to the formula 12 and combined with Table 3, the data of Table 4 can obtain the error precision of the calculation results of the two methods. In combination with the statistical principle function, there are different nonlinearities in different places, or there are multiple extreme points, then the polynomial, especially the low-order polynomial, is used to complete the fitting. It is very inappropriate, so this article uses a smooth spline to fit the curve. The smooth spline is based on the $\sum_{i=1}^{N}\{y_i - f(x_i)\}^2$ by adding a $\lambda \int\{f''(t)\}^2 dt$, introducing the λ smooth factor formula mean square error[15]:

$$\text{RSS}(f,\lambda) = \sum_{i=1}^{N}\{y_i - f(x_i)\}^2 + \lambda \int\{f''(t)\}^2 dt \tag{13}$$

Where $f(x) = \sum_{j=1}^{N} N_j(x)\theta_j$, for the above formula, in the form of a matrix:

$$\text{RSS}(\theta,\lambda) = (y - N\theta)^T(y - N\theta) + \lambda\theta^T\Omega_N\theta \tag{14}$$

Then the idea of θ is equal to 0, that is, the least squares is:

$$\hat{\theta} = (N^TN + \lambda\Omega_N)^{-1}N^Ty \tag{15}$$

Bringing the θ we get into the original fitting function gives:

$$\hat{f}(x) = \sum_{j=1}^{N} N_j(x)\hat{\theta}_j \tag{16}$$

Figure 5, 6, 7 respectively after fitting of each rod long error curve, according to the method as shown in figure obtained by error and error quaternion method compare the fluctuation amplitude is small, and in the X axis rotate - over the Angle of 40 ° to 40 °, as a result of the 3 RPS parallel mechanism makes its own structure characteristics of analysis method is far less than the length of error quaternions is obtained. As a whole, the results obtained by parametric method are more stable than those obtained by quaternion method.

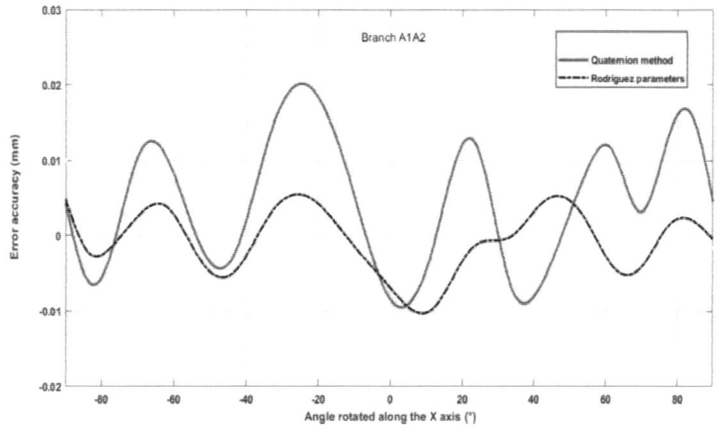

Figure 5. Branch chain A_1A_2 error accuracy

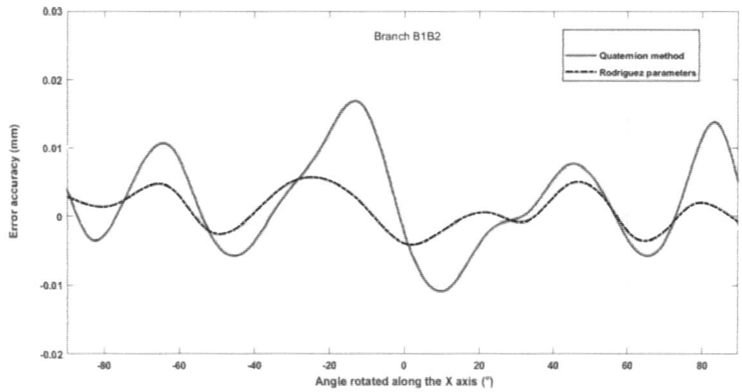

Figure 6. Branch chain B_1B_2 error accuracy

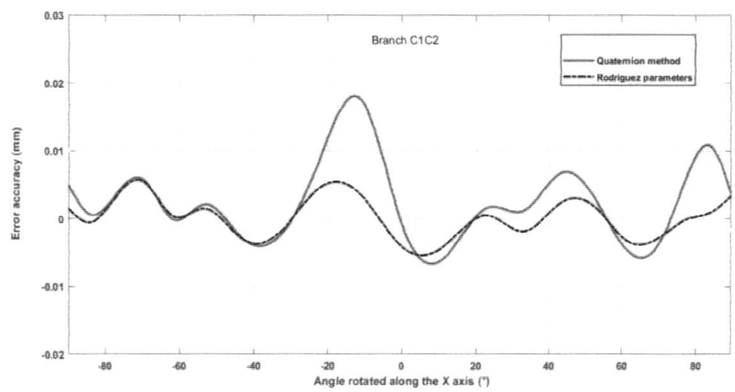

Figure 7. Branch chain C_1C_2 error accuracy

5. Conclusion

By analyzing the quaternion method and Rodriguez parameters described in 3-RPS parallel mechanism gesture differences, analyses the kinematics model established by the two methods in calculating, the results show that the quaternion method in terms of the amount of calculation and efficiency is superior to Rodriguez parameter method, but Rodriguez parameter method in computing precision and computing stability is better than the quaternion method.

For 3-RPS parallel mechanism, it is better to adopt Rodriguez parameter for attitude control if the range of rotation Angle of end-effector is small, and it is better to use quaternion method for control if the 3-RPS parallel mechanism is applied in the situation of high real-time performance. The results provide reference for attitude control and trajectory planning of 3-RPS parallel mechanism.

References

[1] Yang Gaowei, Zhang Jianjun, Yin Ling, Li Weimin. Singularity Analysis of a 3-RPUU Parallel Mechanism [J]. Mechanical Science and Technology for Aerospace Engineering, 2018, 37 (04): 510-518.

[2] CHEN Ji-zheng, YUAN Jian-ping, FANG Qun.A pose estimation algorithm based on Rodrigues parameters[J]. Acta Aeronautica Sinica,2008(04):960-965

[3] URIZAR M, PETUYA V, AMEZUA E ,et al. Characterizing the configuration space of the 3-SPS-S spatial orientation parallel manipulator[J]. Meccanica ,2014, 49(5): 1101-1114.

[4] Kong X. Reconfiguration analysis of a 3-DOF parallel mechanism using Euler parameter quaternions and algebraic geometry method [J]. Mechanism and Machine Theory,2014,74: 188-201.

[5] Wu Yuyao. Attitude control Based on Modified Rodrigues Parameters for Flexible Spacecrafts [D]. Harbin Institute of Technology, 2018.

[6] Tian Xin, Ge Xinsheng. Three-dimensional rigid body attitude control based on modified Rodriguez parameters[J].Mechanics and Engineering,2015,37(03):361-366.

[7] Wang Shunli. Research on 6-3 platform parallel motion machine tool [D]. Nanjing University of Aeronautics and Astronautics, 2010.

[8] Takeda Y., Inada S., Matsuura D., Hirose K., Ichiryu K. Orientation Capability of a 3-RPSR Parallel Mechanism for a Movable-Die Drive Mechanism of Pipe Bender. In: Lenarcic J., Husty M. (eds) Latest Advances in Robot Kinematics. Springer, Dordrecht.2012

[9] Cheng Gang. Study on Structure Design and Dynamic Performance of the Parallel Bionic Robot Leg [D]. China University of Mining and Technology, 2008.

[10] Rodrigues O. Des lois gé ométriques qui régissent les déplacements d 'un système solid dans l'espace, et de la variation des coordonnées provenant de ses déplacements consideéré s indépendamment des causes qui peuvent les produire [J] .Journal des Mathématiques Pures et Appliquées , 1840, 5: 380-440

[11] Cayley A. On the Motion of Rotation of a Solid Body[J]. Cambridge Mathematic Journal, 1843, 3:224-232

[12] Zhou Jianghua, Miao Yuhong, Wang Minghai. Rodrigues Parameter Description of Attitude Motion[J]. Journal of Astronautics,2004(05):514-519.

[13] Wang Yongjun, Qin Yongyuan, Yang Bo. Comparative Study of Quaternion and Rodrigues Parameters in Satellite Attitude Solution[J]. Chinese Space Science and Technology, 2007(03): 18-23.

[14] Huang Zhen, Zhao Yongsheng, Zhao Tieshi. Higher Space Institutions [M]. Higher Education Press. 2014

[15] Yuedong Wang. Smoothing Splines: Methods and Applications [M]. Chapman and Hall/CRC, 2011

IOP Conf. Series: Materials Science and Engineering **616** (2019) 012028 doi:10.1088/1757-899X/616/1/012028

The Multi-Varieties Small-Batch Gas Monitor Flexible Digital Job Shop Design and Integration Technology Research

Kun Wang[1], **Fei Zhang***[,1], **Weihao Wang**[1] **and Yiwu Qian**[1]

[1]Hefei Institute for Public Safety Research, Tsinghua University, Hefei, 230601, China

[*]Corresponding author's: zhangfei@tsinghua-hf.edu.cn

Abstract. With the large-scale construction of city gas pipeline, increasing natural gas pipeline leakage accident disasters caused by the gas monitor is an important on-line measurement instrument for the safety operation of the network of natural gas, which has the characteristics of multi-varieties, style complex, complex procedures, automatic production, low digital level by making multi-varieties small-batch gas monitor as the main the mode of production. In order to solve that the traditional digital manufacturing job shop cannot meet the modern multi varieties and small-batch production of gas monitor manufacturing requirements, based on the complex production process of digital equipment, digital control, digital key technology, it proposes a gas monitor job shop solution based on intelligent manufacturing, construction of the implementation of the integrated framework and implementation activities of information flow model to realize the bottom equipment layer, design and manufacture of middle management and intelligent control layer. It has proved that multi-varieties small-batch of digital manufacturing job shop manufacturing process and development has a good application effect and can provide a reference model for the multi varieties small-batch manufacturing precision instruments in the production, greatly saving the production casting time cost, production cost and resource cost.

1. Introduction

Natural gas pipeline called "urban lifeline"- worked as a crucial public infrastructure in modern urban-brings natural gas to every corner in urban to improve the convenience for public and to support the urban general operation with the development of mass construction of city gas pipe network. However, the natural gas inside the pipeline has the characteristics of flammable and explosive, toxic, spread fast compared with other kinds of pipeline such as water supply, heat supply pipeline. In addition, as the gas detectors have the characteristics of variable-style, complex, fast changes and due to the variety of working principle, production technology and product structure, the production scale is changed from mass production to multi-variety and small-batch production mode which requires a high-improvement for the production quality. With the popularization and promotion of social safety and smart city, the current production mode cannot meet the multi-varietiesdemand and posts an urgent requirement to explore. and discover a new production mode to fulfill the manufacture demand of diversification and small-batch production.

In the current processing of manufacturing of "Industry 4.0", on the basis of intellectualization digital network, the manufacture industry combined IOT and service network to make an intelligent manufacturing system which own the functions of perception, decision-making, control and execution

[1-3]. Currently, domestic and foreign scholars had done a lot of researches and engineering tests on the processing and manufacture of diversification and small-batch production. In 2001, SON Y J, WYSK R A put forward the method of automated building simulation model to solve the problem of real-time scheduling control in complex internal and external environment of job shop[4]. In 2002, Z H Yu had studied on the similar process analysis method based on similar elements and the process similarity analysis application on statistical quality control [5]; In 2005, LEON V J made the components switching and posting time into shortest by modifying the group switching strategy element both similarity and tooling PCB geometric similarity[6]; In 2008, Y Gao had realized the visual monitoring technology of multidimensional vector points in job shop which solves bad real-time performance and small volume of monitoring under traditional production mode as well as the problem of information break from director layer to management layer due to real-time field response[7]. In 2006, S.Y.Nof G discovered control system with the characteristics of coordination, integration, integration, network and distributed decision support to solve the problem in automatic control of the production logistics system[8]; In 2006, Monostori Lstudied the random scheduling mode of the flexible manufacturing system[9]; In 2010, X L Wu had done research on the multi-varieties of small-batch scheduling algorithm of flexible job job shop and proposed a multi-objective hybrid genetic algorithm to solve the assignment problem and process scheduling problem in flexible job shop[10]; In 2011, Obweger H proposed the state management framework based on ECA(E-Business Collaboration Architecture) which followed the default logic rule rule and matched the simple event collected by perceptual nodes to produce the new complex event[11]; In 2013, F Liu proposed to construct multi-varieties small-batch digital production job shop included digital manufacturing foundation support layer, business infrastructure software platform, data collection and information interaction platform, function layer and user interaction layer by doing research on the networking manufacturing course of multi-varieties small-batch production job shop[12].

Based on the analysis above, the production mode had transferred from traditional single species mass production mode to the multi-varieties small-batch production mode. Therefore, on the consideration of transition of the resource scheduling and processing mode, the traditional production cannot meet the demand of production requirements of modern enterprise. In the paper, on the basis of gas detector processing technology and manufacturing processes, it proposed a digital processing and manufacturing methods of multi-varieties small-batch gas monitor which adopted the distributed centralized control with component technology, to solve the problem of low degree of automation and to modify the resource scheduling mode in the processing production.

2. Processing Craft Procedure

2.1 Overall processing craft procedure

Combustible gas detector assembly is consisted of internal temperature and humidity sensor, methane sensor shell cover, air pump, trachea, probe, antenna and battery parts to complete the procedures of preassembly of combustible gas detector. Then complete machine assembly according to the predetermined assembly process and the assembly process is shown in figure 1.

2.2 Main Circuit board manufacturing procedure

Main circuit board which can be connected to and control each part of the sensors is key part of Combustible gas detector to realize the function of data acquisition, filtering and signal processing. The manufacturing procedure includes welding, inspection, testing, and three-proofing validations of the temperature &humidity sensor and main circuit board as shown in figure 2.

3 Digital assembly job shop technology framework

Multi-varieties small-batch combustible gas detector flexible digital manufacturing job shop is on the direction of product production process, on the basis of production information for each stage to integrate equipment, product and manufacturing information into centralized control platform and

IOP Conf. Series: Materials Science and Engineering **616** (2019) 012028 doi:10.1088/1757-899X/616/1/012028

implement the product courses of processing, assembly and test. And the information of each stage is on the basis of professional database to do the process of instructions to send/receive information and also to share the information of each stage on the centralized unified database management platform.

Centralized control platform is based on the underlying hardware and software platform as the basement, the intermediate data standard format conversion process as key point to connect the Industrial Ethernet and all the production instructions and to do the preservation, processing [13], modeling and analysis courses by unifying the information distributed to each data information management servers. In addition, it realized the rapid information transmission and data sharing by constructing newly DNC(Distributed Numerical Control) network to improve the operation efficiency and reliability[14].

The centralized control system of multi-varieties small-batch combustible gas detector flexible digital manufacturing job shop includes user layer, data layer, functional layer, network layer and execution layer as shown in figure 3.

The user layer of job shop is responsible for the staff of processing and constructing job in the job shop to manage the order, acceptance, plan and execution of production task and job shop visual dynamic monitoring task according to each duty.

The functional layer is to achieve the rapid path planning of the actuator in processing and assembling by calling underlying data structure and fast modeling data driven module. Therefore, it can finish the fast modelling course of plan issued product and send message to execution layer.

The data layer is to save the underlying information, upper layer information, plan task orders, and product information of production job shop and to provide an analysis of the data and information traceability queries for functional layer.

The execution layer is responsible for the production, processing and assembly of the combustible gas detector of the action of execution units to realize the operation of machining process by receiving the executed command complete fixed action.

The network layer is the crucial part for the digital job shop information transmission. It realizes the reliability of data transmission and designs the transmission of Ethernet ring structure.

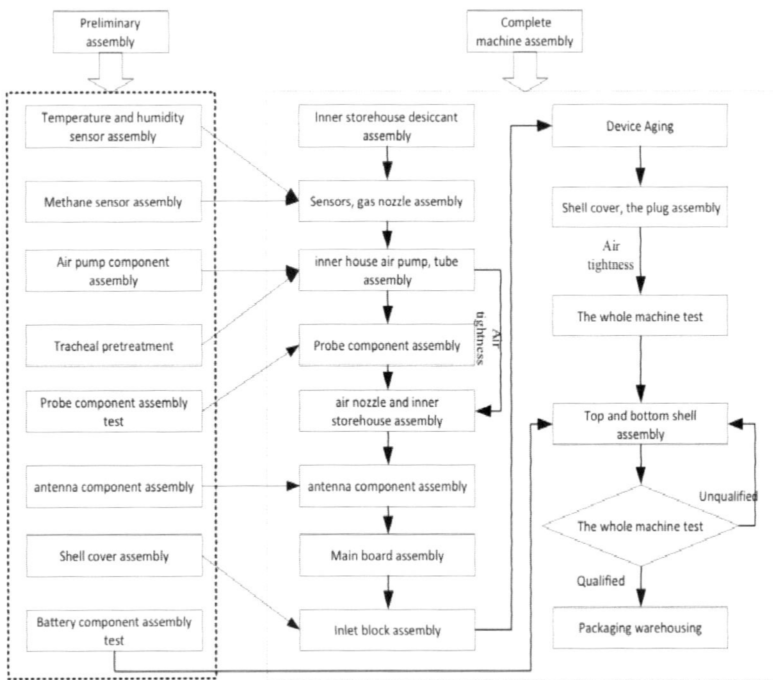

Figure 1. Combustible gas detector assembly process

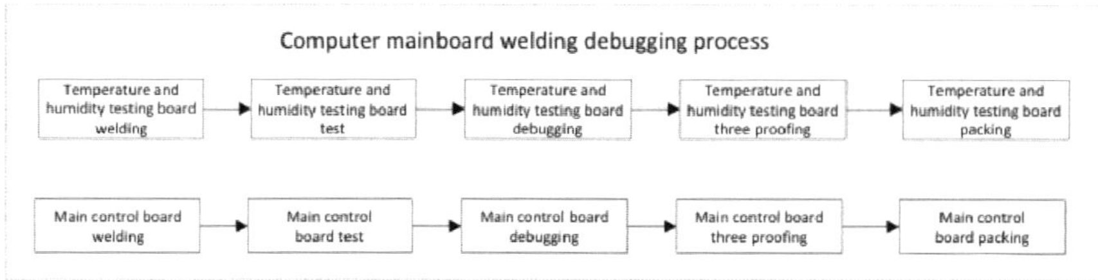

Figure 2. Main circuit processing technological process

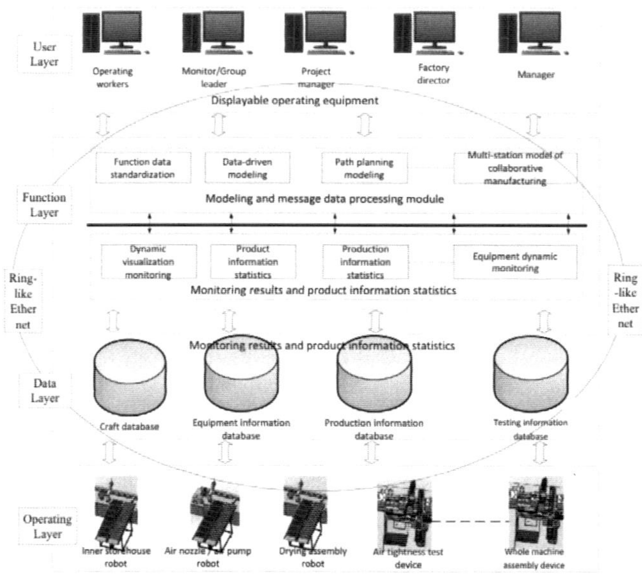

Figure 3. Combustible gas detector digital job shop system structure

4. The construction of the digital job shop

4.1 Circular Industrial Ethernet Structures

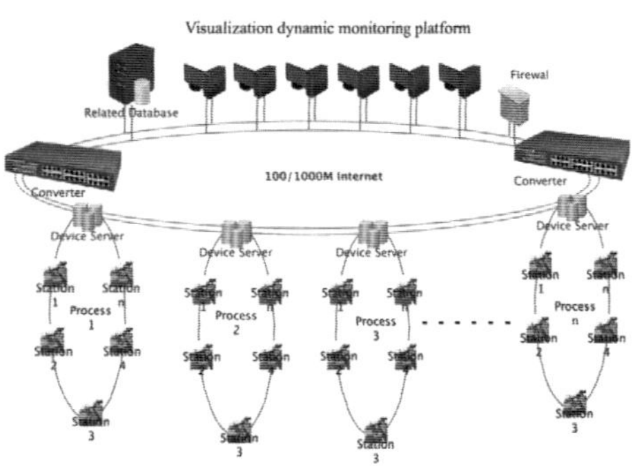

Figure 4. Combustible gas detector system platform

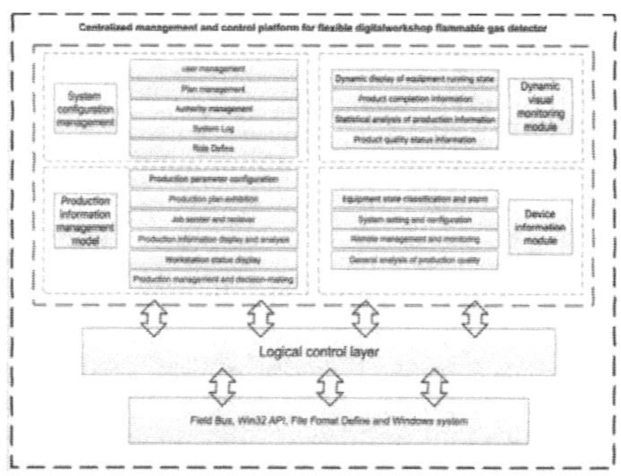

Figure 5. The system function structure logic diagram

The main systems are the field bus control system, distributed control system and network control system in the current computer control. Among the systems, the NCS(Network Control System) is a system that uses form of feedback control system through the network and connects the controlled objects with the controller and the drive through a common network platform[15]. The network control system, worked as a typical spatial distributed system, has the advantages of easy information resource sharing, easy expansion, easy connection, high efficiency, easy maintenance, reliable performance and so on. In the paper, the redundant ring-like Industrial Ethernet control system made full use of the advantages of network control systems to realize the field control of multi-varieties small-batch combustible gas detector flexible digital manufacturing job shop. The top layer of the system is the circular redundant Ethernet control system and bottom layer adopts the different fieldbus types of DCS system to realize structured control mode of "workstations - fieldbus - smart meters" as shown in figure 4.

4.2 The system function structure

The multi-varieties small-batch combustible gas detector flexible digital manufacturing job shop adopts the C/S structure. The server and client communication, control software stay on the shop controller and station controller respectively. Each layers and centralized control platform send and receive two-direction information reciprocally to realize the different functional requirement. And it can be divided into four parts-the system management module, the production information management module, the equipment module and the dynamic monitoring module according to system functions.

The system management function realizes the system access management to different users to meet the demand of job shop condition order placement, receiving, task execution and supervision. And also, users in different permissions to the job shop information management are also different.

Production information management module is responsible for the implementation of production and material plan release, production scheduling, maintenance and management of production line, process flow and organizational scheme of processes. It also evaluates the gas detector quality and scheme adjustment to guarantee the production processing quality and the adjustment of equipment resources, schedule and to optimize production resources by using different types of expert knowledge bases.

The basement of visual production dynamic monitoring implementation is the real-time data acquisition of dynamic assembly job shop. It constructs the combustible gas detector assembly process model based on the working process, scheduling model based on the craft processing plan and the real-time monitoring model based on the dynamic assembly data to realize the function of visualization dynamic monitoring from production and processing to the assembly.

The system realizes the connectivity by defining the form of communication and interface data driven of production assembly equipment, the implementation of the data sharing by driving module, Win32 API function, survival and reading the XML file, dynamic data read between equipment and storages with numerical control equipment. The whole system function structure is shown in the figure 5.

5. Combustible gas detector tooling test

5.1 The overall design of processing job shop

Based on the research achievement of multi-varieties small-batch combustible gas detector flexible digital manufacturing job shop, it designs the flexible digital manufacturing plant to solve the problem of controlling actual production progress, the low degree of visual monitoring process, the bad real-time performance, the collaborative weakness between processes and the adverse impact of unreasonable resource scheduling. The design effect is shown in the figure 6.

The multi-varieties small-batch combustible gas detector flexible digital manufacturing job shop includes detector assembly procedure in storehouse, the sensor assembly process, air tightness testing process, drying equipment assembly process, the bottom shell component assembly, testing, and the whole machine assembly process and so on. After the completion of the cycle, it is expected to increase the delivery rate of the goods by 80%, decrease waste rejection rate by 10% and decrease the finished product inventory rate by more than 30%.

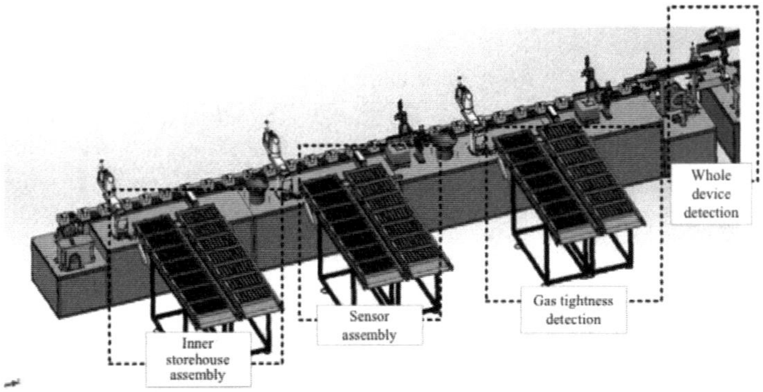

Figure 6. Global design of job shop assembly

5.2 Combustible gas detector performance testing

To verify actual usage and quality problems of the combustible gas detector flexible digital machining manufacturing plants, it tests and verify processed products in the real field. The effect of measuring gas monitoring is shown in the figure 7 and that of the temperature and humidity monitoring is shown in the figure 8, 9.

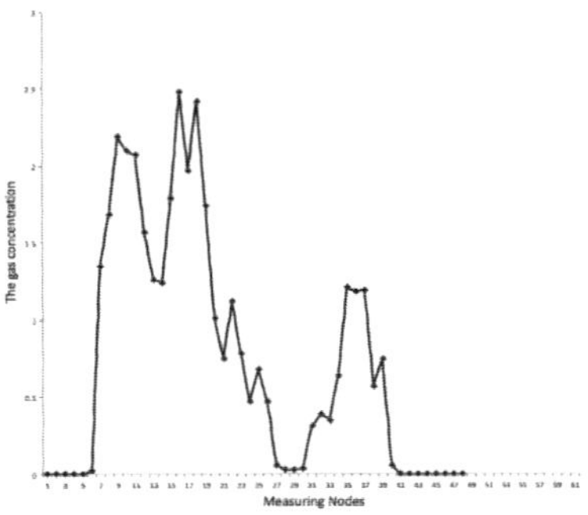

Figure 7. The testing curves of gas sensor concentration

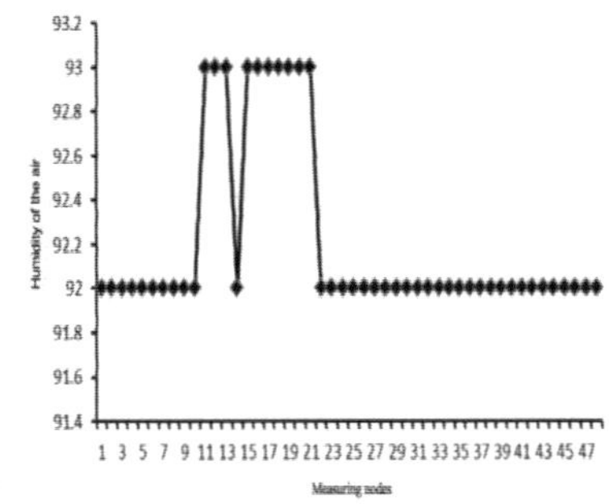

Figure 8. The testing curves of humidity sensor of the air

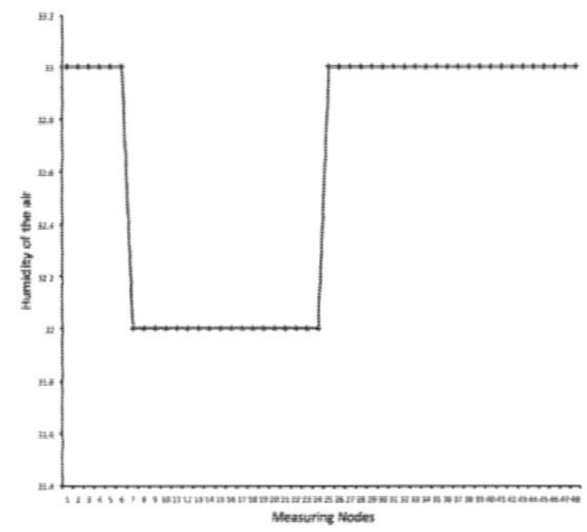

Figure 9. The testing curves of air temperature

6. Conclusion

(1) With the propositions of advanced manufacturing system such as the precision manufacturing, the intelligent manufacturing system, the distributed centralized control system, the concurrent engineering and the omnipotent manufacturing system, the modern manufacturing is developing towards the direction of digital, multiple, flexible manufacturing, integration and intelligence. It realizes the real-time dynamic tracing and collaborative applications of the material in the process of production, equipment, products and personnel to greatly save the production cost and manufacturing time.

(2) In this paper, for the current insufficient situations of the visual monitoring and scheduling ability of multi-varieties small-batch gas monitor processing job shops during work operating and on the basis of analysis of the international and domestic digital plant, it completes the combustible gas detector design and builds the flexible digital manufacture job shop with the system integration from the bottom up of equipment states and the integrity of the product information acquisition. It realizes the implement course of the manufacturing job shop production plan scheduling, the resource

optimization configuration and the visualization of dynamic monitoring during production and manufacturing. In addition, it solves the problem of the manufacturing information island, improves the quality of the products, reduces the production cost, shortens the normal cycle of the castings and improve the market fast response ability of the new product.

(3) The construction of the digital manufacturing job shop accumulates hardware and software implementation experience of multi-varieties small-batch gas monitor processing job shop. The implementation of the system satisfies the need of enterprise production and quality requirements as well as playing a positive role model in the application of digital manufacturing industry in the world.

Acknowledgments

The work is supported by "The National Key Research and Development Program of China(Project No.2018YFC0809304).

References

[1] X Liu, B Xu, J Peng, AS Guanting. Model and Solution of Job-shop Scheduling for Parallel Processes[J]. Journal of Computer-Aided Design & Computer Graphics, 2012 , 24 (1) :120-127.

[2] Ca tfolis Thier ry , Kur t M eat. Hybridiza tio n and Specialization o f Real Time Recurrent Learning-Based N eural N etwo r k. Connectio n Science, 19979 (1): 97-110.

[3] O Gholami,YN Sotskov.A fast heuristic algorithm for solving parallel-machine job-shop scheduling problems [J].The International Journal of Advanced Manufacturing Technology,2014, 70(1):531-546.

[4] SON Y J, WYSK R A.Automatic simulation model generation for simulation-based, real-time shop floor control[J].Computers in Industry,2001,45(3):291-308.

[5] Z H Yu, J J Yin, Z T Wu. Similarity analysis between the process and its application in SPC[J]. System Engineering Theory and Practice, 2002(11):6-11.

[6] LEON V J, JEONG I J. An im proved group setup strategy for PCB assembly[C]//Proceedings of the ICCSA 2005. Berlin, Germany: Springer-Verlag, 2005:312-321.

[7] Y Gao, Z Q Jiang, J Hu, M X Zhang, M Z Liu. Research on visual monitoring of workshop based on multi-dimensional vector points [J]. Modern Manufacturing Engineering ,2008 , 29 (9) :108-108

[8] S.Y.Nof,G Morel, L.Monostori,A.Molina.F Filip. From plant and logistics control to multicenter prise collaboration[J].Annual Reviews in Contr. 2006, 30(1): 55-68 .

[9] L.Monostori,J.Vfincza,T.Kis. Real-Time. cooperative enterprises:Requirements and solution approaches[J]. Information Control Problems in Manufacturing. 2006,30(6):561-566.

[10] X L Wu, S J Li, Y H Du.Research on Batch Scheduling Problem in a Flexible Job Shop[J].China Mechanical Engineering. 201021(4):424-429.

[11] Obweger H,Schiefer J,SuntingerM,etal.Lecture Notes in Computer Science[M]. Springer Berlin Heidelberg, 2011.

[12] F Liu, Q L Wang. Energy Efficiency Assessment of Mechanical Manufacturing System:Characteristics, State-of-the-art and Future Trends. China Mechanical Engineering. 2013,11:1550-1556.

[13] Monostori.L,B.Pfeiffer,Towards adaptive and digital manufacturing[J].Annual Reviews in Control.2010.10(2): 2007-2018.

[14] Obweger H,Schiefer J,SuntingerM,etal.Lecture Notes in Computer Science[M]. Springer Berlin Heidelberg, 2011.

[15] P Bogdan, R Marculescu. Towards a Science of Cyber-Physical Systems Design[J]. IEEE/ACM International Conference on Cyber-physical Systems. 2011:99-108.

IOP Conf. Series: Materials Science and Engineering **616** (2019) 012029 doi:10.1088/1757-899X/616/1/012029

Research on the Three-Dimensional Process Design Method of Shipbuilding Based on MBD Technology

Y W Huang[1], Z Y Tang[2, *], X H Zhang[1], H Zhang[1] and J F Liu[1]

[1]Science and Technology Management Department, Shanghai Waigaoqiao Shipbuilding Co.,Ltd, 3001 Zhouhai Road, Shanghai, 200137, China
[2]School of Naval Architecture, Ocean & Civil Engineering, Shanghai Jiao Tong University, 800 Dongchuan Road, Shanghai, 200240, China

[*]tang970131@163.com

Abstract. From two-dimensional engineering drawings, Computer-aided Design (CAD) technology to today's three-dimensional process design technology of Model Based Definition (MBD) based on feature representation and control, great changes have taken place in human engineering technology. With the applications of MBD technology in aerospace, machine-tool and other fields, it's also being actively explored in shipbuilding industry on engineering applications of the whole process of ship design and production under the background of intelligent transformation. After analysing the problems existed in process design of domestic shipbuilding enterprises, this paper firstly introduces a three-dimensional digital model of shipbuilding based on MBD technology. Then, a three-dimensional process design method is proposed in order to provide theoretical basis for the realization of three-dimensional process design of shipbuilding.

1. Introduction

With the great development of human engineering technology, the closely related information control in design process and the production management mode have changed fundamentally. Although three-dimensional (3D) design software have been widely and deeply applied in shipbuilding field, two-dimensional (2D) engineering drawings are still used in its production process [1]. The efficiency of information transmission between design and production personnel is low, which makes the shipbuilding cycle longer. Under the background of intelligent transformation in shipbuilding industry, the 3D process design method based on MBD technology gradually becomes a key research direction.

2. Background

2.1 MBD technology

MBD technology is called 3D annotation technology [2]. It is not only an essential way to develop 3D design, but also the core technology for 3D models replacing 2D engineering drawings to be the unique data source in shipbuilding. MBD technology originated in aerospace field for shortening aircraft production cycle and reducing costs, which was first proposed by Boeing in the 1990s [3]. Now, it is mostly used in aerospace and machine-tool manufacturing industries or other fields. In shipbuilding industry, MBD technology is also applied to change the ship design and production mode. For example, digital simulation assembly based on MBD is applied to the design of a submarine to solve problems that may occur during the actual assembly process in advance. Hyundai Heavy Industries (HHI) also uses MBD technology to make the 3D model run through all stages of design, production and management, enabling it to be seen by all design and production personnel, administrators and shipowners [4]. What's more, a MBD-based data structure and a management method for shipbuilding are established to raise shipbuilding efficiency by using the MBD-based 3D model as unique data source. The traditional process is improved based on MBD technology [5].

Ship design and production technology based on MBD uses the 3D digital model as the unique data source to integrate all necessary information during ship design and production, to realize 3D process design of shipbuilding and promote collaborative design, production and management across all departments of an enterprise. Meanwhile, the MBD-based 3D digital model can also be the data source of Enterprise Resource Planning (ERP), Manufacturing Execution System (MES) and other information management systems to support knowledge reasoning, process simulation and assistant decision-making, which will improve the digital and intelligent level of ship design and production.

2.2 Process design of shipbuilding

Process design is the general term of process planning design and process equipment design, it mainly refers to the design for process plans and tools. And process design of shipbuilding refers to the design of process methods and specific process procedures for shipbuilding. At present, the process design of domestic shipbuilding enterprises mostly stays at the stage of "3D designing, 2D drawing". There's a certain gap compared with foreign advanced shipbuilding enterprises and other fields such as aerospace and machine-tool. 3D and 2D mixed mode makes it necessary for designers to transform a large amount of 3D and 2D information, which easily leads to information deviation and causes problems in every segment. At the same time, the data source of ship design and production under this mode is not unique, information resources among different departments are difficult to share and not uniform. What's more, it's hard to achieve data integration management since 3D models cannot run through the whole process of ship design and production, which results in data redundancy and heavy workload to design, production and management personnel.

In summary, it is imperative for domestic shipbuilding enterprises to develop 3D process design of shipbuilding. With the development of Internet and information technology, the emergence of MBD technology makes it possible to realize 3D process design of shipbuilding.

3. Establishment of the 3D digital model of shipbuilding based on MBD

3.1 Classification of ship design and production information

MBD technology is based on the unique data source, it needs to establish a uniform 3D digital model which covers a large amount of information. The information can be classified into design information, production information and management information, as shown in figure 1.

Design information mainly includes 3D entity models, annotated information, material properties, etc; production information mainly refers to process information, tool information and technical requirements involved in the production procedure; management information contains production properties, Bill of Material (BOM), management documents, etc.

According to the various stages of ship design and production, BOM can be specifically divided into Engineering BOM (EBOM), Plan BOM (PBOM) and Manufacturing BOM (MBOM). EBOM refers to the detailed list of parts and materials required in the process of ship design and production. Based on the process design level of shipbuilding enterprises, PBOM is a list that describes the manufacturing relationship of parts and components, such as the manufacturing location, manufacturers, material suppliers and so on. MBOM represents the specific processing method and procedure of parts and components.

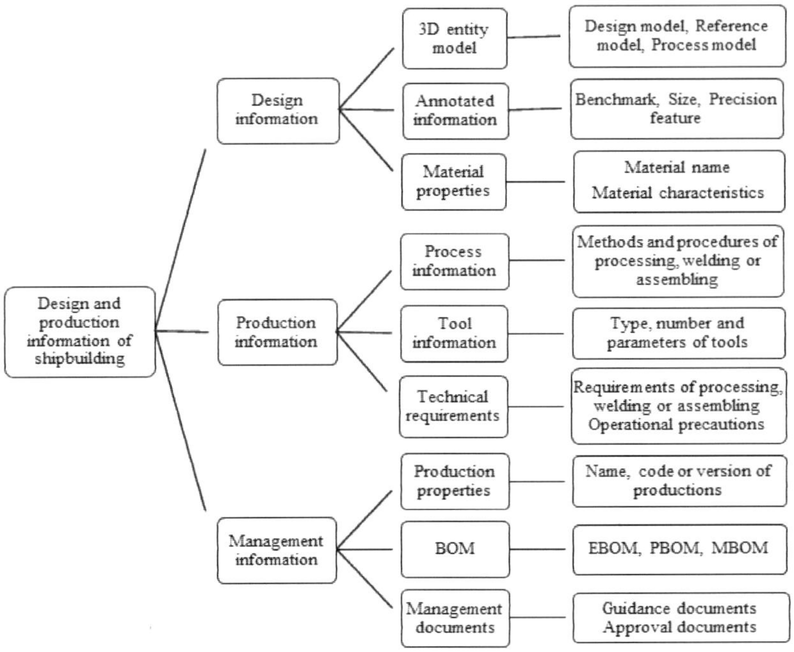

Figure 1. Classification of ship design and production information

3.2 Data management for ship design and production

Shipbuilding is a typical discrete production mode. The types of 3D models, the number of drawings and documents involved in its design and production process are enormous, and the information used at different stages is also various.

Based on the principles of digitalization and integration, the MBD-based data management mode for ship design and production is established, as shown in figure 2. Design department, process department, production department and management department import design information, production information and management information into MBD data management system respectively, thus, a 3D digital model based on MBD technology has been generated. Information transmission and sharing among departments can be carried out through the MBD data management system.

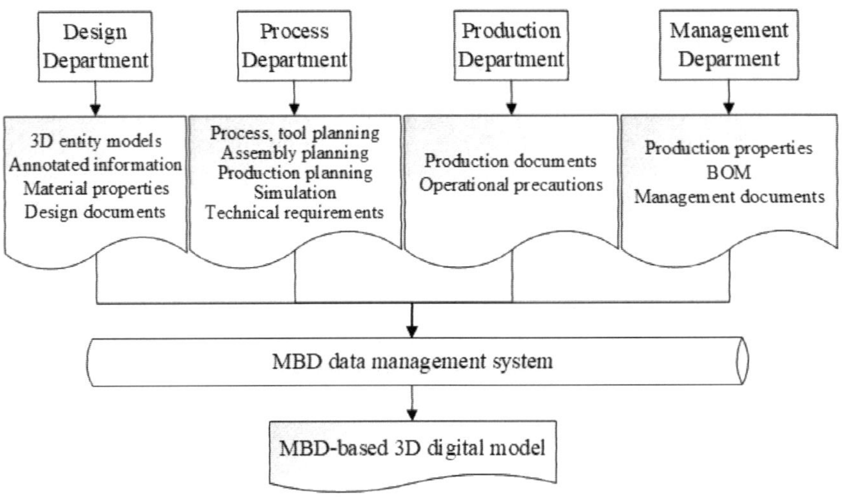

Figure 2. MBD data management system

IOP Conf. Series: Materials Science and Engineering **616** (2019) 012029 doi:10.1088/1757-899X/616/1/012029

3.3 Representation of the 3D digital model of shipbuilding based on MBD

MBD technology realizes 3D model design and real-time transmission of digital information by taking design, production and management information of the model as the core of organization and analysing constraint relationships between various types of information. Technicians can acquire information at every stage of ship design and production by means of the 3D digital model, as shown in figure 3. The 3D digital model truly becomes the only information carrier for ship design and production, which lays the foundation for realization of the 3D shipbuilding process design method based on MBD technology.

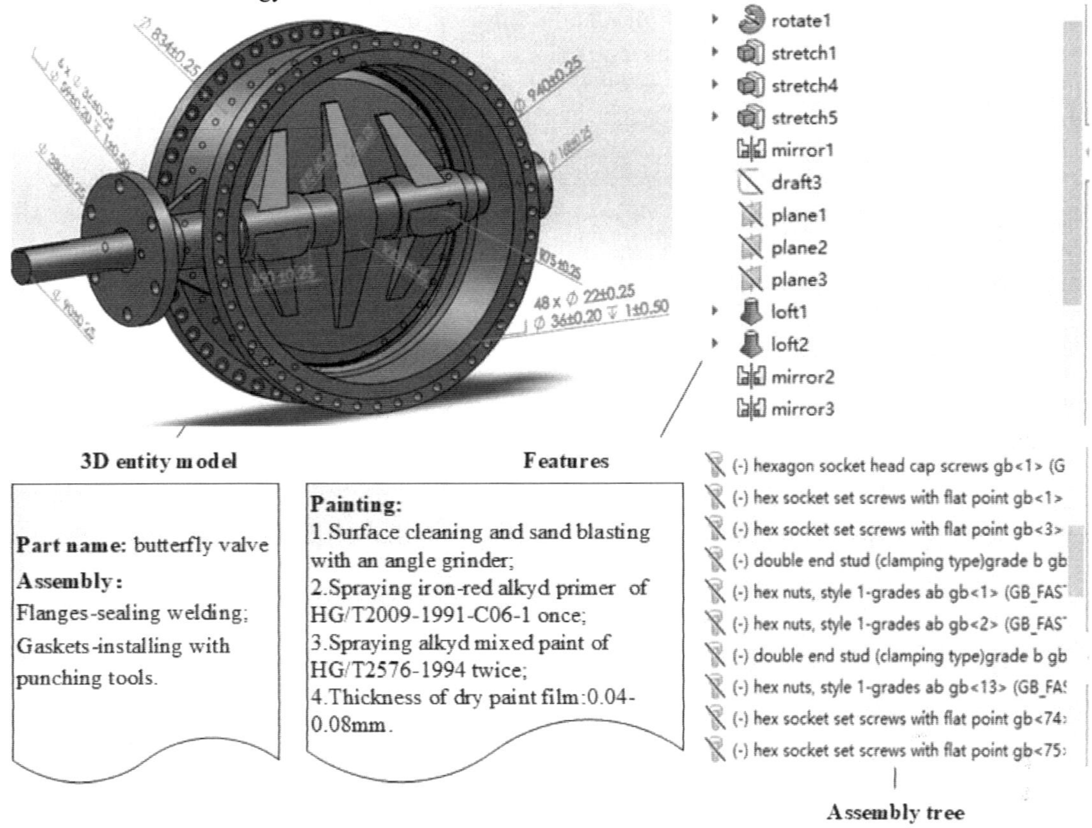

Figure 3. The 3D digital model of shipbuilding based on MBD

Usually, some simulation and analysis work will be done based on the 3D digital model. The exported model needs to be lightened. In the case of ensuring the consistency of design, production, and management information data structure, the complexity of the 3D digital model can be reduced by using corresponding digital definition elements to represent. Firstly, according to the actual demands of production and management, necessary information of different stages (design, production and management) is extracted, reorganized and documented. Then the extracted information files are compressed and imported into MBD data management system to form the 3D digital model. On the basis of meeting the requirements of field production and enterprise management, the accuracy of geometric models is reduced appropriately compared with the design stage, and redundant information is thoroughly filtered, so that the 3D digital model and its information structure are simplified, and the efficiency of file transmission and storage is improved.

4. The 3D process design method of shipbuilding based on MBD

As shown in figure 4, the traditional design and production procedure contains a lot of 2D information which greatly increases the difficulty of information conversion, storage and management, and

IOP Conf. Series: Materials Science and Engineering **616** (2019) 012029 doi:10.1088/1757-899X/616/1/012029

seriously affects the progress of ship design and production. In order to improve the efficiency of design and production, a 3D process design method of shipbuilding based on MBD technology is proposed, which is shown in figure 5.

ERP system is a management platform to provide decision schemes for enterprises with systematic management thought based on information technology. For shipbuilding enterprises, it can cover all kinds of design and management software in ship design and production, and optimize the sharing and utilization of enterprise resources from the supply chain. MES is a kind of monitoring and feedback of ERP system in the field production stage, which can promptly report the problems in the production process and quickly respond through information transmission, and guide the production management with accurate data. The MBD data management system integrated with ERP and MES covers all information in the whole life cycle of ship design and production. It can promote the coordination and unification of design, production and management platforms and make information transmission more timely and accurate.

Meanwhile, the MBD-based 3D process design method of shipbuilding incorporates Knowledge Engineering, which is a new subject that acquires knowledge and information through modern science and technology with high efficiency and large capacity. It studies how to use computer to represent knowledge and solve problems automatically, which is the development of artificial intelligence in conducting knowledge and information. For a long time, due to the lack of information carrier and effective representation mode throughout the whole life cycle of ship design and production in domestic shipbuilding enterprises, knowledge engineering has been difficult to apply to the actual production process. MBD technology is characterized by the unique data source and 3D representation mode, which will make it possible to integrate Knowledge Engineering into ship design and production. Knowledge Engineering acquires knowledge from the inside and outside of systems, verifies its accuracy, and forms a solid knowledge base. Based on professional software, it enables computers do knowledge reasoning, process simulation and assistant decision-making for specific problems, making the research of shipbuilding process design method more intelligent.

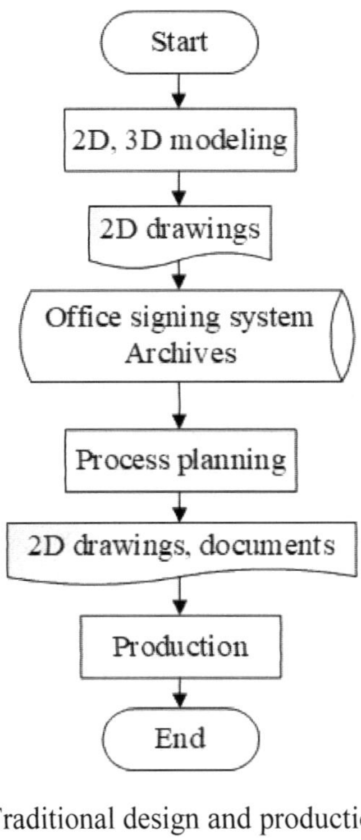

Figure 4. Traditional design and production procedure

IOP Conf. Series: Materials Science and Engineering **616** (2019) 012029 doi:10.1088/1757-899X/616/1/012029

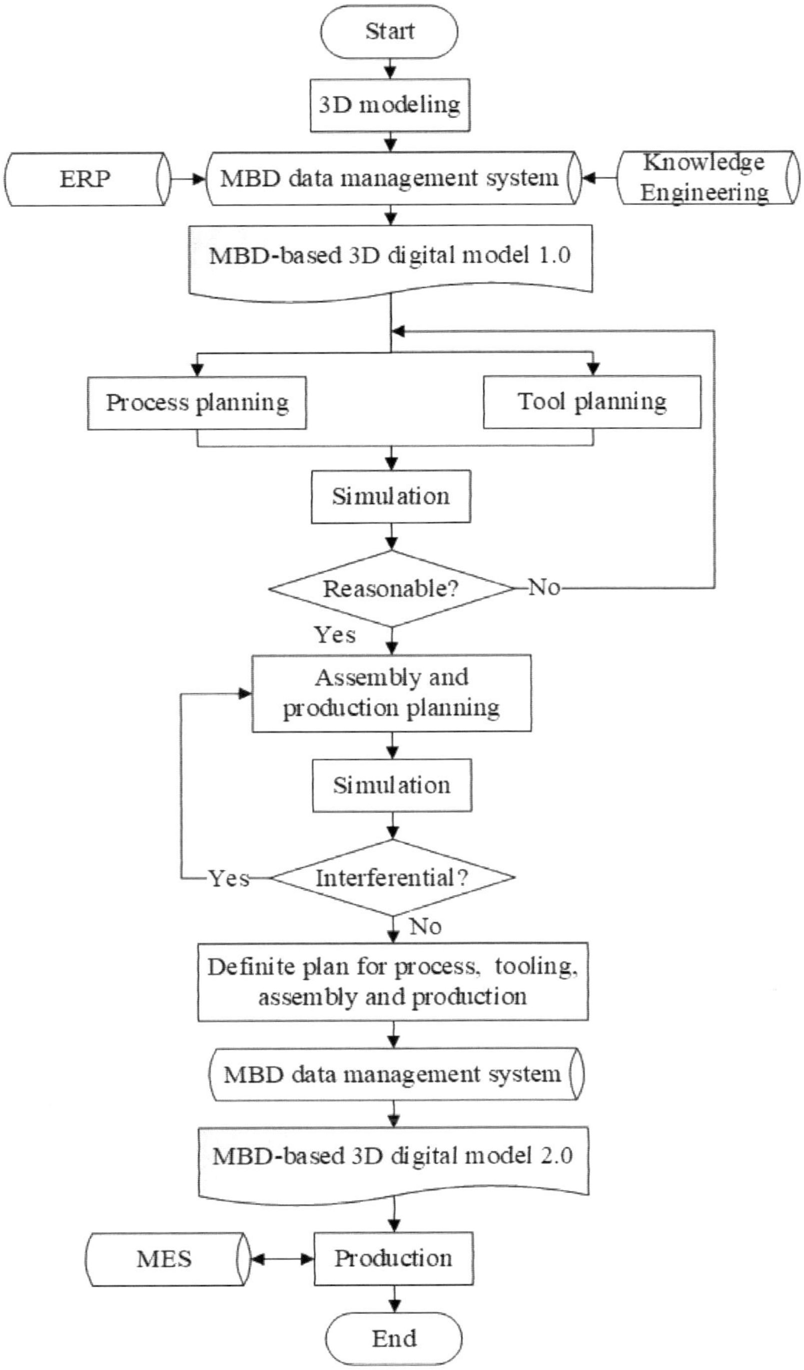

Figure 5. The 3D process design method of shipbuilding based on MBD

5. Conclusion

MBD technology is a key technology for manufacturing industry to realize intelligent transformation. However, nowadays, MBD technology is in the stage of theoretical research in shipbuilding industry which needs continuous development to break through the traditional method and to be applied in the actual design and production of shipbuilding enterprises. And it's worth mentioning that the technical advantages of MBD technology in ship design, production and management have been initially reflected. Based on the existing problems of process design in domestic shipbuilding enterprises, this paper discusses the applications of MBD technology in ship design and production. The 3D digital model and process design method of shipbuilding based on MBD technology are proposed to provide theoretical basis for the realization of 3D process design of shipbuilding. In the future, the research of Model Based Systems Engineering (MBSE) will be carried out for the whole life cycle of ship design and production.

References

[1] Quintana V, Rivest L, Pellerin R and Kheddouci F 2012 *J. Computers in Industry* **63** 79-90
[2] Quintana V, Rivest L, Pellerin R, Venne F and Kheddouci F 2010 *J. Computers in Industry* **61** 497-508
[3] Sharma K J and Bowonder B 2004 *J. International Journal of Manufacturing Technology and Management* **6** 254-64
[4] Kim H W and Yoo W S 2013 *J. International Journal of Non-Linear Mechanics* **53** 55-62
[5] Li M Y, Xu W, Zhang W J and He J Y 2015 *J. Ship Science and Technology* **37** 85-9

IOP Conf. Series: Materials Science and Engineering **616** (2019) 012030 doi:10.1088/1757-899X/616/1/012030

Optimal Design of Stiffened Panels Considering Torsion of Supporting Stiffeners

Jingle Cao*, Yuhong Shi, Huiping Wang, Ruijuan Yao and Shufan Wang

Beijing Institute of Astronautical Systems Engineering, Beijing 100076, China

*caojingle@outlook.com

Abstract. Aiming at the optimal design and analysis of stiffened panels, this paper proposes a new algorithm considering the torsion of supporting stiffeners on axial load carrying capacity. By defining three dimensionless factors: relative bearing capacity, relative transverse elastic stiffness and relative torsional elastic stiffness, the response surface of load-bearing capacity versus two kinds of elastic stiffness is obtained. Based on genetic algorithm, a lightweight optimization model for structure layout and section parameters is established. A certain stiffened panel is optimized on this basis, and the matching design relationship between two kinds of stiffeners is summarized.

1. Introduction

Stiffened panels and shells are widely used as main load-bearing structure elements in aerospace, ship and other fields [1][2]. Buckling is the main failure form of this kind of structures, so structural stability is the main restricting factor of load-bearing capacity [3]. In the analysis of structural stability, stiffened panel is usually simplified to crossed beam system composed of longitudinal stiffeners which bear external load directly and supporting stiffeners which provide elastic support to longitudinal stiffeners. The skin is equivalent to the attached web of the reinforcement. Relevant design standards provide a simplified engineering empirical formula on bearing capacity [4]. However, it ignores the torsion of supporting stiffeners, which makes the results conservative and has great limitations. Accurate stability analysis and optimization design of stiffened panels and shells are mostly based on finite element analysis [5-7]. The finite element analysis based on explicit dynamics algorithm can obtain relatively precise load-bearing capacity which fits well with the experimental results, but it is very time-consuming and sometimes unaffordable. Hence, the engineering empirical formula which can forecast the load-bearing capacity quickly is still of great significance in the initial stage of structure design and analysis.

From the perspective of theoretical analysis, this paper improves the traditional engineering algorithm, and derives the formula for calculating axial load carrying capacity of stiffened panels considering the torsion of supporting stiffeners. On this basis, an optimization model on section parameters and structure layout of stiffened panels is established based on genetic algorithm, and the matching design relationship between longitudinal stiffeners and supporting stiffeners is summarized.

2. Theoretical analysis

2.1. Basic analytical model

Decouple longitudinal stiffeners and supporting stiffeners. When subjected to axial compression load, supporting stiffeners will not only bend but also twist, which can provide both transverse stiffness and corner stiffness. Therefore, longitudinal stiffeners can be regarded as a multi-span compressive column supported by a series of transverse elastic supports and torsional elastic supports, as shown in figure 1.

By loading concentrated force (or torque) at the joint of the supporting stiffener, the elastic stiffness is the ratio of the concentrated force (or torque) to the deflection (or rotation angle) of the joint.

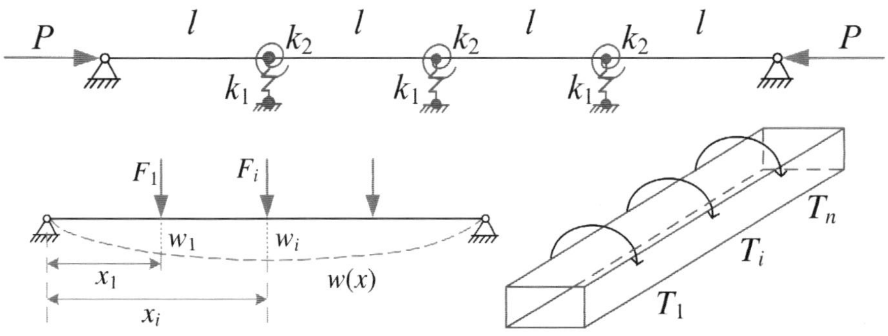

Figure 1. Theoretical analysis model.

2.2. Buckling of compressive column under two kinds of elastic supports

The transverse force caused by transverse elastic supports and the bending moment caused by torsional elastic supports are both regarded as external loads, as shown in figure 2. The differential equation of the deflection curve on compressive column is

$$\frac{d^2y(x)}{dx^2}+\alpha^2y(x)=\frac{M(x)}{EI} \tag{1}$$

where coefficient $\alpha^2=P/EI$, EI is the bending stiffness of longitudinal stiffeners, P is axial compression load, $y(x)$ is the deflection and $M(x)$ is the bending moment on cross-section.

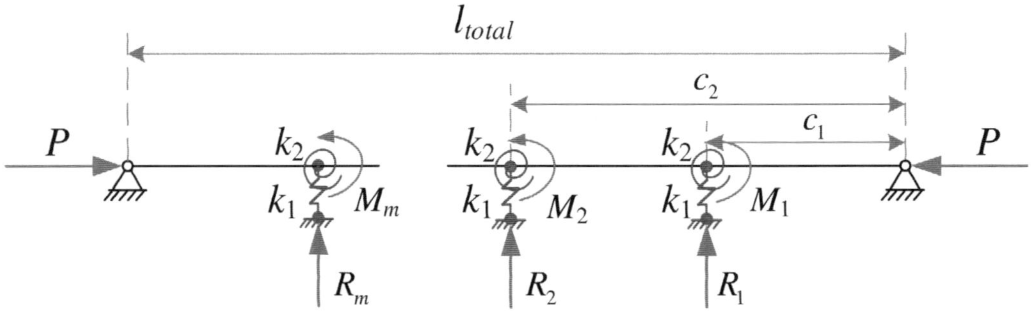

Figure 2. Compressive column under two kinds of elastic supports.

Consider the case that only one transverse force Q applied at a distance c from the right end. The function of deflection curve is

$$y_Q=\begin{cases}\dfrac{Q\sin\alpha c}{P\alpha\sin\alpha l}\sin\alpha x-\dfrac{Qc}{Pl}x & x<(l-c)\\ \dfrac{Q\sin\alpha(l-c)}{P\alpha\sin\alpha l}\sin\alpha(l-x)-\dfrac{Q(l-c)}{Pl}(l-x) & x>(l-c)\end{cases} \tag{2}$$

While the function of rotation angle can be described as

$$\theta_Q = \begin{cases} \dfrac{Q\sin\alpha c}{P\sin\alpha l}\cos\alpha x - \dfrac{Qc}{Pl} & x < (l-c) \\ -\dfrac{Q\sin\alpha(l-c)}{P\sin\alpha l}\cos\alpha(l-x) + \dfrac{Q(l-c)}{Pl} & x > (l-c) \end{cases} \tag{3}$$

Consider the case that only one bending moment M applied at a distance c from the right end. The function of deflection curve is

$$y_M = \begin{cases} \dfrac{M\cos\alpha c}{P\sin\alpha l}\sin\alpha x - \dfrac{M}{Pl}x & x < (l-c) \\ -\dfrac{M\cos\alpha(l-c)}{P\sin\alpha l}\sin\alpha(l-x) + \dfrac{M}{Pl}(l-x) & x > (l-c) \end{cases} \tag{4}$$

While the function of rotation angle can be described as

$$\theta_M = \begin{cases} \dfrac{M\alpha\cos\alpha c}{P\sin\alpha l}\cos\alpha x - \dfrac{M}{Pl} & x < (l-c) \\ \dfrac{M\alpha\cos\alpha(l-c)}{P\sin\alpha l}\cos\alpha(l-x) - \dfrac{M}{Pl} & x > (l-c) \end{cases} \tag{5}$$

According to the deflection of each joint equals the displacement of each transverse elastic supports, and the rotation angle of each joint equals the corner of each torsional elastic supports, superimpose formula (2)-(5) to obtain the linear system of equations

$$\begin{aligned} \frac{R_m}{k_{1\_m}} &= \sum y_Q + \sum y_M \\ \frac{M_m}{k_{2\_m}} &= \sum \theta_Q + \sum \theta_M \end{aligned} \tag{6}$$

Formula (6) can also be described in the form of matrix

$$[A](R_1 \quad M_1 \quad L \quad R_n \quad M_n)^T = (0 \quad L \quad 0)^T \tag{7}$$

The buckling load can be obtained in case of forcing the determinant of matrix coefficient $[A]$ to be zero.

Assumed the elastic supports are equally spaced and the length of each span is l. According to Eular's formula on buckling load of compressive column with two hinged ends $P_{cr} = \pi^2 EI / l^2$, defining three dimensionless factors: relative bearing capacity $\overline{P} = P / P_{cr}$, relative transverse elastic stiffness $\overline{k}_1 = k_1 / (\pi^2 EI / l^3)$ and relative torsional elastic stiffness $\overline{k}_2 = k_2 / (\pi^2 EI / l)$. Now coefficient matrix $[A]$ is only related to $\overline{P}$, $\overline{k}_1$, $\overline{k}_2$ and span number m, so $|A|=0$ can be described as

$$\overline{P} = \Phi(\overline{k}_1, \overline{k}_2, m) \tag{8}$$

For each span number m, the response surface of the relative bearing capacity versus two kinds of relative elastic stiffness is obtained, as shown in figure 3. Then the buckling load of each compressive column can be easily calculated according to the response surface.

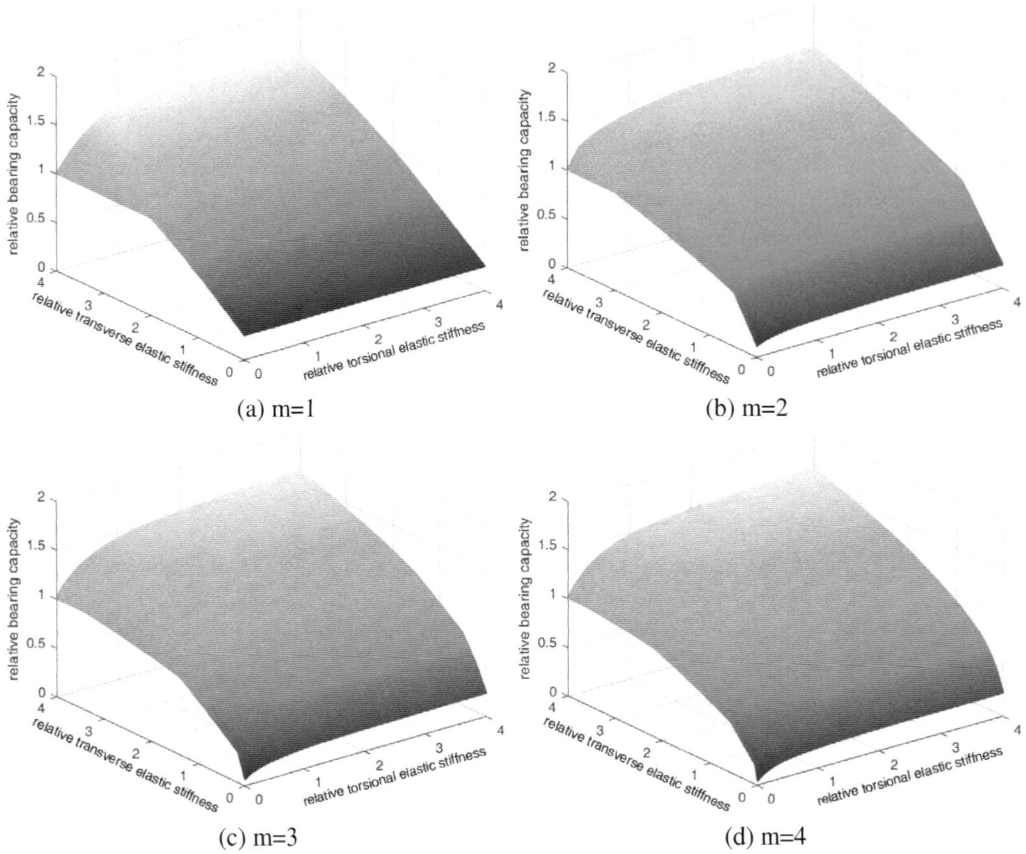

Figure 3. Response surface of the relative bearing capacity varying two kinds of relative elastic stiffness.

2.3. Stiffness of elastic supports

2.3.1. Stiffness of transverse elastic supports. Apply concentrated load F_i to the joint x_i of the supporting stiffener. The differential equation of the deflection curve on a supporting stiffener is

$$EI_s w'''(x) = Q(x) \tag{9}$$

where $w(x)$ is the deflection, $Q(x)$ is the shearing force on cross-section, EI_s is bending stiffness of the supporting stiffener. Then the stiffness of transverse elastic support is the ratio of concentrated force to the deflection on the joint $k_{1\_i} = F_i / w(x_i)$.

2.3.2. Stiffness of torsional elastic supports. Apply concentrated torque T_i to the joint z_i of the supporting stiffener. According to Umansky theory, the differential equation of the rotation angle on a thin-walled stiffener with closed section is

$$\frac{1}{v} E_1 J_{\bar{\omega}} \theta'''(z) - GJ_d \theta'(z) = -L(z) \tag{10}$$

where $\theta(z)$ is the rotation angle, $L(z)$ is the torque on cross-section, GJ_d is the free torsional stiffness, E_1 is reduced elastic modulus, $J_{\bar{\omega}}$ is the fan-shaped moment of inertia, v is a parameter reflecting the warping degree of the section. The analytical solution of equation(10) can be expressed as $\theta = \frac{1}{\gamma} T \frac{l}{GJ} g(\frac{l}{b})$, where parameter b described the section dimension and function $g(l/b)$ is related to the constrained torsional term which can be ignored when ratio l/b is large enough. Then the differential equation can be simplified as

$$GJ_d \theta'(z) = L(z) \tag{11}$$

Then the stiffness of torsional elastic support is the ratio of concentrated torque to the rotation angle on the joint $k_{2\_i} = T_i / \theta(z_i)$.

3. Optimal design

3.1. Optimization model

According to the theoretical analysis above, section parameters and the number of stiffeners have a great influence on load carrying capacity. Reasonable matching design of two kinds of stiffeners can reduce structure weight greatly.

In this paper, lightweight design of stiffened panels with fixed number of longitudinal stiffeners is carried out. That is to get the minimum value of the structure weight under the condition of meeting the requirements of a given bearing capacity. In case of the quantity of design variables and the existence of discrete variables such as the number of supporting stiffeners, this paper intends to use genetic algorithm for optimal design. The mathematical expression can be described as

$$\begin{cases} Find \quad a_i, b_i, n \\ Min \quad W(a_i, b_i, n) \\ s.t. \quad P \geq P_{\min} \\ \underline{a_i} \leq a_i \leq \overline{a_i}, \underline{b_i} \leq b_i \leq \overline{b_i} \\ n = 2, 3, 4... \end{cases} \tag{12}$$

where a_i and b_i are section parameters of two kinds of stiffeners, n is the number of supporting stiffeners, W is structure weight, P and $P_{\min}$ are the calculated and limited bearing capacity.

3.2. Optimization model

The optimal design of a stiffened panel with two longitudinal stiffeners is carried out. T-section longitudinal stiffeners are 2400mm length which are simply supported at both ends while box-section supporting stiffeners are 1200mm length with two fixed ends, as shown in figure 4.

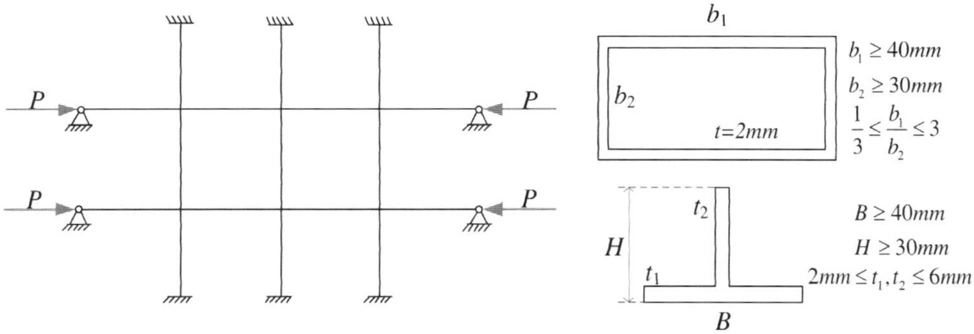

Figure 4. Structural model.

IOP Conf. Series: Materials Science and Engineering **616** (2019) 012030 doi:10.1088/1757-899X/616/1/012030

Given a set of initial parameters (column 1 of Table 1), the bearing capacity is 141339N and the structure weight is 8.763kg. Taking this bearing capacity as the minimum critical bearing capacity, the optimization iteration is carried out respectively, for the case that having different number of supporting stiffeners. Each optimization iteration process is shown in figure 5 and the optimal results are rounded up to engineering reasonable values, as shown in Table 1.

It can be found out that the optimized structure weight in each case is significantly lower than the initial one. The width B and thickness t_1 of longitudinal stiffeners reduce to be nearly the minimum limited values in all four cases. This is because comparing with flange, web provides a lower contribution to the moment of inertia in T-section.

Meanwhile, the dimensions of the optimized supporting stiffeners are much smaller than the initial ones. According to the response surface shown in figure 3, when the relative supporting stiffness reaches a certain value, the variation trend of the bearing capacity against the stiffness of elastic support tends to be smooth. That is, the contribution of the increase of supporting stiffeners section dimensions to the bearing capacity is obviously reduced. Therefore, the trend of optimization iteration is increasing the cross-section size of the longitudinal stiffeners and reducing the cross-section size of the supporting stiffeners.

Furthermore, when containing different number of supporting stiffeners, the optimized weight shows a great difference and the value reaches the minimum in case 2 (having two supporting stiffeners). This shows the great influence of the matching layout of two kinds of stiffeners on structure lightweight design.

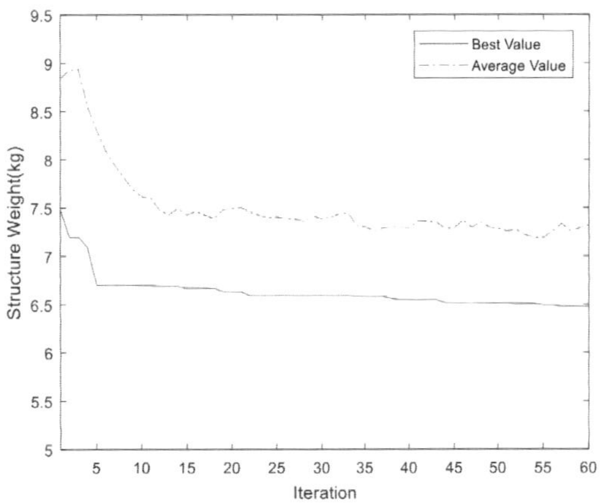

Case 1. one supporting stiffener

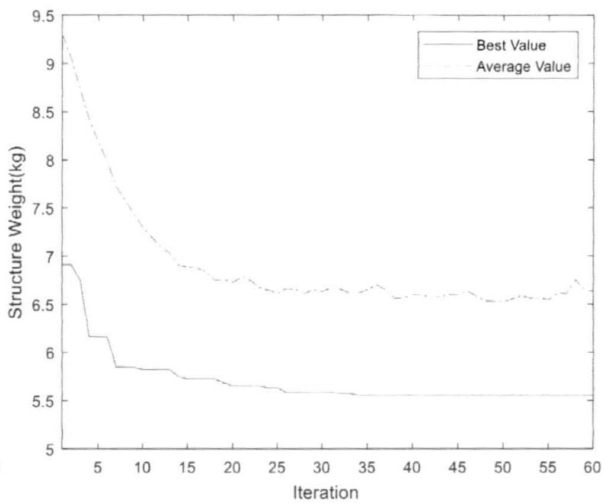

Case 2. two supporting stiffeners

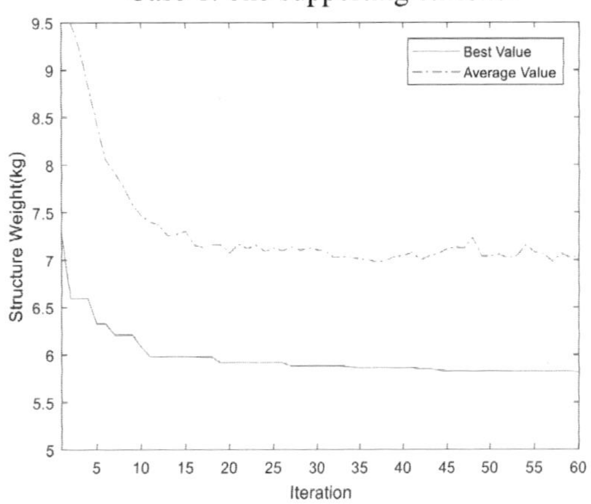

Case 3. three supporting stiffeners

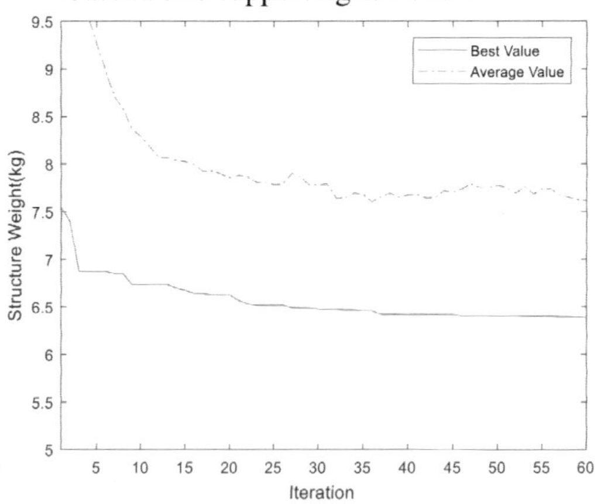

Case 4. four supporting stiffeners

Figure 5. Optimization iteration process.

Table 1. Optimization Results.

	Initial	Case 1	Case 2	Case 3	Case 4
n	3	1	2	3	4
b_1 / mm	60	40	40	40	40
b_2 / mm	60	30	32	35	33
H / mm	30	58	45	38	35
B / mm	50	42	40	40	40
t_1 / mm	4	2	2	2	2
t_2 / mm	4	5.9	4.5	3.9	3.6
P / kN	141.3	144.3	141.8	142.0	143.6
W / kg	8.763	6.647	5.504	5.824	6.381

4. Conclusion

This paper presents a method for calculating the axial load carrying capacity of stiffened panels considering the torsion of supporting stiffeners. The longitudinal stiffener is regarded as a multi-span compressive column supported by a series of transverse elastic supports and torsional elastic supports. By defining three dimensionless factors: the relative bearing capacity, the relative transverse elastic stiffness and the relative torsional elastic stiffness, the response surface of the bearing capacity versus two kinds of elastic supports is obtained.

Based on genetic algorithm, a lightweight optimization model for structure layout and section parameters is constructed. A certain stiffened panel is optimized on this basis. The results show that

a. In the case of having strong supporting stiffeners, the trend of optimization iteration is increasing the cross-section size of the longitudinal stiffeners and reducing the cross-section size of the supporting stiffeners.

b. When the structure contains different number of supporting stiffeners, the weight of the optimized structure has great changes. Compared with the optimization of section parameters, structure layout matching of two kinds of stiffeners has a greater impact on structure lightweight design.

References

[1] Noor A K, Venneri S L, Paul D B, et al. Structures technology for future aerospace systems. *Computers & Structures*, 2000, 74(5):507-519.

[2] Houston G, Quinn D, Murphy A, et al. Design rules for stiffened panel buckling containment features. *Thin-Walled Structures*, 2017, 116:69-81.

[3] Lanzi L. A numerical and experimental investigation on composite stiffened panels into post-buckling. *Thin-Walled Structures*, 2004, 42(12):1645-1664.

[4] Wan Q, Wang F H. Longitudinal Stability Analysis and Optimum Design of Supportless Long-span Deck Structure. *Shipbuilding of China*,2011,52(01): 17-25.

[5] Badran S F, Nassef A O, Metwalli S M. Y-stiffened panel multi-objective optimization using genetic algorithm. *Thin-Walled Structures*, 2009, 47(11): 1331-1342.

[6] Zhang G F, Duan S H, Wu C L. Method research for post-buckling analysis of stiffened structure base on FEM. *Structure & Environment Engineering*,2013, 40(02): 43-50.

[7] Kim S H, Kim C G. Optimal Design of Composite Stiffened Panel with Cohesive Elements using Micro-Genetic Algorithm. *Journal of Composite Materials*, 2008, 42(21):2259-2273.

IOP Conf. Series: Materials Science and Engineering **616** (2019) 012031 doi:10.1088/1757-899X/616/1/012031

Design of Laser Marking Machine Product Family Based on Ethnography

Ziyao Zhang[1] and Jinglian Chen[1,*]

[1] College of Arts and Design, Beijing Forestry University, Beijing 100044;

*Corresponding author, e-mail: chenjl@bjfu.edu.cn

Abstract. The study analyzed the product family design elements of laser marking machine, and was aimed to improve the family image and market competitiveness of laser marking machine. From the perspective of ethnography, using the product family design method, this paper analyzed and summarized the psychological experience of different groups of people on the family design of laser marking machine products, and used the results as an important design reference in the design and development of laser marking machine. On the basis of this, this paper made a conclusion that the ethnographic method can effectively analyze the design flaws and future development direction of products for different groups of people, and provided guidance for product family design.

1. Introduction

Product family design refers to what gives similar or even identical styling characteristics to different products produced by the same enterprise, so that makes it has a family-based identification factor, which results in a unified and coordinated effect between different products [1]. And product family, is considered to be an effective way to enhance product image and market competitiveness to increase product added value and brand effect. Laser marking machine products have a wide range of products under one brand because of their diverse use requirements, so family in the laser marking machine industry has become a key competitiveness in this field [2]. Product family not only strengthens the brand image of the product, but also shortens the design cycle, reduces development costs, and makes the product more holistic.

The family of products is often formed by having the same or similar characteristics of the same series of product design elements [3]. Like modeling situation, parts, colors, human-machine interface, etc. However, this feature creates a different psychological feeling among the technicians, salesmen and designers of products (expert group) and the expected users of the product (amateur group). Therefore, considering the needs of the two groups of people, the products can attract new users while being recognized by the regular users, which is beneficial to the product family for more diverse market demands and design iterations and innovations.

2. Ethnography and product family design

Ethnography is a group culture phenomenon that describes a common cognitive system of a certain group or a certain class in a society [4]. Different groups have different cognitive effects on the same thing because of their internal shared knowledge. In recent years, the methods of studying population culture and cognitive style in group culture have begun to be borrowed from product design. The different ways of life for different cultural groups enable products to meet diverse needs; the integrated perception of each group enables products to target a broader market. Discovering the gap between the design of existing products and user perceptions can discover new design cues and requirements. Product family design also introduced the concept of biological genetics, so a group of products with the same or similar DNA naturally form a product family. The current research is mainly based on the extraction and expression of product family DNA, and the use of perceptual engineering methods combined with the emotional image of the brand and user needs for the next step of design. The focus is mostly in the expression stage, but there can be an analysis process from the extraction to the expression. Due to the unique family characteristics and market environment of the laser marking machine, the existing analytical methods have limitations for it.

This paper adopts the ethnographic perspective and the method of questionnaire survey to further investigate the extract of the laser marking machine product family DNA. And cognitive differences were analyzed by different groups' recognition of the various components of the product. In combination with the design suggestions of the laser marking machine using habits and methods, the design elements of the laser marking machine are refined to provide reference for the design of future laser marking machines and similar products.

3. Laser marking machine and product family design

The concept of product family first appeared in engineering design, mainly researching the development platform of generalization, modularization and standardization to better meet the needs of customers for mass customization, and the same way can be applied to industrial design. There are different series of products in the product family, and the products have the same or different parts due to the difference in combination and function. The parts are composed of components and the components are composed of elements. The product shape generally has a lot of component shape features, so it is impossible to make all the component shape features become the part that attracts the user [5]. However, due to the high recognition degree and important functionality, certain form factors can become a visual focus.

Laser marking machines are not well-known electronic products. As an industrial digital product, there are relatively few existing researches. Also researches usually pay more attention to the design results. But the laser marking machine is a relatively simple industrial product with only a total of nine main components. Through the analysis of functionality and usability, combined with the product family theory, the combination of the nine main components and the family products can be divided into three parts: individual parts, general parts and overall shape. The overall shape includes the modeling ratio of the laser marking machine and the similarity of the product family. General parts form factors are generally not easily changed as outsourced parts or parts that are less popular in the product family. They are: display, table, lifting shaft and galvanometer. The individual parts of the components is functional and plastic, and it is an important part that can reflect the family characteristics. For this reason, the focus of this paper is also on these components. They are: optical cavity, laser head, chassis, control button, vents. Figure 1 shows the representation structure of the laser marking machine product family.

4. Group division

The semantic matching conveyed by product design is an important indicator for evaluating the quality of product design. However, different groups of people have different cognitive for the same product since their background cultures and values are rather different. By analyzing the cognitive differences between groups, the comprehensiveness of design can be improved; and high-quality designs with high cognitive matching are selected efficiently. For the industrial digital products such as laser marking machines, the popularity of the public is very low, so the polarization of their cognitive differences will be very obvious. Therefore, the laser marking machine research group can be divided into two categories: one is the factory technicians, operators, sales personnel, designers, etc. who have contacted the products in detail; the other is the general consumers who have never contacted the laser marking machine, who are called the expert group and the amateur group respectively. These two groups have a considerable gap in their understanding of the product, and there will be a gap in the subjective feelings of the laser marking machine.

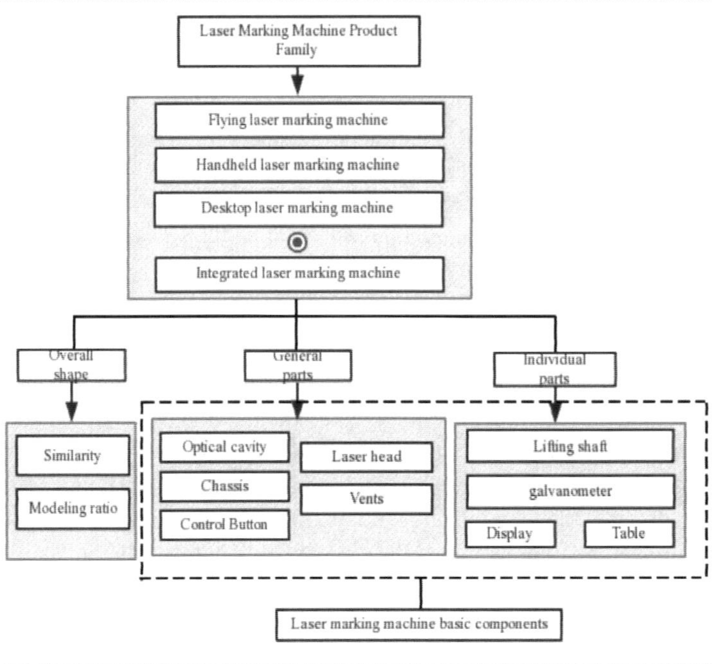

Figure 1. Structure of the laser marking machine product family.

5. Research methods

5.1. Ethnography research method

The ethnographic research methods are generally used in the initial user research phase of product design. The results of design surveys using this method are often more in-depth and intuitive than traditional research methods. It focuses on the users' existing dissatisfaction and future expectations about the product style. It not only studies the behavior and emotions of social groups, but also studies the attitudes, motivations and reasons behind it. Therefore, the design rules are further found in a diverse group, and through finding the difference and concerning between the group culture and the physical product will thereby generate the evolution of product design.

5.2. Questionnaire research method

This study used a questionnaire survey method to investigate the differences in the perception of product family among different groups by laser marking machine.

Experimental preparation: Search for images of different brands of laser marking machines from the internet, collect a total of 19 brands and 156 pictures, and then remove images that are too weakly family-based and have low quality. Finally, a total of 15 brands of 91 pictures were chosen, then the same brand of different types of marking machine is summarized into one picture. Each picture is a product family, which gives 15 product family pictures (Shown in Figure 2). The five-point scale consisting of individual parts, the expert and amateur group is scored by 1-5 based on the degree of family influence of each component on each set of images. At the same time, some subjective questions are combined to verify the accuracy of the data results.

Figure 2. Product family pictures.

6. Analysis of research results

6.1. Questionnaire analysis

A total of 30 product family design questionnaires were received, 15 for each of the expert group and the amateur group. The acquired data is processed by spss25.0. Firstly, the reliability analysis is performed on the two sets of data respectively. The result of the reliability analysis of the expert group is 0.74; the amateur group is 0.89, so the data of both sets are greater than 0.7, which proves that the data has internal consistency and credibility.

Each product family diagram will get a five-part score matrix. Figure 3 implies that there is no strong difference from the average, but the difference between individual samples is strong. Sample 9 has the strongest difference, and sample 15 and sample 8 have the highest composite score.

Difference sample analysis: The sample 9 has a large degree of family-based, and the difference between the series products is relatively large. The main family characteristics are vents and metallic shells, which means that the amateur group has low recognition of family products with the same material texture. Preference sample analysis: Sample 15 is most consistent with the current cognition of laser marking machines, indicating that the popularity of laser marking machine products is relatively low, and the general perception of this product is relatively old. The sample 13 has a higher score because the body color is composed of white and blue, which are the most satisfied colors of the expert and the amateur group.

6.2. Analysis of cognitive differences

Average and standard deviation analysis of each component of each group was shown in Figure 4. The average has different results, the average score ordering of the expert group is: optical cavity>laser head>control button>vents>chassis. The amateur group is: optical cavity > laser head > chassis > control button > vents. The data of the amateur group is more in line with the experimental expectations of this article, but the data of the expert group has the lowest score on the chassis. The Interview analysis of expert group shows that the chassis has the highest importance for the 60% of expert group. That means the chassis is given the highest expectations for the expert group but only the lowest satisfaction. And the average ordering of chassis, control buttons, and vents is not gradual. It also shows that there is a difference between the expert group and the amateur group.

In terms of standard deviation, it is obvious that the amateur is larger than the expert group. The reason is that the expert group will be more conservative in considering the functional importance while the score gap is therefore more conservative.

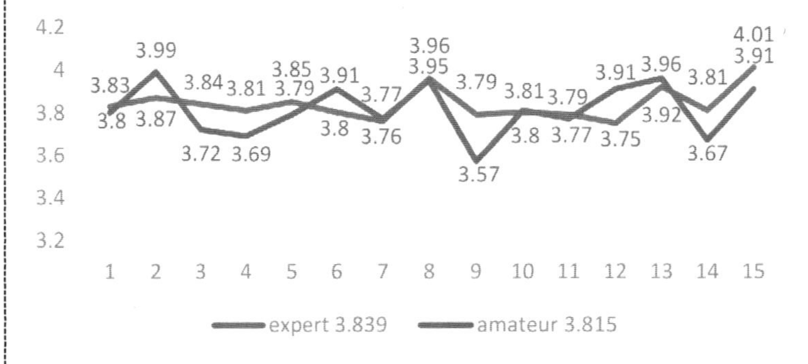

Figure 3. Matrix average score chart.

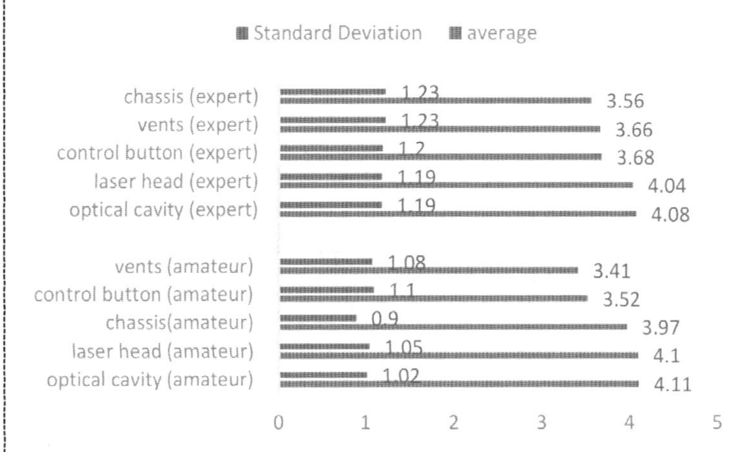

Figure 4. Average and standard deviation of each component.

6.3. Ratio difference analysis

The Wilcoxon test in the non-parametric test of the two related samples is used to compare and check. For the five components, Table 1 shows the two components of the chassis and the vents are smaller than the significant level value sig<0.05, which indicates that there is a significant difference between the amateur group and the expert group in the preference of the vents and chassis in the importance of the laser marking machine product family. Among them, the difference in cognition for the chassis is most obvious.

Table 1. Wilcoxon test.

expert-amateur	Optical Cavity	Laser Head	Chassis	Control Button	Vents
Significant difference	0.64	0.8	0.006	0.35	0.02

7. Discussion and conclusion

7.1. Discussion

Based on the above analysis, the design elements of the laser marking machine product family can be summarized. First of all, the chassis is the most indispensable part of the laser marking machine product family, but the most promising components. And for the chassis, the expert group and the amateur group have the most significant difference, so the design should be more comprehensive, considering the multi-faceted crowd experience. The design of the vents should be aesthetically pleasing on the basis of satisfying the function and the cognition difference. Secondly, the optical cavity and laser head are the most important product family components, both experts and amateur group have a common understanding. In summary, the sequencing of the family design of laser marking machines is: optical cavity, laser head, chassis, control buttons, vents. However, the design of the laser marking machine should start from the optical cavity but not just the optical cavity. White and blue are the most recognized colors, but the design still has to be combined with its own brand features to avoid product homogeneity. Finally, the design should not only be further improved in family, but also the new products should be able to update the cognition of each group. In the future, different groups can feel the evolution of product DNA.

7.2. conclusion

This paper used the application of ethnographic methods in the product family design, analyzed the different feelings of the expert group and amateur group for the same product, on this basis, summarizes the cognitive differences of different groups of people on the various parts of the product, and obtains the design essentials of the further design of the laser marking machine. However, this study only studies the evolution of product family DNA through two groups of laser marking machine products, so the follow-up work will be based on this, combined with emotional imagery, brand value and other factors to design the laser marking machine, improve this Design method.

References

[1] Shijian Luo, Wenjie Li 2016.9 M. *Product Family Design DNA.*
[2] Xin Hao, Jinglian Chen. *Modeling Optimization Design of Laser Marking Machine.* International Conference on Arts 2015.
[3] Di Cai, Jing qian 2007 *J. Machinery.* **37(5)** 45-48
[4] Xiaotong Chen, Xiang Gu 2014 *J. Packaging Engineering.* **20** 33-37
[5] Lu Liu, et al 2017 *J. Journal of Agricultural Engineering.* **17**

IOP Conf. Series: Materials Science and Engineering **616** (2019) 012032 doi:10.1088/1757-899X/616/1/012032

A Test Scheme Design for High Speed Active Cabin Separation

Wei Feng [1,2,a], Changxun Li [1], Fanwei Meng [1,2], Qiwei Chen [1,2], Wangmin Yi [1,2], Shengyu Zheng [1,2], Linzheng Wang [1], Chenxing Wang [1] and Lili Zhang [1]

[1] Beijing Institute of Spacecraft Environment Engineering, Beijing 100094, China
[2]Beijing Engineering Research Center of the Intelligent Assembly Technology and Equipment for Aerospace Product, Beijing 100094, China

[a] Fengwei_bisee@163.com

Abstract. In Mars explorations, the EDL (Entry, Descent and Landing) process is autonomously controlled, called "black 7minutes". In order to verify the EDL process of the Mars probe cabin (the heatshield facing the Mars surface and the aeroshell back to the surface of Mars) high speed separated, this paper proposes an active motor pull separation scheme and controlling method of axial force, normal force and torque multi-coupling, solves the technological problem of the force applied , followed and safe recycling cabin in short time and distance, establishes a test system and completed the separation test of the heatshield and the aeroshell by real product. The results show that the test success criteria are met and the test scheme is successful. This method and experimental system can be used not only in the separation experiment of Mars probe, but also in the separation or servo simulation of other detectors or general mechanisms, can be widely used or used for reference. This paper explains the general scheme and key technology, followed by other papers on ADAMS Simulation, influence on air-resistance and so on.

1. Introduction

In Mars explorations, The EDL (Entry, Descent and Landing) process is autonomously controlled, called "black 7minutes". In order to verify the separation design of the heatshield and the aeroshell in the EDL process of the Mars explorations, and to solve the problems of the deviation and uncertainty of the simulation verification, it is necessary to carry out the test on the ground to fully verify the design and the key technology.

In Flight test, During the development of the US Viking Parachute, 4 times separation tests of heatshield were carried out[1][2]. Before the Curiosity, there was no more high-altitude supersonic flight test[3]. The separation test of the heatshield (including on-orbit flight) of chinese manned spacecraft has been carried out 16 times. European ExoMars conducted a wind tunnel test for the process of separation of heatshield. MER (Mars Exploration Rover, in the US) used the POST2 software system to carry out the simulation analysis of the separation of heatshield[4][5]. EXOMARS(in Europe) carried out simulation of the separation of heatshield and aeroshell and establish the entire EDM model[6]. China Harbin Institute of Technology Deep Space Exploration Basic Research Center carried out Monte Carlo six-free numerical simulation verification of the full error model. According to the requirement of ground test for separation test, combined with simulation verification, this paper carries out the design of ground active separation test scheme.

IOP Conf. Series: Materials Science and Engineering **616** (2019) 012032 doi:10.1088/1757-899X/616/1/012032

2. Introduction

The biggest difference between ground test and on-orbit is that the gravity of the earth is 1g and that of Mars is about 0.33g, which requires compensation from ground test system. The weight of test equipment, such as test hangers, which increase the degree of freedom of motion of the specimen, will be added to the specimen, resulting in the difference of mass and moment of inertia. In order to make the ground test as consistent as possible with the on-orbit state, the equivalence of the stress state should be considered.The actual force state of the cabin in the short-term separation process (about 1 s) is shown in figure 1.The force state of the cabin on ground is shown in figure 2.

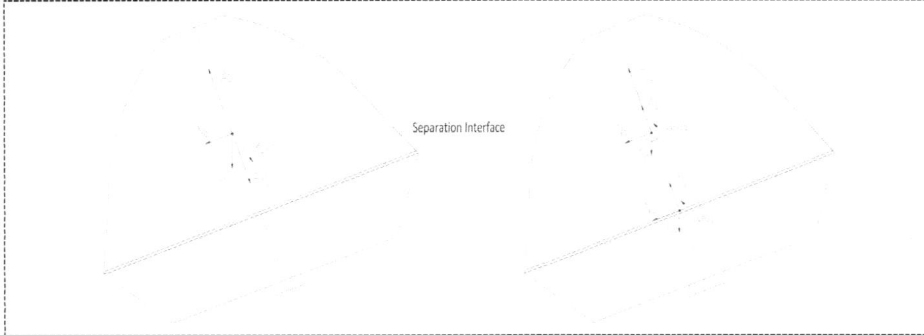

a) Before separation (in orbit) b) After separation (in orbit)

Figure 1. Force state of cabin separation on orbit

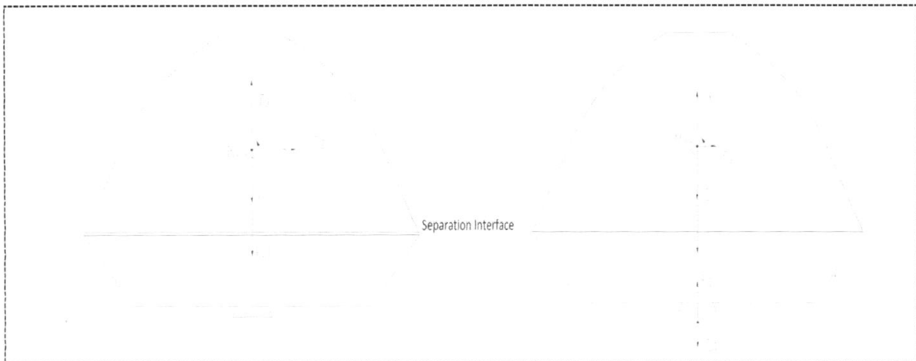

a) Before separation (ground)) b) After separation (ground)

Figure 2. Force state of cabin separation on orbit

Before separation:

$$Fx - G - F_{ab} = G_b \sin\theta - A_b - F_{ab} \quad (1)$$

$$Fz = N_b + G_b \cos\theta \quad (2)$$

$$My = C_{bmzg} \quad (3)$$

After separation:

$$Fx - G = G_b \sin\theta - A_b \quad (4)$$

$$Fz = N_b + G_b \cos\theta \quad (5)$$

IOP Conf. Series: Materials Science and Engineering **616** (2019) 012032 doi:10.1088/1757-899X/616/1/012032

$$My = C_{bmzg} \tag{6}$$

From the formula, when the earth gravity of the aeroshell is compensated by Fx, the axial force Fx-G, the normal force Fz and the torque My are all consistent with the on-orbit force state.

According to the existing conditions and test requirements, the suspension plan is put forward including Fully passive Test Scheme , Fully passive Test Scheme and Fully passive Test Scheme. Fully passive Test Scheme: Axial acceleration is less than 1g, not a constant value, not applicable to this working condition. The Test Scheme is applicable only when the acceleration accuracy is not strictly required. Hybrid Test Scheme: Detonation time and control time are difficult to synchronize, not applicable to this working condition. It is used when the requirement for accuracy is medium. Fully active Scheme: It is used when the requirement for accuracy is high Recommended, but Servo control system is complicated. Considering fully, only the active scheme can meet the test requirements.

3. Technical Scheme

3.1. Test technical scheme

According to the stress analysis, the stress realization scheme of the specimens is shown in the figure 3. The specimens are subjected to axial force, normal force and torque, in which the axial force is controlled by servo motor, and the normal force and torque are mainly controlled by springs. The test technical scheme is shown in Figure 3.

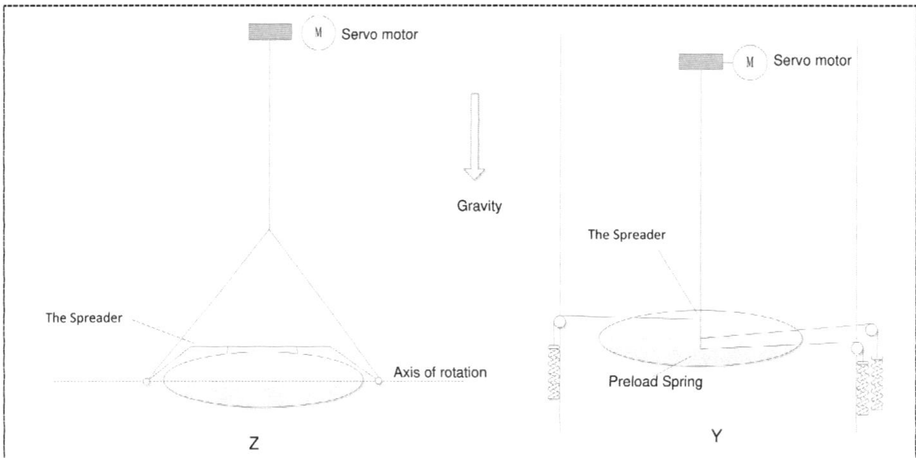

Figure 3. Test technical scheme

3.2. Difficulties and solutions

According to the above test characteristics, the main design difficulties of ground test system lie in force servo and cabin safety at high speed.

The normal force and the rotational moment of the aeroshell are applied at high speed conditions. There is no simple linear coupling between the normal force, the rotational moment and the axial force of the back cover. The normal force pulling spring assembly is used for the application of the normal force (horizontal direction) of the heatshield (the aeroshell), and the normal force under spring is shown in Figure 4. The torque force pulling spring assembly is used for the application of the torque (horizontal direction) of the heatshield (the aeroshell),and the torque normal force under spring is shown in Figure 5.

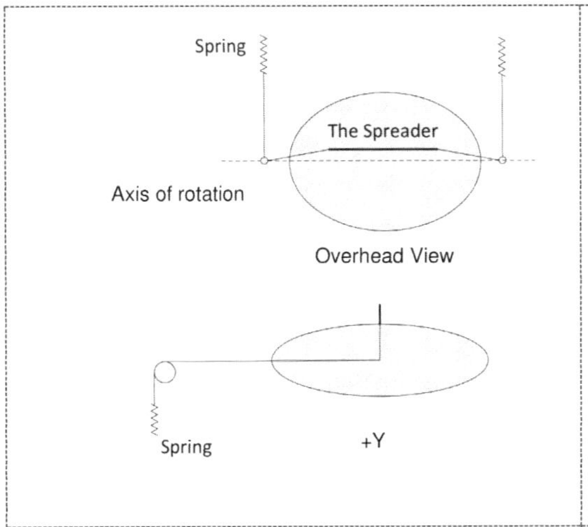

Figure 4. Normal force under spring

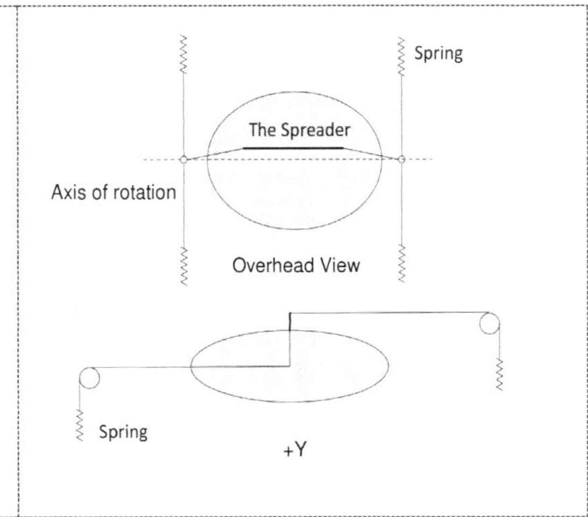

Figure 5. Torque under spring

Safe recycling of the cabin: After the test data is measured, the high-speed moving cabin needs to be decelerated to a standstill for recycling and retesting.

- A safety net is placed 8m from the ground.
- In the case of separation, to avoid the product bumping into the test stand, connect the calculated length of the protective rope at the connection part of the spreader and the product.
- Use balancing weight design to prevent malfunctioning products such as motors from falling off the ground.

4. Test verification

4.1. Components of Test System

The test system consists of six parts: the test platform, the test spreader system, the shock test system, the high-speed photogrammetry system, the product simulation part and the test auxiliary system.

Test platform: providing axial force application, normal force and rotational moment application for the separation process of the hanging cabin, and measuring the force applied in all directions. The stand size can be matched with the size of the test piece. Test spreader system: the separation and lifting of the hanging cabin under two kinds of separation conditions, the spreader and the hanging cabin connecting, the lifting point passes through the center of mass; the test spreader is connected to the test platform through the lifting interface. Impact measurement system: Measuring the shock response of the separation surface of the hanging cabin and the fixed cabin and other measuring points, and carrying out data processing; the strain measuring system measures the strain of key parts of the system. High-speed photogrammetry system: Measuring the separation speed, angular velocity and attitude angle of the suspended cabin; recording separation process. Test Auxiliary System: The auxiliary tooling and equipment used in the test process, which should meet flip requirement of the product as shown in the figure. Test Product Simulation Part: simulation parts with similar quality characteristics and force characteristics with real products, used to test the system.

4.2. Simulation

Through ADAMS simulation analysis, the following conclusions are given:

- Under the test conditions, no collision occurred within 0.1s of the heatshield and 0.5s of the aeroshell;
- Axial tension , air resistance compensation and spring axial component compensation value;

- The simulation results show that when the normal tension is along the -Z direction, and the torque is applied along the +Y, it is the limit condition, and the axial force of the heatshield of the motor system under this condition is given.

4.3. Experiment results

The test data of the main axial cable force, the upper torque force, the lower torque force, and the normal force are obtained by sensors under the aeroshell nominal conditions, and the curves are shown in figure 6, figure 7, figure 8 and figure 9.

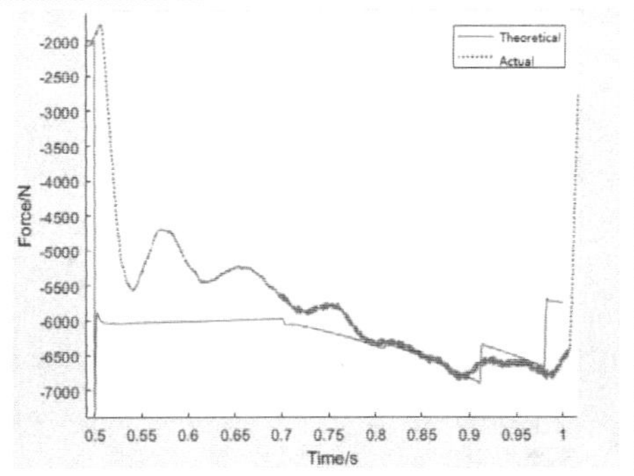

Figure 6. Main cable force of standard conditions

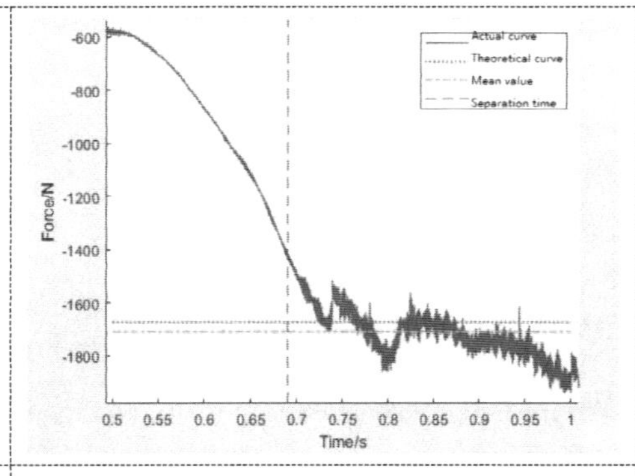

Figure 7. Upper torque force of standard conditions

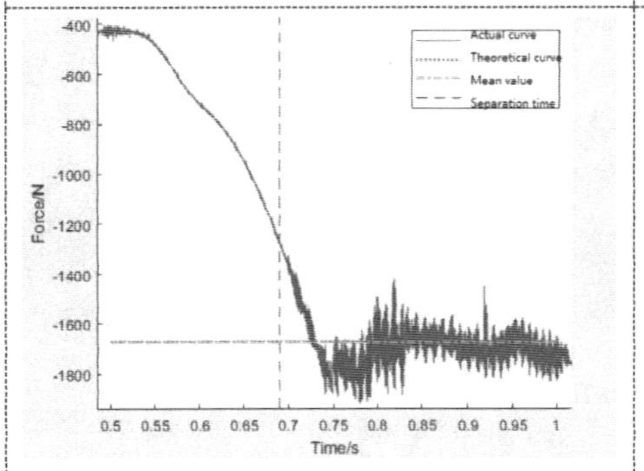

Figure 8. lower torque force of standard conditions

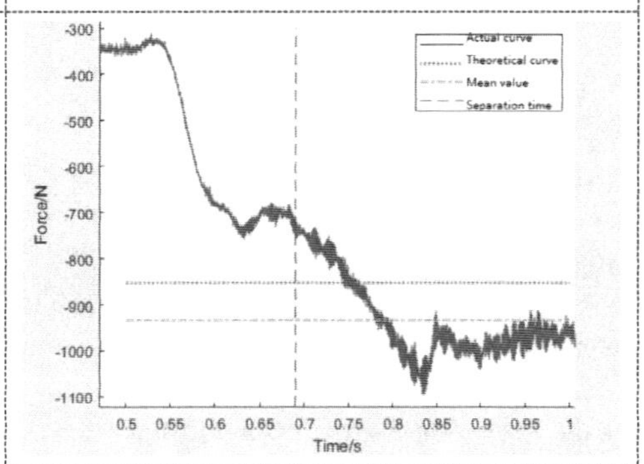

Figure 9. Normal force of standard conditions

The analysis of the results of the nominal condition test of the aeroshell meets the requirements. The comparison of actual and theoretical data is shown in Table 1.

Table 1. Result of nominal working condition of the aeroshell

	upper torque	Lower torque	normal	main cable
Theory	1672	1672	852	
Measured	1708	1674	933	6.46%

5. Conclusions

According to the EDL Process of the Mars probe cabin separated index, this paper proposes an active motor pull separation scheme. The test system is established, the Adams simulation analysis is carried out, and the ground validation test is completed. The following results are obtained:

IOP Conf. Series: Materials Science and Engineering **616** (2019) 012032 doi:10.1088/1757-899X/616/1/012032

- An overall scheme of active high-speed separation based on servo system and air spring is proposed.
- Adams simulation analysis is carried out to provide compensation values for axial tension, air resistance compensation and spring axial component.
- An experimental system has been established. The experimental results show that the experimental data meet the criterion of success and the experiment is successful.

This method and experimental system can be used not only in the separation experiment of Mars probe, but also in the separation or servo simulation of other detectors or general mechanisms, can be widely used or used for reference.

References

[1] Reginald R. Lundstrom etc. Flight Tests of Viking Parachute System in Three Mach Number Regimes, NASA Langley Research Center Hampton, Va. 23665, 197410

[2] J. R. Lang, D. A. Sarver, Experimental aerodynamic Characteristics of the Viking Lander Capsule During Aeroshell Staging, Martin Marietta Report TR-3720301, September 1972 (under contract NAS1-9000)

[3] Juan R. Cruz, Aerodynamic Decelerators for Planetary Exploration: Past, Present, and Future, NASA Langley Research Center, Hampton, Virginia, 23681-2199,2006

[4] Behzad Raiszadeh, Mars Exploration Rover Heat Shield Recontact Analysis, NASA Langley Research Center, Hampton, Va., 23681,2006

[5] S. A. Striepe, D. W. Way, A. M. Dwyer, and J. Balaram, MARS Smart Lander Simulations for entry, Desent, and Landing, AIAA Atmospheric Flight Mechanics Conference and Exhibit5-8 August 2002, Monterey, California

[6] Maurizio Capuano, Stefano Portigliotti , Franco Gilardi, Olivier Bayle, Thales Alenia Space – Italia, Strada Antica di Collegno 253, 10146 Torino – Italy, ESA/ESTEC, Noordwijk, The Netherlands, EXOMARS Mission 2016 EDM Design and Delvelopment Plan, IAC-2011.A3.3A.9

IOP Conf. Series: Materials Science and Engineering **616** (2019) 012033 doi:10.1088/1757-899X/616/1/012033

Test Research on Airdrop Shock Vibration of Armored Vehicles

Jun-fang HOU, Guang-chun YU, Wei-hua ZHANG, He-ping Wang and Yuan-yuan LIANG

Proving Group of Ground Vehicles, Beijing, China

Email: houjf1234@163.com

Abstract. In view of the airdrop adaptability of a lightweight high-mobility wheeled armored vehicle, the shock vibration of the vehicle's onboard electronic equipment under simulated airdrop was tested and the shock characteristic of the equipment was analyzed. The test results show that the onboard equipment of the airdrop vehicle is mainly subjected to three stages of shock vibration, which includes the crash of the lifting rope, shock of the cushion airbag and the rebound shock. The cushion airbags can effectively reduce the shock acceleration peak of the onboard electronic equipment of the light wheeled armored vehicle, and the shock peak of the onboard computer and power supply is about 9-10 g with the shock duration of 0.15-0.2 s. While compared with the shock of the base, the buffering effect of the isolators to the airdrop shock vibration is not satisfied. Resonances of the isolation system are caused by the airdrop shock and the shock peaks of the tested electronic equipment are higher than that of the base.

1. Introduction

With the development of military science and technology, the ground wheeled armored vehicles are developing towards the direction of lightweight and high-speed, and the high mobility is becoming a key development direction. The rapid delivery of ground vehicles to the battlefield by air transport has become a typical mobility mode of modern warfare, which has attracted much attention from the armies of various countries. Although the airbag cushion system can greatly reduce the shock at air landing, the vehicle body and the onboard equipment will still bear large shock vibration. Therefore, in the development stage of the equipment, shock vibration laboratory test is needed to test the impact resistance of the equipment. In GJB150.18A-2009[1], it is recommended to use the measured shock vibration for bench test, and the shock load will be performed according to the typical shock peak and duration given in the standard if no measured data are available.

Due to the disadvantages of high cost and long test period, the simulated airdrop is often adopted to replace the practical airdrop in the adaptability test of the vehicle equipment, so as to reduce the test cost and shorten the test period. Domestic and foreign scholars have researched on the shock test method and shock spectrum data processing method. The finite element model of airbag and vehicles was studied by Huangjie HONG[2]. Yibing HU[3] studied the filtering method of shock vibration data and verified the application of wavelet analysis in airdrop shock signal processing. Xiaobo LEI[4] analyzed the shock response spectrum of aircraft when landing, and obtained the digital filter coefficient of shock spectrum by using the ramp response invariability method. Shoujun LIU[5] simulated the airbag landing process based on ANSYS and compared with the test results. Xin Xu[6] simulated the dynamic landing process of the airdrop and obtained the changing stress.

Aiming at the airdrop adaptability of a light wheeled armored vehicle, the shock vibration of the main onboard equipment in the vehicle was tested by simulated airdrop experiment, and the shock vibration characteristics of the onboard equipment are analyzed.

2. Airdrop Cushion System

The airdrop cushion system for vehicle equipment and cargo mainly includes parachute, cargo platforms and cushion airbag. The airdrop process is shown in figure 1. During the airdrop of vehicles, the main parachute is drawn out by the traction parachute system, so that the vehicle maintains falling at a constant speed. At the same time, the conformal balance lever keeps the cushion airbag inflated by the air inlet under its own gravity. During landing, the airbag compresses the air under the shock load, and pushes the air outlet to open. The air in the airbag is discharged at a specified flow rate, the falling speed of the vehicle is reduced at the same time. Finally, the airbag lands at a safe speed, which owes to the cushion effect of the airbag.

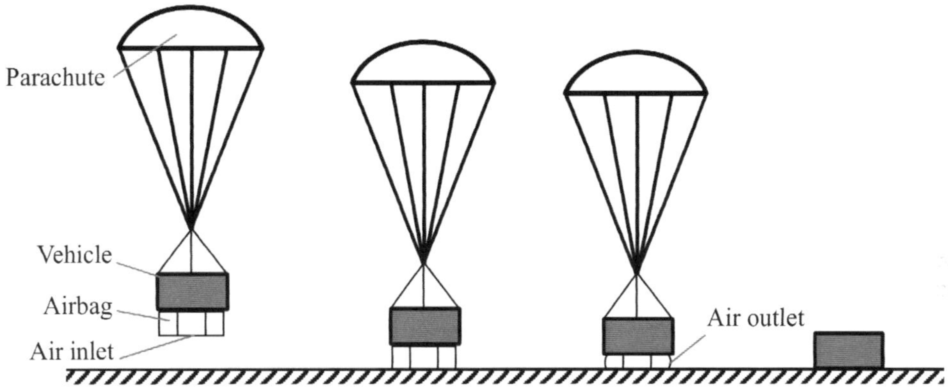

Figure 1 Airbag cushion process of vehicles

The structure of the airdrop cushion airbag is shown in figure 2. The overall structure adopts the cushion airbags without cargo platform, and two supporting beams are connected to the bottom of the vehicle body. There are four air bags connected under the supporting beams with volume of about 12m^3. Two conformal balance lever are connected with the bottom of the airbags, which can keep the airbags filled with air when landing.

Figure 2 Structure of the cushion airbag

3. Test System

The schematic diagram of the shock vibration test system is shown in figure 3. The test system for shock vibration of simulating airdrop is composed of multi-channel data collector, shock acceleration sensors and the acquisition software. The single channel sampling frequency of LMS multi-channel data collector can reach 200kHz, which can meet the requirements of shock acceleration test with high response frequency. The frequency response range of shock acceleration sensor is 0.2Hz-25khz and the measuring range is ±5kg, which can meet the peak test requirements of shock vibration.

IOP Conf. Series: Materials Science and Engineering **616** (2019) 012033 doi:10.1088/1757-899X/616/1/012033

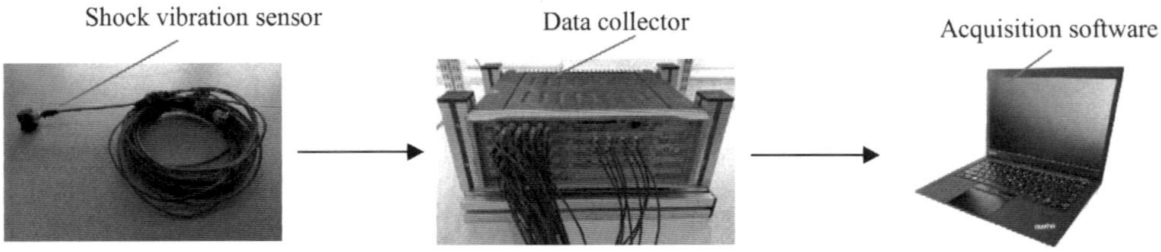

Figure 3 The shock vibration test system

4. Test Results and Data Analysis

4.1 Simulating Airdrop Test

The landing speed of vehicle equipment during airdrop is generally 7-9m/s, and the drop height of the simulating airdrop can be calculated according to the landing speed of 9m/s, which is 4.1meters. Firstly, the shock acceleration sensor was installed on the base and body of the onboard electronic equipment, as shown in figure 4. The data cable is connected to the data collector outside the vehicle, and the sampling frequency is set to be 100kHz. The airdrop cushion system was fixed at the bottom of the tested vehicle, and then the tested vehicle was lifted by a crane with airdrop ropes by 4.1 meters above the ground. Then the rope was suddenly released, allowing the tested vehicle to fall freely to the ground, and the shock acceleration signal during the landing was tested and recorded.

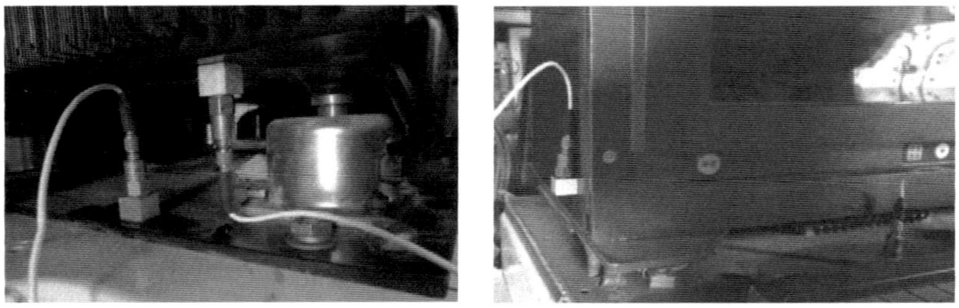

Figure 4 Position of the shock acceleration sensors

4.2 Data Processing and Analysis

Literature [7] pointed out that filter should be avoided to apple in the shock acceleration measurement due to the continuous spectrum with unlimited bandwidth of the shock vibration in the frequency domain, however, the shock vibration of the actual movement is often characterized by complex time domain waveform, which requires low-pass filter to eliminate the mechanical and electrical interference, and to filter the resonance and high frequency noise of the accelerometer. These noise signal superimposed on the shock acceleration waveform, affected accurate interpretation of the peak acceleration. In the shock acceleration measurement, low-pass filtering is usually required, and the cut-off frequency of the low-pass filter is generally not less than 10/T Hz, where T is the pulse duration of shock acceleration.

According to the preliminary analysis of the measured shock acceleration signal, the duration of the main pulse is about 0.15-0.2s, so the cut-off frequency of low-pass filtering should not be less than 70Hz. In this paper, the time-domain signal of impulse acceleration tested was processed by a fourth-order Butterworth low-pass filter, and the cut-off frequency of the filter was set to be 200Hz. Figure 5 shows the original shock vibration signal and filtered signal of the communication device and control terminal.

It can be seen from the shock signal of the communication device and control terminal that the shock signal mainly includes three shock periods. According to the recorded video in the simulating airdrop, it can be analyzed that the first impact is caused by hitting of the lifting rope to the top deck of the vehicle after release due to elastic force. Its time range is short, and the devices connected to the top deck will have a larger peak value. The second shock signal is the main shock signal, which lasts for a long time and has a large peak value. It is the sudden shock generated from the airdrop cushion airbag touching the ground to the first landing of the vehicle's bottom until the air in the bag fully exhausted. Due to the buffering effect of the cushion airbag, the duration of the shock is prolonged and the peak acceleration is effectively reduced. The third shock signal has a small peak value and a short duration, and there are several small peaks in the base position of the onboard equipment, where the peak value is caused by multiple bounces after hitting the ground.

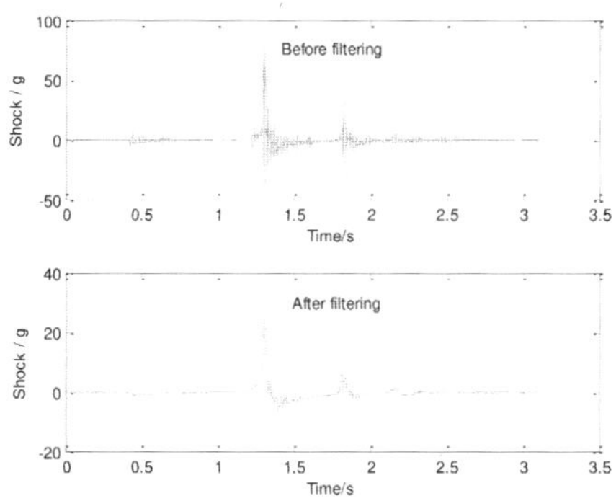

Communication device

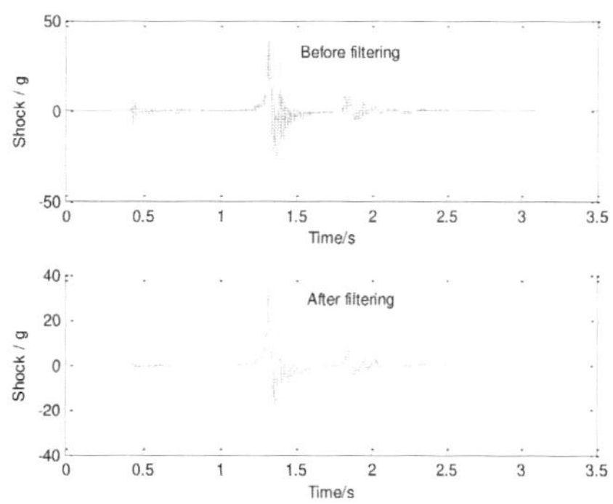

Control terminal

Figure 5 The shock vibration signal before and after filtering

Figure 6 and figure 7 is the filtered shock signal of the base and body of the onboard computer and power supply. it can be seen from the tested shock vibration that the cushion airbag can effectively reduce the landing shock peak of the onboard electronic equipment of the light wheeled armored vehicles, and prolong the shock duration period. The peak acceleration of the two equipment is about 10-15 g with shock duration about 0.15 0.2 s. While the buffer effect of isolators of the electronic equipment is not obvious, and the shock peak value of the equipment is higher than the impact of the installation base, which is attributed to resonance of the isolation system induced by the shock vibration.

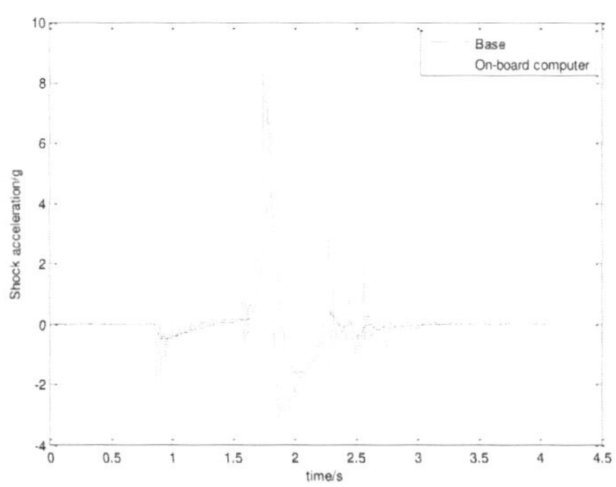

Figure 6 Shock signal of the base and the onboard computer

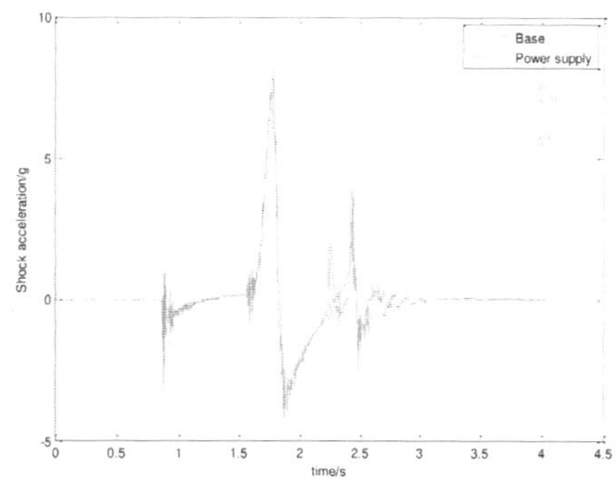

Figure 7 Shock signal of the base and the power supply

5. Summary

Through the test and analysis of the impact vibration of the electronic onboard equipment of a wheeled armored vehicle during airdrop, the test results show that the onboard equipment of the airdrop vehicle is mainly subjected to three stages of shock vibration during the simulating airdrop, which includes the impact of the lifting rope, the shock of the cushion airbag and the rebound landing shock. Cushion

IOP Conf. Series: Materials Science and Engineering **616** (2019) 012033 doi:10.1088/1757-899X/616/1/012033

airbag can effectively reduce the shock acceleration peak of the onboard electronic equipment of light wheeled armored vehicles. The peak impact of the equipment is about 10-15g, and the impact duration is about 0.15-0.2s. However, the buffer effect of isolators of the electronic equipment is not obvious, and the shock peak value of the equipment is higher than the impact of the installation base, which is attributed to resonance of the isolation system induced by the shock vibration. Compared with the rear peak sawtooth wave of the ground equipment recommended in GJB150.18A-2009, the peak value of airdrop shock vibration of the measured equipment is relatively low with prolonged duration.

References

[1] GJB 150.18A-2009. Laboratory environmental test methods for military materiel - Part 18: Shock test.
[2] Huangjie HONG, Hongyan WANG. Research on Assessment Method of Airbag Cushion System for Airdrop Equipment. Journal of Mechanical Engineering, 2015, 51(4):148-155
[3] Yibing HU, Jie ZHANG. The Filtering Method of Shock Test Data[J]. Journal of Test and Measurement Technology, 2013, 27(6):484-489
[4] Xiaobo LEI,Xiaobao DENG. Analysis of Shock Response Spectrum for Aircraft Landing. Science Technology and Engineering, 2014,14(34):298-301
[5] Shoujun LIU. Simulation and experimental study on ramjet airdrop airbag cushion performance. North University of China, 2015
[6] Xin Xu. study on of the landing process Simulation of ram-type airdrop cushion airbag. Mechanical Engineering & Automation, 2017, 2(201):59-63
[7] Yiqing CAO, Shanming LI. Effect of the Low-pass Filter Cut-off Frequency on the Shock Acceleration Calibration. Review of Science and Technology, 2008,26(16):81-84

IOP Conf. Series: Materials Science and Engineering **616** (2019) 012034 doi:10.1088/1757-899X/616/1/012034

A Discussion on Furniture Rental Service for Urban Migrants in China——From the Gain of a Lifestyle Design Research in Hong Kong & Beijing

Y Du*

Department of Industrial Design, Dalian University, No.10 Xuefu Street, Dalian, Liaoning, China

*Ruby830619@icloud.com

Abstract. This article describes the results of a study on lifestyle design research in relation to the consumption of space & furniture of the urban migrants in China. The methodology adopted was based on online investigations of life patterns and residential situations of the straits and desires concerning with furniture and space. The results showed that Chinese urban migrants might be a suitable audience for renting furniture service, while the persona 'Artistic Youth' seems to be the most potential consumer group. The study found that an adequate service should be considered with Functional categories, Styles, Grade, Price and Accessory products, then a viable furniture rental model was initially conceived. The study also proposed the different emphases on furniture rental service between Hong Kong and Beijing: HK's would focus on practical function and full use of space, BJ's should pay more attention to whether it can improve the quality and taste of living.

1. Background

In 2016, there were approximately 245 million migrant workers in China, accounting for 1/6 of the total population, whose average age is less than 30 (29.8) [1]. In Beijing, 40% of the resident population are urban migrants (8.08 million) [2]. In Hong Kong, 10404 mainland talents have come in through a 'Quality migrant admission scheme' and a total of 57,000 mainland people settled in this city within the same year [3].

China's urbanization begins in the early 1980s, which is developing rapidly but at a low level [4]. It has brought about tremendous changes to municipal environment and lifestyles. In order to understand the living conditions of the urban migrants, HK Polytechnic University launched a lifestyle research project in Hong Kong and Beijing to investigate the consumption of space & furniture for them. This was a design research project which aims at investigating daily utilization habits and primary requirements of space & furniture. On the chance to provide deeper insights of spatial and furniture consumption preferences to formulate appropriate design ideas for them.

2. Lifestyle design research

Lifestyles are important because the ways of living can be transformed into consumption patterns, while specific lifestyles showed certain rules of consumer goods, we can also comprehend ways of living by studying the consumption patterns [5]. Lifestyle research can provide richer information than simple demographic data because: (a) Lifestyle is a group phenomenon; (b) Lifestyle covers all aspects of lives; (c) Lifestyle reflects individual's core life interests; (d) Lifestyle shows differences in various demographic variables [6-9]. The term has a more restricted sociological meaning in reference to the distinctive style of life of specific status groups, with contemporary consumer culture it connotes individuality, self-expression, and a stylistic self-consciousness [10-13].

Based on the lifestyle space framework (**Figure 1**) of Benny Ding Leong in Asian Lifestyle Design Lab (ALDL) in HK Polytechnic University which is designed for China and comparable areas of Chinese society and culture, the research focus and tool selection were determined from four aspects including context (structural factors), behavior (material consumption), organization (consumption pattern) and ideology (attitude). It discusses the living mode, residential property, cultural background, consumption habit and ideology of urban migrants from two dimensions of space and time.

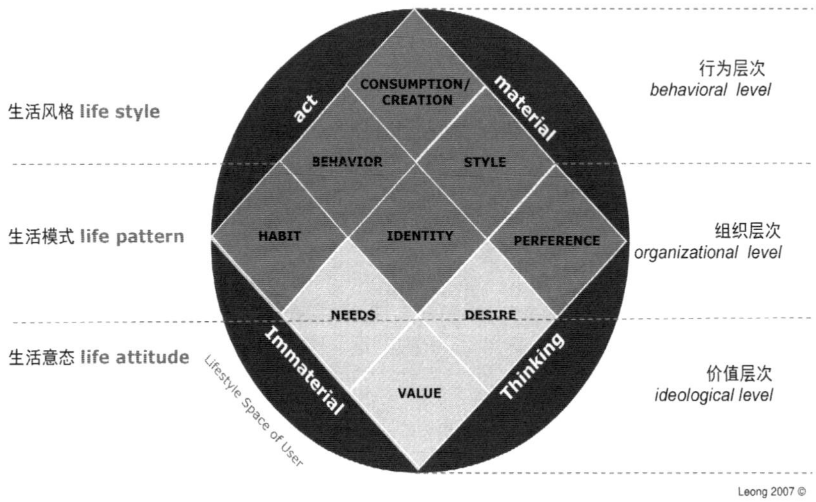

Figure 1. Lifestyle space (Leong, 2007)

3. Research process review

Given the diversity of respondents, a pilot survey was conducted online to identify the target population. It shows that our audiences have many characteristics in common: young, highly educated, under a transition stage from college to society, engaged in mental work, currently lack of wealth reserves and living in rental houses, etc. After classification and filtering, 20 participants (10 from HK, 10 from BJ) were recruited as target samples. We sent each of them an electronic document package which included a personal demographic card, a series of questions on new house purchasing & home decoration, a VALS [14] questionnaire, a set of photo journey tasks, and a pair of lifecycle diagrams of workdays and holidays. After receiving their feedbacks, we organized and analysed their responses and conducted in-depth interviews immediately. The research process is shown in **Figure 2**.

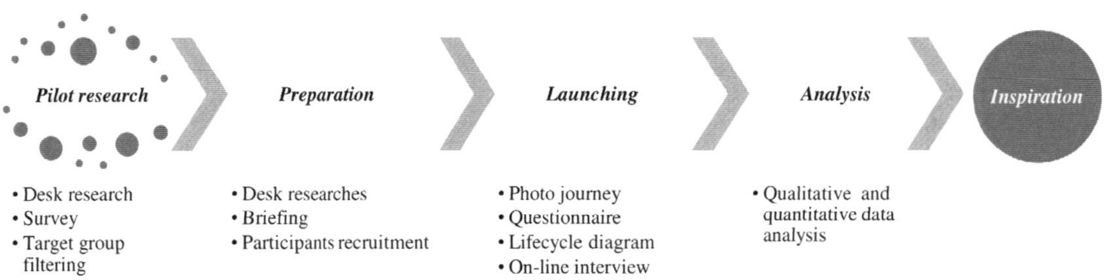

Figure 2. Project research process

IOP Conf. Series: Materials Science and Engineering **616** (2019) 012034 doi:10.1088/1757-899X/616/1/012034

4. Findings

Our participants showed great enthusiasms, many of them expressed hopes on following up the project. On the basis of information and desires they shared, we began to understand their hidden ideas and to sum up the main straits and gaps faced by their lives, and finally found some research directions which could be explored further. For instance, a good modal of furniture rental service would be a helpful choice for them.

4.1. Why may urban migrants need furniture rental services?

It is found that most of new urban migrants are living in rental houses. All our respondents have changed their residence during last three years. Migrants described their current living conditions as 'Temporary Arrangements'. Under such circumstances, they did not want to spend too much energy or money on shelters. Many of them would rather have a rough and miserable living condition than frequent moving and disposing of old furniture and necessaries.

Although many migrants complained about the extremely bad quantity and quality of the existing furniture, they would be very careful to consider buying a new piece unless absolutely necessary. This is because many landlords forbid tenants to leave their belongings when they left, and the process of transportation and disposal is extremely troublesome.

4.2. Which type of urban migrants in China may ask for the service of rental furniture?

4.2.1. Traditional conceptual limitations. The Chinese character '富' represent the meaning of 'wealthy'. It has a radical '宀' on the top symbolizes 'houses', a '一口' in the middle symbolizes 'male labor forces', and a '田' at the bottom symbolizes 'fields'. More than 2000 years farming economy generated a traditional Chinese wealth view: Thinking highly of the tangible asset.

Compared with other nations, Chinese attach more importance to ownership than usufruct. They tend to accumulate material wealth for future generations. Under the impact of entrenched indigenous value, it may not be that easy for traditional Chinese to accept an occupying form as renting.

4.2.2. Persona classification. With this concern, we asked our participants 'Will you consider renting furniture if the quality is guaranteed and the price is reasonable?' Beyond our expectation, half of the 20 participants expressed a willingness on trying. Although we have not done extensive research all over the country, this rate is conceivable higher than an average proportion.

In order to explore what type of urban migrants may ask for the service, we clustered all the interviewees into 3 types of persona: Pragmatist, Materialist and Artistic Youth. Each persona has his/her typical character tags. (**Figure 3**)

From the classification results, persona 'Artistic Youth' shows a most willing to try furniture rental services. It accounted for 9 of all 20 participants. Most of them are around 30 years old, have stable relationships or just stepped into marriages.

They are vanguards, usually have busy work with nice income, have independent minded and clear pursuit of future. They hold avant-garde values, comprehensive abilities and spirits of exploring. They would love to build a simple but quality life, while the current living environments still leave big gaps from their ideal states. Therefore, they seem to be the potential consumers of an early furniture rental service. And with the development of the economy, the proportion of such cluster will continue to increase.

IOP Conf. Series: Materials Science and Engineering **616** (2019) 012034 doi:10.1088/1757-899X/616/1/012034

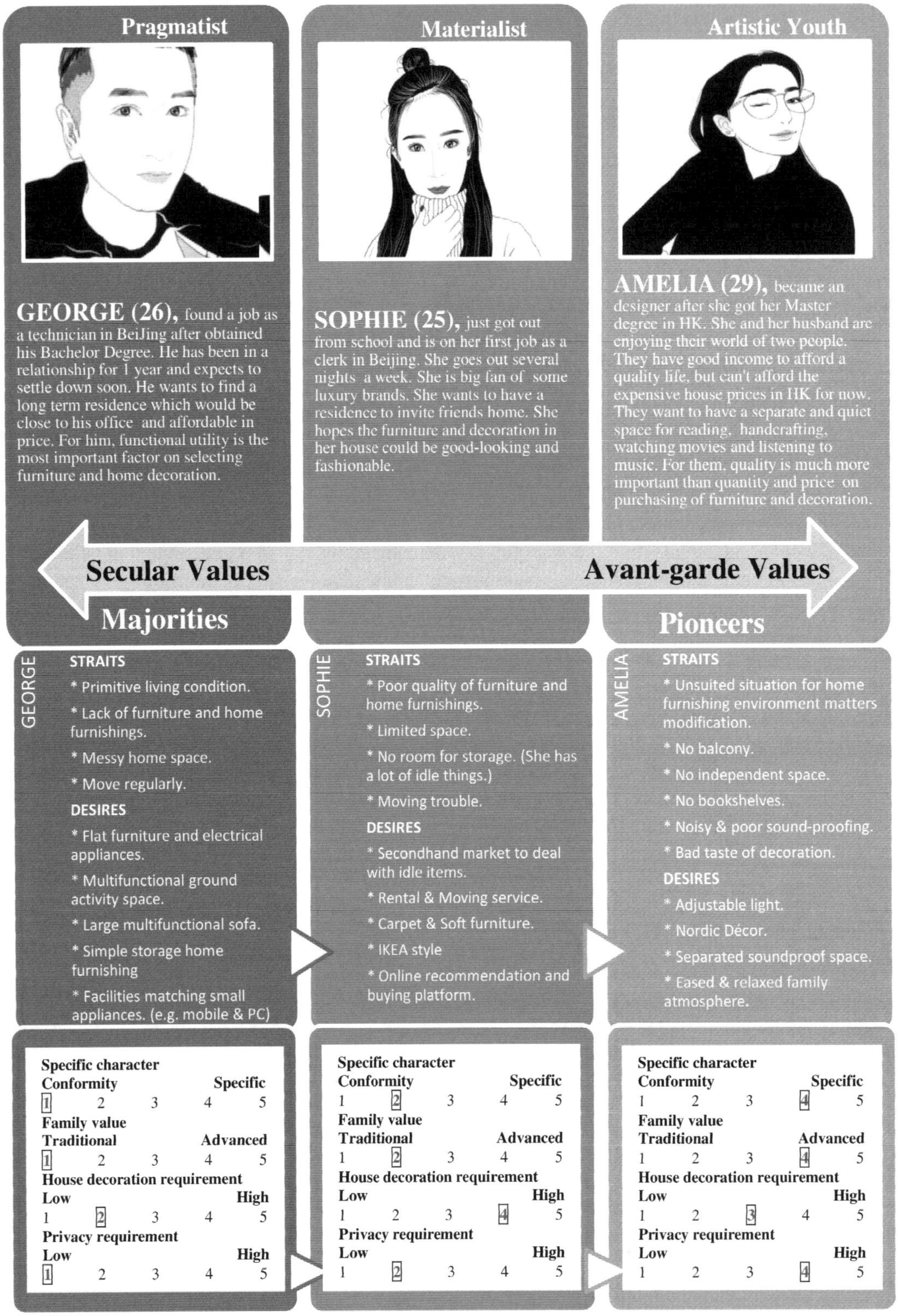

Figure 3. Persona profile

4.3. Which types of furniture might be the most popular ones?

4.3.1. Functional categories. From the feedback of lifecycle diagrams and in-depth interviews, we know that urban migrants seldom cook at home. They spend much time on PCs and smartphones. Their longest stays are in bed or in sofa. Hypothetically, if someone's willing to customize a piece of furniture for them, what they want is a large and soft couch or a comfortable and spacious bed. Besides, many of them desire to have a set of terminal furniture which can meet all requirements of relaxing, working and dinning at home.

4.3.2. Styles. Nordic Style is the highest frequency word mentioned by the interviewees. From the pictures they chose for their ideal home decoration, the following keywords can be used to describe their favorite home style: simplicity, modern, warm, bright, soft, wooden.

4.3.3. Grade & Price. 10 respondents mentioned IKEA during interviews. This brand wins itself a place in its stylish design and affordable price. People described IKEA as a good choice for rental phases, but they would love to choose other brands of better quality when decorating their own houses. Regards IKEA as a reference, at its price, consumers will tend to buy rather than rent. For a rental mode, a more upscale and expensive category might be more suitable.

4.3.4. Accessory products. Since some interviewees said they like to do activities on the floor, a soft carpet will be sought-after. Since some respondents wondered a romantic family atmosphere, an adjustable light may be a good helper. Since some others wanted to invite friends over for weekend, a set of home theater equipment will be popular. Decorative accessories (as rugs, mirrors, artworks), household appliances (as DVD, televisions, hi-fi equipment) can also be added into the list to meet the needs of users and to improve their quality and taste of livings.

4.4. What're the keys to a successful furniture rental service?

The interviewee who refused to try furniture rental service declared two major concerns: **Fear of damage to rented furniture** and **Dislike using a second-hand furniture.** It appears that a value and outstanding rental service should both provide high quality products and professional supporting services. Firstly, there should be dedicated service structures guaranteed for furniture's newness and cleanliness. Secondly, there should be a varied and appropriate price setting with short and long term plans available to suit different budgets. Simultaneously, there should be professional supporting service to take care of delivery and setting up from the beginning till removal of furniture in the end. **Figure 4** shows an ideal furniture rental process here.

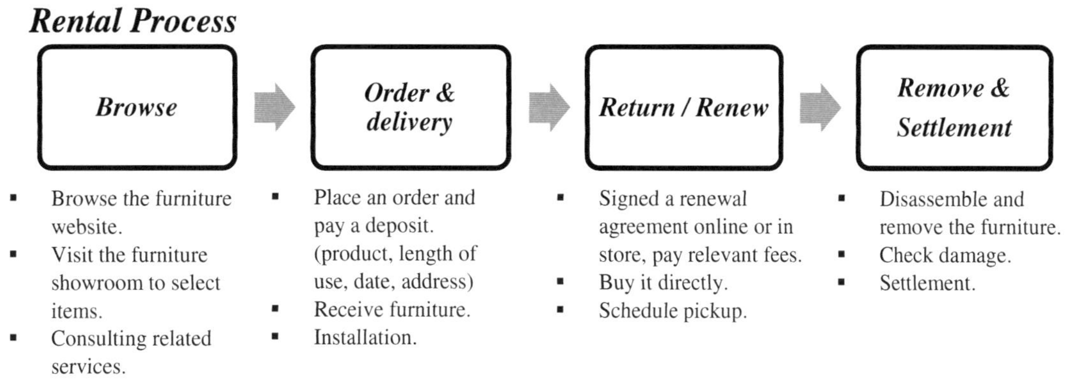

Figure 4. An ideal furniture rental process

5. Discussion

5.1. Different emphases on furniture rental service between Hong Kong and Beijing

The primary objective of furniture rental service for urban migrants is to meet functional requirements and improve the user's quality and taste of life. Due to the social system and cultural differences, the living conditions and matching house rental model vary greatly between Hong Kong and Beijing, which inevitably affects the implementation of furniture rental services and design priorities.

Frankly speaking, any rental service in Hong Kong must encounter a very embarrassing situation - there has no extra space can be used to exploit. In Hong Kong, with the outrageous housing prices and extremely limited living spaces, no material product is as valuable as disposable dwelling space. Customarily, no matching furniture is prepared for a rented house, in addition, the landlord requires the tenant to dispose of his own furniture when he left. This may provide a market for renting practical necessities furniture (such as beds) for some short term tenants. From our interviews, Hong Kong respondents wanted furniture to be maximize functionality and saving space.

Meanwhile, furniture rental service shows a better prospect in Beijing. Most Beijing migrants are living in rental houses with shabby facilities. They've got certain financial capabilities, but are resigned to the present states. Beijing's living space is not as cramped as imagined, many rental rooms still left plenty of idle space.

Furthermore, urban migrants in Beijing hold relatively higher requirements for the quality and taste of living environment. They are more sociable. (Of the 10 participants interviewed, 8 said they were very gregarious and wanted to invite friends to their homes.) They pay more attention on face-saving. Compared with Hong Kong interviewee, they engage in more leisurely work and spend more time at home.

From all these points of view, migrants in HK and BJ would have different priorities when renting furniture: HKs would focus on practical function and full use of space, BJs would pay more attention to whether it can improve the quality and taste of living. **Figure 5** illustrates some of the design entry points for migrants in Hong Kong and Beijing.

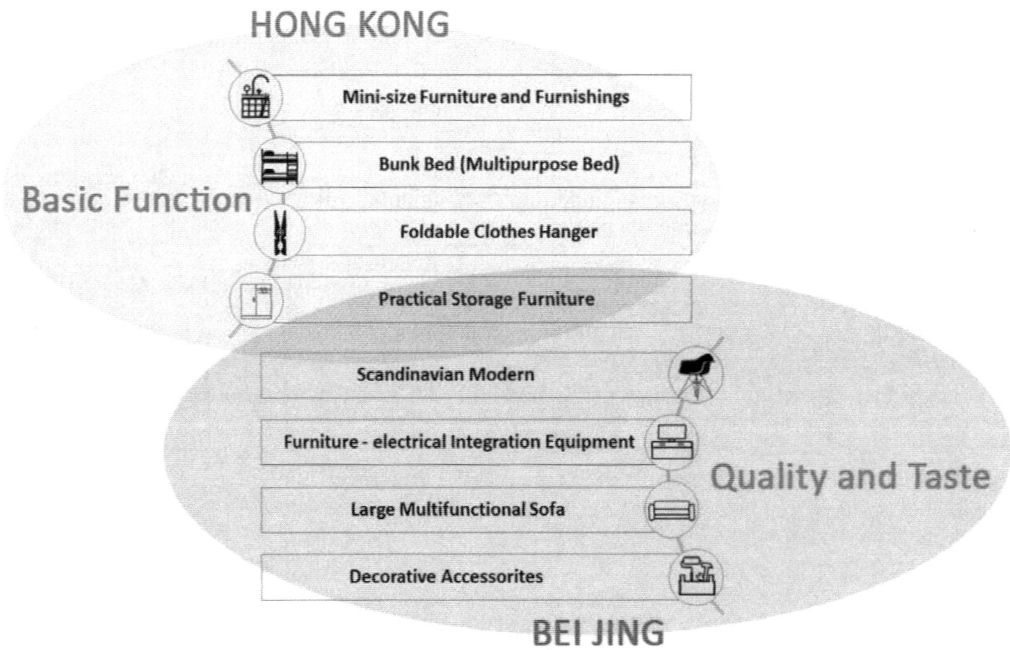

Figure 5. Emphasis of furniture leasing service in Hong Kong & Beijing

5.2. Limitations and directions for future research

Research on rapid urbanization and floating population has been a pop topic for years, and the project presented here could be regarded as a preliminary exploration of Lifestyle Research on it. As with other qualitative research work, there are still limitations should be noted.

Firstly, although this study was conducted in two mega cities with representativeness of different regions, cultural developments and social systems in China, the results may not be generalized to the entire population of urban migrants in Hong Kong and Beijing. The screening of the study objects also has limitations restricted to the channel and platform of participant recruitment.

In addition, the methodology employed in this work were all carried out in online forms. The validity of the migrants' self-reported data on lifestyle patterns is also a concern. And there was not much opportunity to control the seriousness of the participants and the authenticity of the feedback information.

Besides these limitations, the current research provides very specific starting points for future research. For instance, a group of urban migrants are found as potential consumers of a compatible furniture rental service, which might serve as an independent valuable subject for further practicality study. Integrating the correlations between consumer habits and lifestyles, we found it is well worth the effort to make product (service) design relevant to daily life-pattern analysis, environmental psychology variables and core value induction, which may expand the research model to a wider group of audience and improve the understanding of consumer demands.

Reference

[1] National Health and Family Planning Commission of the PRC 2017 *China's Migrant Population Development Report*
[2] Beijing Municipal Bureau of Statistics 2017 *Beijing Statistical Yearbook* (Beijing: China Statistics Press) 3-2
[3] Census and Statistics Departiment Hong Kong Special Administrative Region 2017 *Demographic Trends in Hong Kong 1986-2016* (Website of the Census and Statistics Department www.censtatd.gov.hk)
[4] Fan C C 2012 *China on The Move: Migration, The State, And The Household*（London: Routledge） 15–22
[5] Leong B D Lee B 2017 *The internet of life: Changing lifestyles and sustainable values in fast-developing China and India*. Routledge Handbook of Sustainable Design (London: Routledge) 275-276
[6] Anderson W T Golden L L 1984 *Lifestyle and psychographics: A critical review and recommendation* Advances in Consumer Research **11(1)** 405–411
[7] Gronow J Warde A (Eds.) 2001 *Ordinary consumption* (Hove: Psychology Press) 2
[8] Miller D 2001 *The dialectics of shopping* (Chicago: University of Chicago Press)
[9] Chaney D 1996 *Lifestyles: Key Ideas* （London: Routledge）33-49
[10] Weber M 1968 *Economy and Social* (Berkeley and Los Angeles: University of California Press)
[11] Sobel E 1982 *Lifestyle* (London: Academic Press)
[12] Rojek C 1985 *Capitalism and Leisure Theory* Tavistock
[13] Featherstone M 1987 *Lifestyle and Consumer Culture* Theory, Culture & Society **4(1)** 55–70
[14] The US VALS™ Survey (Website of STRATEGIC BUSINESS INSIGHTS http://www.strategicbusinessinsights.com/vals/presurvey.shtml)

IOP Conf. Series: Materials Science and Engineering **616** (2019) 012035 doi:10.1088/1757-899X/616/1/012035

Research on the Effectiveness of Service Design Based on the Quantitative Method of Kansei Engineering—Taking the LANNA Coffee Service Process as an Example

Zhengyu Wang*, Meiyu Zhou and Xiangyu Liu

East China University of Science & Technology,No.130,Meilong Road,Xuhui District,Shanghai,P.R.China

*E-mail: 753501966@qq.com

Abstract. The increasing enrichment of industrial products has made it difficult to meet the growing consumer demand for the material functions of products. This has caused the non-material or mental function of the product to be strengthened, promoted product design to gradually expand from the material dimension to the non-material service design, and accelerated the transformation of the manufacturing industry into a service-oriented product manufacturing. Service design has gradually become a new hot academic research field in product design. As a traditional industry, how to promote the transformation of coffee shop from "product" to "service" has become the research direction of this study. After investigation, we find some problems in coffee shop service, such as single focus, unbalanced proportion distribution of internal structure of service design, as well as vague and uncertain user demand. This paper attempts to use the forward quantitative inference method of Kansei Engineering to analyze consumers' perceptual demands for coffee shop services through questionnaires, the overall process of service and the weight of service elements. We also try to set up the service elements and multivariate linear regression equation between the service effectiveness, put forward the service design effectiveness evaluation method, and on this basis, provide the analysis of the service system to serve the interests of the parties with complimentary method to make the adjustment and evaluation of service design more intuitive.

Key words: service process; Kansei engineering; effectiveness

1. Introduction

With the development of economy, the progress of science and technology and the improvement of social production efficiency, the consumer's consumption concept also add other dimensions to the product change. With the enhancement of consumers' overall perception of products, product functionality which meets users' basic needs is no longer a decisive purchase reference for consumers, Instead, consumers' perceptual demand for products has increased. Therefore, the dimension of product design presents the characteristics of complexity and multi-coupling, and gradually develops to non-materialization. Product design changes from single function satisfaction to product service satisfaction. The involvement of Kansei Engineering provides research tools, methods of collecting Kansei information [1] and data processing support for service design. For example, In 1999, Rostlinger and Goldkuhl [2] tested the service of washing machines using kansei engineering. In the same year, Nishino [3] applied kansei engineering to explore Internet services and found wider applicability. In 2011, Zhang ST [4] put forward the research and application of Kansei Engineering and Genetic Algorithms to the intelligent design system of product shape, 2012, Hartono [5] proposes a comprehensive framework including the kano model and QFD is applied in a luxury hotel services, thus, experts and scholars at home and abroad for the combination of perceptual engineering and design services have conducted extensive research, But this type of research has a smaller proportion than other types of research, Most of them focus on the design of physical products, such as automotive interior [6] ,camera [7],mobile phone [8], etc. However, its value to the correctness of the service hypothesis, the "dynamic virtuous circle" of each "contact" in the service process, the evaluation of the service structure, and the combing of user needs are not taken seriously.

In previous studies, kansei engineering pays more attention to physical products and focuses more on research methods and tools. This study attempts to use the kansei engineering theory to discuss the "effectiveness" of service design. Through sorting out the whole process of service design and analyzing the structure of service contacts, the quantitative method of Kansei Engineering is introduced to study the validity relationship between service contacts and users' feelings. We aim to find and identify the key contacts that affect the service quality, optimize the original service contact structure, and establish the evaluation criteria of service system effectiveness through quantitative research, and use the quantitative engineering inference method for service design effectiveness research. The linear regression equations of service design effectiveness are fitted to construct an evaluation system based on the design effectiveness of perceptual engineering services.

2. Effectiveness of service design

Service design is a kind of interdisciplinary knowledge, and its definition is constantly changing with its own development. Most discussions on service design involve the five principles of service design proposed, namely user-centered, co-creation, order, object and integrity [9]. These five principles explain the effectiveness of service design in different dimensions.

1) User-centered. In the process of service design, the user is the object of service design concern. Therefore, the effectiveness of service design is reflected in users' experience of service process and evaluation of service results.

2) Co-creation. The needs of different stakeholders are included in the design considerations, so it is reflected in the effectiveness of the service process.

3) Order. The procedural characteristics of service design lead designers to take "time" as an important consideration when designing services. Therefore, in the discussion on the effectiveness of services, the consideration of order becomes relatively important.

4) Objects. In order to improve the quality of service, some service processes are designed to appear in a form of "invisible", which will change users' expectations in a positive or negative way. Therefore, the effectiveness of service design is also reflected in the enhancement of users' sense of gain through materialization.

5) Integrity. The integrity of service design means that cooperation and support between different disciplines can be achieved through service design to improve customer experience and employee satisfaction. This should be considered in the initial phase of service design.

3. Factors influencing the effectiveness of service design

The empirical research method of service design involved in this study is an improved method based on the "iterative process" method [9]."Iteration"research methods, can be summarized as four key steps, explore, create, rethink, implementation), the four steps are repeated, is building the basic introduction to design complex design process [9], but its research for service only provides a rough framework, and did not set a specific research specification. So in this paper, combined with the study of specific effectiveness problems, the original iterative method is improved into a new research method, namely: discovery, demonstration, design and implementation.The following will give a brief overview of the research tools used in this paper to study the effectiveness of service design, and take LANNA coffee as an example to describe the four key steps of the "iterative process".

Step one, discovery. The research tools in this part mainly focus on information collection and problem discovery. Understand and clarify the real motivation of user behavior to ensure the effectiveness of the problem itself. The main research tools of this study at this stage are: Stakeholders map, Journey map and shadowing practice.

(1)Stakeholders map

The function of the stakeholders map is to clearly represent all participants in the service and list their importance and influence on the service, as well as to present visually or physically by means of the design diagram. It is beneficial for participants and designers of service design to reach interest consensus quickly and help them allocate resources more effectively. The stakeholders map will help the study to complete the basic information about the effectiveness of the service design process.

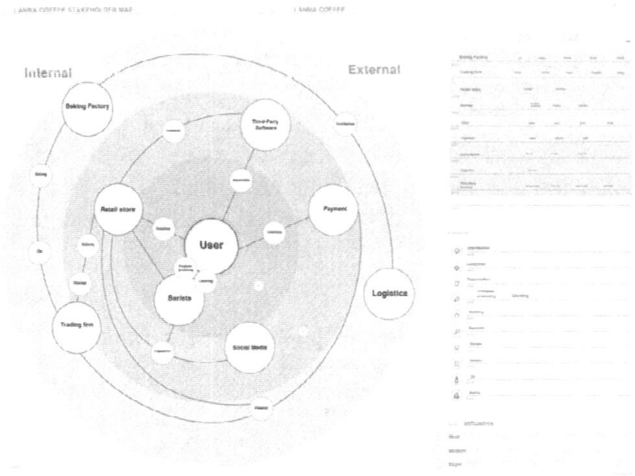

Figure 1. Stakeholders map.

According to the stakeholders map of LANNA coffee (figure 1), based on the weight that affects the user experience (the closer to the user, the greater the weight that may affect the user), internally, the main stakeholders that affect users are baristas and stores, while externally, the main stakeholders are third-party supported software.

From the perspective of controlling variables, external third-party software, due to its stable and controllable nature, can almost be regarded as a constant in the dimension of effectiveness discussion, and will not be discussed here.

It is not difficult to see from the map that the main factors affecting the user experience are stores and baristas. The main interest concerns of stores are profit and supply chain adjustment, with profit prioritized, and the second is to ensure the normal operation of coffee shops. However, the interests and concerns of baristas are different from those of the stores. The main interests and concerns of baristas are to establish a good relationship with users and create a comfortable dining environment. The second is to ensure the quality of coffee and deliver the core demands of stores.

(2)Journey map

The purpose of service journey is to list the whole process of service in a linear manner, and to clearly present the related actions of stakeholders in the whole service through the design drawing. Its main purpose is to find the "broken points" in service design which make users feel bad and improve the effectiveness of the overall service.

Taking LANNA's service process as an example (figure 2), the elements related to service are vertically listed: elements (materials), contacts, foreground behaviours, consumer behaviours, background behaviours and users' mood. Horizontal listing of the different stages of the service process, vertical and horizontal intersection grid shows that in the process, the service behaviour occurs, and this is called a focus moment. As in the previous interview, it has been found that the user experience of three service processes in the focus of bad moments, which are the broken points in the figure. The research and presentation of the effectiveness involved in this study will also be carried out based on this.

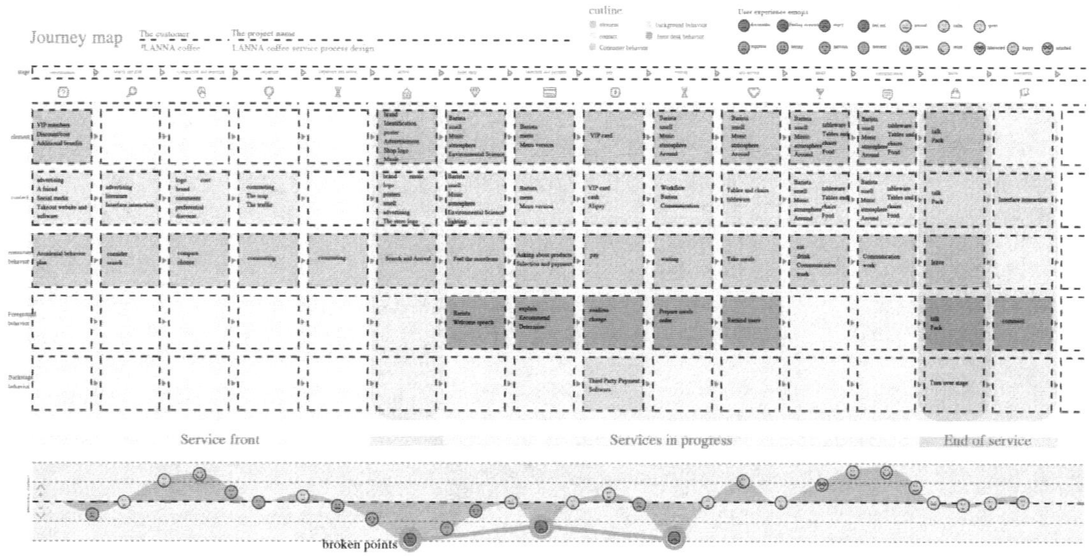

Figure 2. Journey map.

(3)Shadowing practice

The shadowing practice method is similar to the participatory observation method in kansei engineering, which is to observe users without their knowledge. Shadow exercises will play a key role in the research on the effectiveness of user experience in this topic.

The evaluation of the effectiveness of the first phase depends on various "satisfactory" feedbacks. By listing the degree of influence of different stakeholders on the service, the priority of different degrees of satisfaction will be sorted out, and the source of the problem and relevant demonstration methods will be fully assumed as far as possible, so as to provide corresponding mature hypotheses and relevant feasibility demonstration methods for the demonstration in the second stage.

The second stage is argumentation. The stage of argumentation is the stage of concept formation in the whole process. The main purpose of this stage is to identify the problems in the service process, list the feasible Suggestions, put forward the relevant demonstration process according to these Suggestions, and complete the design briefing. In this stage, the three main research tools used to discuss effectiveness are business model canvas , 4PS and SWOT.

(1)LANNA coffee business model canvas

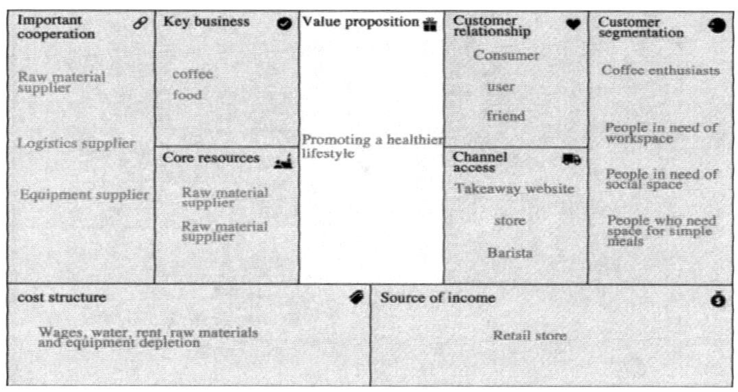

Figure 3. LANNA Coffee Business Model Canvas.

The Business Model Canvas(BMC) is designed to help researchers identify which key activities, resources and partners are used to communicate the core proposition of an enterprise to users. It includes the cost structure of the enterprise, the way of profit, the relationship between the expectation and users, and the specific delivery channels. In this study, BMC can help researchers improve the accuracy of questions and reduce speculation, so as to ensure the effectiveness of demonstration.

In the business model canvas of LANNA coffee (figure 3), orange represents the cost block of the business model, green represents the profit block, and the value proposition in the middle is the core of the whole business model. The value proposition is realized through the consumption module, and the profit is realized through the output module. Through the business model canvas, can be clear: first, the overall business model of the core value proposition. Second, it can tease out the relationship between cost (orange) and profit (green).

(2)4PS

4PS refers to Product, Price, Place, Promotion [10]. This tool follows the canvas of business model and subdivides the services provided by the enterprise to users for discussion in these four dimensions. The figure below is the 4PS diagram of LANNA coffee (figure 4).

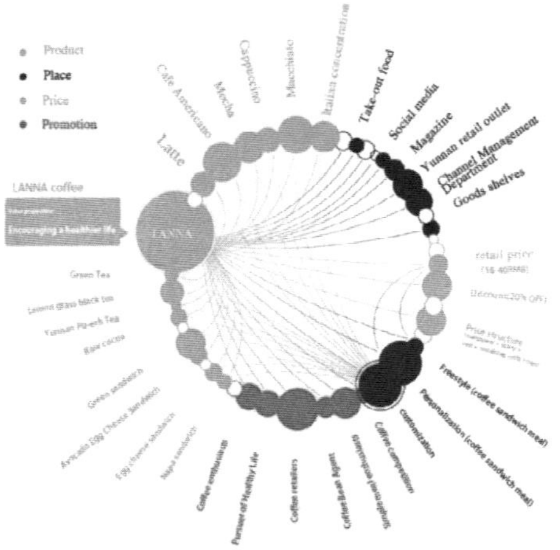

Figure 4. 4PS diagram of LANNA coffee.

(3)SWOT analysis of LANNA coffee

SWOT analysis is mainly used to present the internal and external advantages and disadvantages of the service. Help researchers adjust the proportion of resource allocation to make their service strategies clearer.

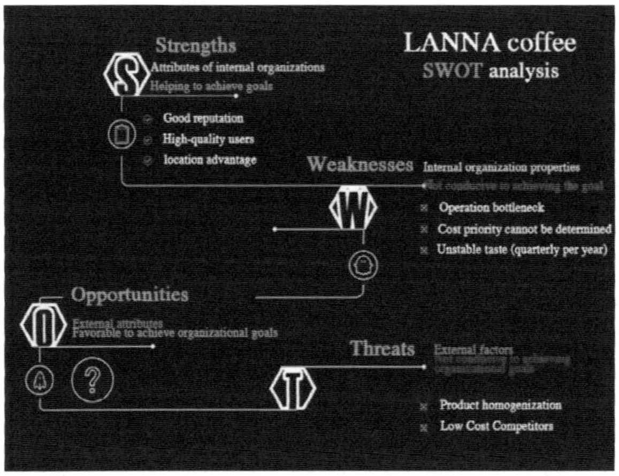

Figure 5. SWOT analysis of LANNA coffee.

Take LANNA as an example. In SWOT analysis, LANNA coffee does not describe its opportunity (mark in figure 5).

Stage three, design. The design phase consists of two phases: design justification and design testing. In the first stage, relevant assumptions should be put forward and demonstrated for the problems found in the previous stage. In the case that it is as real as possible, the "prototype" that can be demonstrated for the overall service research design should be put forward. The effectiveness of this stage is reflected in that it follows the principle of "integrity" in the service design. The main research tools in this stage are: portrait, "If... " hypothesis, storyboard, desktop walkthrough, etc.

(1)Portrait

The portrait mainly helps researchers to comprehensively consider the needs of different dimensions by using the research results of the first two stages in the design stage. The shaping of the portrait is conducive to accelerating the process of service research and ensuring the effectiveness of the design.

(2) "If" hypothesis

"If... " hypothesis belongs to "re-proof" phase of the design phase, in which participants are encouraged to describe what they consider to be an ideal state of service or a more desirable service experience. Explore future service scenarios, which in service design are often a valid way to analyze the whole (or fragment) of a service individually.

(3)Storyboard

The purpose of storyboard is to "materialize" the concept of service and the imagination of service. By describing the context of the service in as much detail as possible, the researcher can help the researcher deeply understand the user experience described by the participants and provide some insights for the service or the prototype of the service. It helps designers to incorporate research findings into the specific design of the service architecture, thus ensuring the effectiveness of the practice.

(4)Desktop walkthrough (service prototype)

Desktop walkthroughs can help researchers reproduce service scenarios , and can also help participants discuss and co-develop "service prototypes" or analyze and tweak existing services.

The effectiveness evaluation in the third stage is similar to that in the second stage, but the effectiveness evaluation in the third stage focuses more on the effectiveness evaluation with user experience as a higher priority.

The fourth stage is implementation. The main purpose of this stage is service implementation and information feedback. In this stage, the effectiveness of the results in service design can be guaranteed by collecting the satisfaction data of users on the service results. The research tool in this stage is the concept interpretation and service blueprint.

(1)Concept interpretation.

Concept interpretation helps service providers to provide more accurate services for users. However, researchers need to further clarify the relevant responsibilities of service participants and the common "interest consensus" through ""concept interpretation", so as to ensure the effectiveness of service results.

(2)Blueprint.

Blueprints are a way to elaborate on each service touchpoint. Service blueprints help researchers identify the most important parts of a service, help them understand how the service performs, and allocate resources efficiently.

4. Research and tools for service design and kansei engineering

From the current service design research and practice, the problems existing in the service design mainly concentrate on the question of "validity" and "availability". Because the "availability" of service design research focuses on service execution level and specific tool level, is based on the specific service environment, which is not the focus of this study, we do not get into details in this paper.

Previous literature exploring "validity" related issues has several findings. First of all, the excessive pursuit of service design user experience has resulted in misleading conclusions, blind designs, reduced operability and the lowered efficiency from the implementation of "iteration" issues. Secondly, the rational analysis of positivism in service design tend to be mostly qualitative. The fuzziness of the qualitative reasoning itself has brought problems, such as in accurate details, and hard-t-define perceptual intentions. Kansei engineering is a translated method that can transform the user's fuzzy visual perception into the details of the product design process[6]. Its intervention provides tools such as reliability analysis and quantitative inference for service design, reveals problems and rules in the process of service design from a more intuitive and datamatic perspective. It also makes up for limitations in qualitative inference, helps designers to accurately grasp perceptual design details, and assists designers in making design decisions.

4.1. Quantitative inference research tool for Kansei Engineering

(1)The quantitative inference of reliability analysis in kansei engineering

The reliability analysis method adopted in this study is the "Cronbach's alpha" coefficient. The calculation formula is as follows:

$$a=(k/k-1)*(1-(\sum Si2)/ST2) \quad (1)$$

Where K is the total number of items in the scale, Si2 is the intra-item variance of the score of item i and ST2 is the variance of the total score of all items. It can be seen from the formula that the coefficient a evaluates the consistency among the scores of each item in the scale, which belongs to the internal consistency coefficient.

Likert scale was used in the questionnaire in this study. which is composed of a group of statements. A likert scale of five or seven points is established for each group of statements, and the subjects are asked to score the items in the scale according to their attitudes. Finally, the attitude scores of the subjects can be obtained by adding them together.

(2)Multiple linear regression quantitative inference in kansei engineering

The regression analysis used in this study is multiple linear regression analysis, and the formula of multiple linear in this study is:

$$y = b0 + b1x1 + b2x2 + \ldots\ldots bkxk + e \quad (2)$$

Where b0 is the constant term, b0 is the regression coefficient, and b1 is x2, x3, x4... When xk is fixed, the effect of x1 on y is increased by one unit, that is, the partial regression coefficient of x1 on y; Similarly, b2 is x1, and when xk is fixed, the effect of x2 on y for each unit added, that is, the partial regression coefficient of x2 on y, and so on.

In this study, multiple regression is mainly used to solve three problems: first, to determine whether there is a correlation between variables; If they exist, find a mathematical expression that expresses their relationship. Second, predict or control the value of one or more variables based on the value of one or more variables. Third, carry out weight analysis, that is, to find the main and secondary weights and their mutual relations among multiple variables whose functions affect a variable.

5. The case study

At present, in the process of accelerating the transformation from industry to service priority, the coffee industry is the first to complete the transformation from "product" to "service". Therefore, as a coffee service provider, it is necessary to pay more attention to the continuous improvement and differentiation of services to maintain competitive advantages. In the normal coffee service, customers always want new services and new experiences. Therefore, the service design of coffee should be open, so as to continuously adjust the service to meet the new needs of customers. Designers need to have an insight into customers' real emotional needs in order to design new services that meet customers' expectations.

Take localized cafes in Shanghai as an example, on the one hand, influenced by European and American countries, local cafe its service form presents the state of coexistence of Europe and the United States has very good out of the cup, has established good customer relations with customers, but on the other hand, due to the different consumption patterns and lifestyles, make Europe and the United States and the Shanghai local cafe cafe in service mode difference, so service design problems that should be reviewing local cafe.

To sum up, the cafe in Shanghai provides a relatively ideal model for this study, both in terms of the effectiveness of the service design process and the service design results.

5.1. Experiment and data processing

In this section, researchers, practitioners in the coffee industry, users and service designers jointly participated in the experimental design, and selected perceptual words related to service experience from the original kansei engineering lexicon. They also constructed original research materials based on perceptual words, and constructed questionnaires for data collection. The focus group interview method was used to screen perceptual vocabulary, and the results are shown in table 1. In order to ensure the efficiency of the questionnaire, 5-order Likert scale was adopted.

Table 1. Kansei adjectives filtering table.

Traditional kansei adjectives	Quick, familiar, professional, timely, high-quality, convenient, friendly, valued, reliable, creative, save, save time, customized, expensive, cheap, complete, direct, in general, real, excellent, trusted, unified, elastic, a new, simple, exquisite, efficient and safe.
Service design effectiveness vocabulary	Quick, familiar, professional, timely, high-quality, convenient, friendly, economical, valued, time-saving, customized, expensive, cheap, general, true, excellent, trustworthy, novel, efficient.
Service design process effectiveness vocabulary	Convenient, friendly, economical, time-saving, customized, expensive, cheap, trustworthy, valued, novel, efficient.
Service design results validation vocabulary	Convenient, friendly, economical, time-saving, customized, expensive, cheap, trustworthy, novel, efficient.

In this study, ten cafés with similar price and location with LANNA coffee were selected as the research samples for the overall effectiveness of coffee service. The coffee brands are: View, Blueking, Manner, Wagas, Roast, Rumas, Carving time, cat's sky city, Seesaw and Huzhazha coffee. After confirming the research sample, the consumers were randomly selected to fill out the questionnaire. A total of 300 copies were handed out, and 253 copies were recollected. Considering the total number of questionnaires actually collected and the number of questionnaires in each cafe, and the comprehensive assessment of the questionnaires by the expert group, each store selected 20 questionnaires to participate in the final analyses. The average scores of 20 questionnaires in each store are shown in Table 2.

Table 2. questionnaire average score table.

QN [a]	**1**	**2**	**3**	**4**	**5**	**6**	**7**	**8**	**9**	**10**	**11**
Q [b]	3.9613	3.7431	3.8477	3.8659	3.5568	3.6023	3.4545	3.5273	3.6	3.4977	3.7477
Cat	4.0952	3.9285	3.9524	4.4048	3.5952	3.6667	3.3095	3.4523	3.6905	3.4286	3.8333
Wagas	3.976	3.8333	3.9523	4.0952	3.5476	3.5	3.3809	3.524	3.7619	3.6667	3.9762
Carvingtime	4.0952	3.8095	4.0476	4.2381	3.5476	3.5	3.2143	3.5952	3.7857	3.9048	4.3571
Rumas	4.0952	3.9285	4.0952	4.0952	3.5238	3.4761	3.2143	3.8809	3.8809	3.8571	4.0238
Manner	3.976	3.3333	3.4524	3.1905	3.7619	4.3809	4.5476	4.0952	4.0952	3.881	4.0714
Roast	3.4047	3.1190	3.3333	3.2143	3.524	3.3095	3.5714	3.5476	3.7619	3.6905	4.0238
Huzhazha	3.5952	3.4285	3.619	3.5714	3.69	3.7143	3.8333	3.5952	3.8571	3.881	4.1905
View	3.9285	3.8333	4.024	4.0952	3.8809	4.1905	4.0714	4.1191	3.9048	4.024	4.381
Blueking	3.6190	3.6904	3.7857	4.0714	3.476	3.619	3.6428	3.7619	3.7857	3.8095	4.0952
Seesaw	3.8095	3.9761	4.024	3.9762	3.4285	3.9048	3.4286	3.9762	4.1191	3.9285	4.0476

[a] Question number.

[b] All questionnaire samples.

The reliability analysis of the statistical results is mainly carried out with the "Cronbach's alpha" analysis tool, and the reliability results are shown in table 3.

Table 3. Reliability Diagram for Survey Data.

Sample	Cronbach's Alpha	Number of terms
All the samples	0.969	220
Cat	0.963	20
Wagas	0.947	20
Carving time	0.96	20
Rumas	0.95	20
Manner	0.951	20
Roast	0.726	20
Huzhazha coffee	0.662	20
View	0.642	20
Blueking	0.837	20
Seesaw	0.917	20

It can be concluded from the above table that the overall data credibility is higher than the basic confidence of 0.6, so the results of data feedback are credible.

Table 4. The equation of linear fitting chart.

coefficient[a]

	Unstandardized Coefficients		typical coefficient		
Model	B	Standard error	Trial version	t	Sig.
1 (constant)	15.775	.000		.	.
2	-11.459	.000	-13.762	.	.
3	11.486	.000	12.735	.	.
4	1.119	.000	1.933	.	.
5	-2.717	.000	-1.553	.	.
6	2.816	.000	3.905	.	.
8	-.771	.000	-.783	.	.
9	-.510	.000	-.299	.	.
10	.470	.000	.328	.	.
11	-3.812	.000	-2.630	.	.

[a]dependent variable: 1

Table 5. The Prediction Equation Accuracy.

Model	R	R-squared	Adjusted R-Square	standard error of estimate
1	.816[a]	.666	.624	.14958

[a]predictive variable: (constant), VAR00003.

5.2. Experimental analysis

The results of linear fitting of data are as follows. From the fitting data, kansei vocabulary 3 is the most concerned valid vocabulary (accessibility), and kansei vocabulary 2 has the least impact on the overall service (looking for convenience). The coincidence degree of the linear whole was 66.6%

(accuracy), exceeding 50% of the low correlation degree, and the experimental results were established. Perceptual vocabulary is moderately correlated with the effectiveness of service experience.

Fitting out multiple linear equations in coffee service design:

$$y = -11.459x1 + 11.486x2 + 1.119\ x3 - 2.717\ x4 + 2.816\ x5 - 0.771\ x6 - 0.510\ x7 + 0.470\ x8 - 3.812\ x9 + 15.775 \quad (3)$$

In the overall weight analysis, looking for convenience (-11.459); Accessibility (11.486); Friendly environment (1.119); Waiting for coffee (-2.717); Coffee taste (2.816); Cost performance (0); Professional degree of service (-0.771); Service efficiency (-0.51); Value perception (0.47); Service feedback (-3.812).

According to the weight analysis, the *Kansei adjectives* “convenient access” takes up the largest proportion in the overall coffee service, showing a relatively large correlation with the effectiveness of the overall service. It reflects users' demand for effective service design results in Shanghai, that is, the priority of “convenience” gradually takes up a larger proportion in the overall service effectiveness perception. Such a weight, on the one hand, explains the rise of cost-effective window stores in Shanghai, and on the other hand, reflects users' demand for efficiency. Simply, the focus of users on the effectiveness of the coffee service is that it is either easily accessible or easily delivered. In other words, the priority of “convenience” is gradually increasing in the recognition of overall service effectiveness.

However, the convenience of finding coffee shops presented in the results shows a weak correlation (-11.459), which reflects the randomness of consumption of coffee services themselves , and confirms the low viscosity of users and coffee shops mentioned above. The low user viscosity is mainly reflected in two aspects. On the one hand, users do not really pursue the taste of coffee and choose to enjoy the service of coffee , but hope to satisfy their own social experience through coffee. Coffee and coffee have gradually separated, and coffee has gradually become a kind of social media. On the other hand, the rise of take-out can meet people's demand for the timeliness of coffee drinks, and at the same time, it is extended through the social places of coffee drinks.

6. Conclusion

This article through to the perceptual technology and service design paper proves that the first, the perceptual design engineering and services can be combined for research, perceptual quantification methods can help services in the engineering design of participants quickly establish service elements of service design priority level, thus effectively control costs, achieve the goal of comprehensive improve service efficiency. Second, this study establishes the design service efficiency and service elements of multivariate linear regression equation, it is concluded that the service effectiveness factors of service design the effect of weight, thus help practitioners in the case of cost must be filtered to the service elements, help them build a service quality effectively, contains effective service process, and the effective service results.

References

[1] Tsuchiya, Toshio, et al. “A fuzzy rule induction method using genetic algorithm.” *International Journal of Industrial Ergonomics* 18.2–3(1996):135-145.

[2] Rostlinger,A,Goldkuhl,G,1999,Expanding the scope:From language to generic practice, in Proceedings *4 th International Workshop on the Language Action Perspective on Communication Modelling* LAP’99,.Copenhagen,Denmark,September 12-13.

[3] Nishino T, Nagamachi M, Ishihara K, Ishihara S, Ichitsubo M, Komatsu K, 1999.Internet Kansei engineering system with basic Kansei database and genetic algorithm. In: *Proceedings of TQM and Human Factors* (Linkoping, Sweden:Centre for Studies of Humans, Technology and Organization), pp 367–372.

[4] Zhou S T, *Research on Product Form Intelligent Design System Based on Kansei Engineering*

and Genetic Algorithm . Diss. Lanzhou University of Technology,2011. (in Chinese).
[5] Hartono, and Markus. "Incorporating Service Quality Tools into Kansei Engineering in Services: A Case Study of Indonesian Tourists." *Procedia Economics and Finance* 4.Complete(2012):201-212.
[6] Nagamachi, M. "Kansei Engineering: A new ergonomic consumer-oriented technology for product development." *International Journal of Industrial Ergonomics* 15.1(1995):3-11.
[7] Yang, Chih Chieh. "A classification-based Kansei engineering system for modeling consumers' affective responses and analyzing product form features." *Expert Systems with Applications* 38.9(2011):11382-11393.
[8] Yang, Chih Chieh, and H. C. Chang. "Selecting representative affective dimensions using Procrustes analysis: An application to mobile phone design." *Applied Ergonomics* 43.6(2012):1072-1080.
[9] Schneider, Jakob. "This is service design thinking." (2011).
[10] Bitner, M. J, and B. H. Booms. "Deregulation and the future of the US Travel Agent Industry." *Journal of Travel Research* 20.2(1981):2-7.